Waste Utilization in Geotechnical Practice

Scrivener Publishing
100 Cummings Center, Suite 541J
Beverly, MA 01915-6106

Publishers at Scrivener
Martin Scrivener (martin@scrivenerpublishing.com)
Phillip Carmical (pcarmical@scrivenerpublishing.com)

Waste Utilization in Geotechnical Practice

Edited by

Sandeep Samantaray
Dept. of Civil Engineering, National Institute of Technology, Srinagar, Jammu and Kashmir, India

Subhashree Samantasinghar
Dept. of Civil Engineering, Odisha University of Technology and Research, Bhubaneswar, India

Soumendra K. Mohanty
SM Consultants, Bhubaneswar, India

Vivek
Dept. of Civil Engineering, National Institute of Technology, Srinagar, India

Abinash Sahoo
Dept. of Civil Engineering, Odisha University of Technology and Research, Bhubaneswar, India

and

Deba Prakash Satapathy
Dept. of Civil Engineering, Odisha University of Technology and Research, Bhubaneswar, India

Scrivener Publishing

WILEY

This edition first published 2025 by John Wiley & Sons, Inc., 111 River Street, Hoboken, NJ 07030, USA and Scrivener Publishing LLC, 100 Cummings Center, Suite 541J, Beverly, MA 01915, USA

For more information about Scrivener publications please visit www.scrivenerpublishing.com.

Wiley Global Headquarters
111 River Street, Hoboken, NJ 07030, USA

For details of our global editorial offices, customer services, and more information about Wiley products visit us at www.wiley.com.

Library of Congress Cataloging-in-Publication Data

ISBN 9781394242184

Front cover images courtesy of Wikimedia Commons
Cover design by Russell Richardson

Set in size of 11pt and Minion Pro by Manila Typesetting Company, Makati, Philippines

Printed in the USA

10 9 8 7 6 5 4 3 2 1

Contents

Preface

The book explores how various waste materials can be effectively utilized in geotechnical construction projects related to soil and foundation improvement, presenting innovative research ideas along the way. Geotechnical engineering involves studying the behavior of earth materials—including soil, rock, and groundwater—and designing structures and systems that interact with them. In this context, waste materials such as construction and demolition debris, industrial byproducts, municipal solid waste, and other normally discarded materials are examined for their potential in applications like soil stabilization, land reclamation, and erosion control.

The book covers topics including the properties of various waste materials, methods for testing and characterizing them, potential environmental impacts, and the design considerations and techniques required to incorporate these materials into geotechnical structures. Additionally, it discusses the economic and regulatory aspects of waste utilization in geotechnical practice, highlighting cost-saving potentials and the protocols governing waste use in construction projects.

Geotechnical engineering is a vital field within civil engineering. The addition of various materials can enhance soil strength and durability, and using waste materials offers significant environmental and economic advantages. However, the large quantities of waste generated often lead to storage challenges. *Waste Utilization in Geotechnical Practice* serves as an informative and practical guide to land-based waste disposal and its potential applications in soil improvement.

This book offers technical guidance on ground improvement technologies that transform marginal soils into substrates capable of supporting all types of structures. Soil improvement involves altering one or more soil properties to enhance its engineering performance and is necessary when existing conditions are inadequate. Although poor soils might be managed by excavation and replacement or by employing deep foundations, it is often more cost-effective to improve the soil in situ through various treatments. With rapid urbanization and industrial growth, addressing poor soil conditions is critical to achieving durable structures at a competitive cost.

An essential guide for both novice and practicing engineers, this reference covers projects involving soil stabilization and admixtures. It examines the utilization of industrial waste and by-products, commercially available admixtures, conventional soil improvement techniques, and state-of-the-art testing methods. The book highlights studies on the potential use of waste materials in geotechnical ground improvement, identifies key elements of various waste utilization methods, and explains their role in enhancing engineering performance. It serves as a valuable resource for researchers in related fields.

The book will appeal to a broad readership, including professionals in geotechnical engineering, environmental scientists, waste management experts, and students in these fields. Finally, we extend our gratitude to Martin Scrivener and the team at Scrivener Publishing for their invaluable editorial support throughout the publication process.

The Editors

1

Interaction of Landfill Leachate with Olivine-Treated BCS: Suitability for Bottom Liner Application

Deepak Kumar Padhi[1], Udit Narayan Bhoi[1], Biswajit Majhi[2*] and Siprarani Pradhan[2]

[1]*M.Tech, Odisha University of Technology and Research, Bhubaneswar, Odisha, India*
[2]*Odisha University of Technology and Research, Bhubaneswar, Odisha, India*

Abstract

Leachate is defined as any contaminated liquid that is generated from water percolating through a solid waste disposal site, accumulating contaminants, and moving into subsurface areas. Landfill leachate is characterized by high organic and inorganic pollutant concentrations and is extremely toxic to the environment. Due to its high toxicity, landfill leachate is a major threat for surface water as well as ground water. Nowadays, due to the rapid growth of population and industries, disposal of solid or industrial waste is a major issue. In landfills, leachate is a liquid that enters the landfill from external sources such as rainfall, surface drainage, etc. The leachate infiltrates through the pores of the soil from surface to ground water and makes the water contaminated. Therefore, to prevent this scenario from occurring, it is necessary to modify subsurface soils as barrier materials in waste containment facilities in order to confine municipal and hazardous waste materials from the surrounding environments as well as to stop landfill leachate from seeping into the nearby hydro-geological system. In this present work, the potential of olivine-treated black cotton soil (BCS) as a bottom liner material in landfills is examined. The BCS is treated with varying olivine percentage of 10%, 20%, 30%, 35% and 40% of the soil mass. Some laboratory tests namely: Index properties, standard proctor test, Modified proctor test, unconfined compressive strength and hydraulic conductive are conducted to evaluate the geotechnical properties of BCS

**Corresponding author*: bmajhi@cet.edu.in

Sandeep Samantaray, Subhashree Samantasinghar, Soumendra K. Mohanty, Vivek, Abinash Sahoo and Deba Prakash Satapathy (eds.) Waste Utilization in Geotechnical Practice, (1–30) © 2025 Scrivener Publishing LLC

treated with olivine. The findings show that olivine having 35% of soil mass used for treatment of BCS give the maximum desirable results.

Keywords: Landfill, liner material, BCS, olivine geotechnical tests

1.1 Introduction

A significant soil type in India is a dark, grey color BCS (BCS). When wet, it experiences high swelling, and when dried, it experiences high shrinkage. As a result, this soil grows and becomes slick throughout the wet season and shrinks throughout the dry time of the year, resulting in serious soil fractures. The BCS contains a lot of clay. Chemically, BCSs consist of lime, iron, magnesium, alumina, and potash; however, they are free of nitrogen, phosphorus, and organic matter. It expands in volume by 20% to 30% of the original volume when pressure is applied. It is quite challenging to maintain the soil because of its swell and drying characteristics [1–4]. Any construction should undergo high stabilization because of its unique properties. BCS is a sedimentary type of soil that is present in its original location; it does not spread from that location. BCS is a result of the particular rock's wear and tear. Black soil can only occur under mild climatic conditions and with igneous or basalt rock as the parent rock. Then, as a result of the igneous rock weathering or breaking and the lava cooling and solidifying, black soil is created. It is also referred to as lava soil because it is made of lava.

Clayey soils are widely used in waste containment facilities as barrier materials to stop the seepage of landfill leachate into the nearby geological and hydrological system as well as to keep hazardous and municipal waste products out of the environment. A barrier material must have low hydraulic conductivity, sufficient strength, and high compatibility with the percolating leachate to be effective. Clayey soils typically exhibit low permeability, but they can be impacted by temperature and moisture content changes, which can cause contraction and the subsequent development of fractures [5–8]. Due to varying settlements or loads that the liner material experiences, tensile cracks may also develop. Desiccation, differential settlements, swell-shrink, or pressures can all lead to the development of cracks, which can drastically alter the clay liners' strength and increase leachate penetration via the liner structure. Based on these limitations, researchers and engineers are looking into practical methods to enhance the technical potential of naturally clayey soils that have strong features that expand and contract before being used as lining materials.

As a result, it is now widely accepted in the building of liner that natural clayey soils can be blended with various compounds to enhance their hydro-mechanical qualities, such as the byproducts, organic compounds, and synthetic substances and prevent/minimize cracking. Many studies [9–15] have looked into the usage of waste products and other byproducts to address the instability of soil liners. Several investigations have also enhanced the hydro-mechanical characteristics of soil liners through the application of various synthetic and natural materials.

To ensure that the barrier material performs as intended during field application, it is imperative to determine whether it is chemically compatible with the leachate or percolating fluid it will likely encounter. This is because the leachate's interaction with the soil can change its hydro-mechanical characteristics. After a long period of leachate percolation, a substance is deemed suitable for the leachate if it retains structural reliability and minimal hydraulic conductivity. Therefore, utilizing tap or distilled water to assess the engineering features of soil barriers is far from realistic of real-world field situations.

To reduce leachate migration in constructed landfills, [16] one has to investigate an optimization technique to assess the marine clay combined with CO_2-carbonated olivine and its hydraulic conductivity. The results demonstrate that, in order to reduce leachate migration in landfills that are intended for that purpose, hydraulic conductivity of marine clay combined with CO_2-carbonated olivine is measured using an optimization technique. The response surface methodology, which was utilized to plan the trials and analyze the data, was implemented to optimize the factors that reduced the hydraulic conductivity of the clay treated with CO_2-carbonated olivine.

He [17] prepared a mixture of pond ash and a sample of BCS, mixing pond ash from various sources. To determine the strength of the mix specimen, for 7 and 28 days, he conditioned a specimen of black cotton blend. He found that when soil samples were treated with 25% class C fly ash (18.98 percent of CaO), swelling pressure dropped by 75% and 79%, respectively, after 7 and 28 days of curing. They [18] examined the impact of expansive soil on its hydraulic conductivity, plasticity, compaction, swelling pressure, FSI, and swell potential. Using fly ash values of 0, 5, 10, 15, and 20% on a dry weight basis, the FSI was reduced by about 50% by adding 20% Fly Ash to the ash-blended expansive soil. They concluded that the qualities of plasticity decrease as the fly ash content increases. Expanding soils coupled with fly ash have lower hydraulic conductivity as fly ash content rises because the maximum dry unit weight rises with fly ash level. Satyanarayana (2004) [19] investigated how adding lime and fly ash might affect expansive soil's engineering characteristics. The researcher discovered that the ideal proportion combination of the elements for building roads and embankments was 70%, 26%, and 4%. He/She [20] used bentonite to create liners for water-retention and waste-containment systems, either on its own or after being altered with natural soils. The study demonstrates how the clay liner's hydraulic conductivity is affected by the coarser fraction's size. It is known that, in addition to clay concentration, the coarser fraction's size affects the liner's hydraulic conductivity at low bentonite values. The hydraulic conductivity rises with the size of the coarser fraction for a fixed percentage of clay. He [21] investigated the stabilization of expansive soils by 0–30% using fly ash from and desulpho gypsum. The mixture of expanding soil, desulpho gypsum, and fly ash had a variable quantity of lime added (0 to 8 percent). The samples were then allowed to cure for seven and twenty-eight days. It was found that as the percentage of stabilizer in the combination increased, swelling percentage decreased by roughly 23 percent and swell rate increased. With the 25% fly ash addition and 30% desulpho gypsum, the swelling percentage was further lowered throughout the curing phase. He [22] studied the effects of fly ash and rice husk ash on soil strengthening. He advised the use of 25% fly ash concentrations to reinforce the expansive subgrade soil and 15% to combine with RHA to create a layer that reduces swell. The byproduct of burning coal in thermal power plants is a sort of industrial waste known as fly ash. According to the test results, compaction and CBR characteristics have significantly improved. Furthermore, it has been discovered that fly-ash is a waste product ineffective for stabilizing expanding soil. They [23] studied the geotechnical characteristics of pond ash samples at the inflow and outflow points of two Indian ash pond sites. To determine the strength properties, samples of both compacted and loose pond ash were subjected to triaxial tests (consolidated drained, and consolidated undrained) with measurements of the pore water pressure under a range of confining pressures. In many ways, the behavior of the ash samples taken from the ash pond inflow point was comparable to that of sandy soils. Although their specific gravity and MDD were significantly lower than the sands, they outperformed Yamana sand in terms of strength. There were notable variations in the values of the ash samples taken from the ash pond area's point of outflow. They [24] found that after doing multiple experiments on pond ash, the compaction energy changed from 357 to 3488 kJ/m^3 and the OMC decreased from 38.82 to 28.09 percent. The compacted specimens exhibited a dry density ranging from 10.90 to 12.70 kN/m^3. Additionally, it is found that decreasing the water content percentage of the OMC results in an increase in the unconfined compressive strength value for a sustained DOS of 13–14 percent before it declines for both standard and modified proctor densities because the ash particles lubricate the surface. The compaction energy and the unconfined compressive strength were found to be linearly correlated. Pond ash becomes more ductile when it is reinforced with fibers. He [25] looked at the crystal chemistry of olivine, as well as the use of particular trace elements as thermometers for garnet and spinel peridotites. They also attempted to infer the genesis of the cratonic lithosphere from olivine compositions. They discovered that the diamond inclusions had high levels of Cr. They [26] examined the Fe^{3+} replacement behavior using atomistic computer simulations and came to the conclusion that in the Si-O-Mg-Fe^{3+} system, bond substitution—in which ferrous iron replaces both Mg and Si is superior to vacancy substitution. They [27] investigated the use of olivine as a diagnostic tool for planetary settings because elements such as V, Cr, Fe, and Ti can occur in varying oxidation states, influencing its partitioning behavior. The partition coefficients across minerals and how they change with temperature and pressure have been the subject of the majority of research that describe trace elements in olivine, with a primary focus on spinel peridotites. The literature review show that more research has been required for the treatment of expansive soil like BCS to increase its strength and stability by adding different admixture and easily available earth mineral which is cost effective and has the ability to change soil behavior and prepare it for building purpose. To increase soil strength or use it as barrier material, various stabilizers are also used and mixed with BCS with proper proportion to make it suitable for liner material.

In this paper, the potential of olivine-treated BCS as a bottom layer material in landfills is investigated. BCS is treated with different percentages of olivine of 10%, 20%, 30%, 35%, and 40% of the soil mass. The geotechnical characteristics of olivine-treated BCS are assessed using a series of laboratory experiments, including index properties, hydraulic conductivity, unconfined compressive strength (UCS), and the standard and modified proctor tests.

1.2 Materials and Methodology

1.2.1 BCS

The texture of BCS is delicate. It is extremely malleable because of the high clay content. Compressing the earth is extremely easy. It frequently exhibits notable swelling and shrinkage. It expands in the presence of water, causing the soil layer to collapse. BCS is the second largest soil group in India.

Figure 1.1 The appearance of materials (a) BCS (b) Olivine and (c) Leachate.

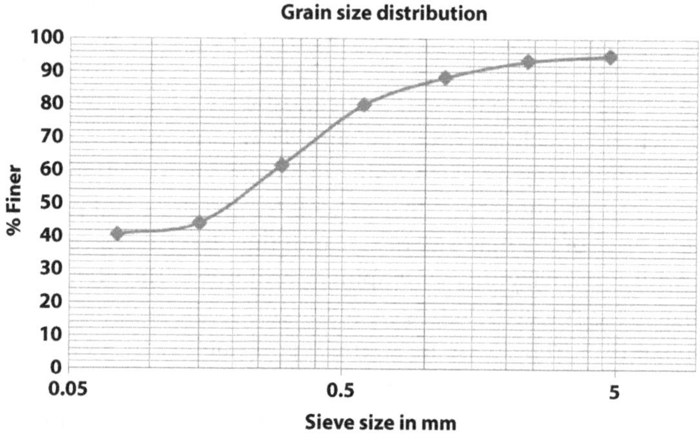

Figure 1.2 Grain size distribution of BCS.

Table 1.1 Physical characteristics of BCS.

Properties	Value
Liquid limit (%)	60.02
Plastic limit (%)	26.33
Plasticity index (%)	33.69
Specific gravity	2.66
Maximum dry density (gm/cm^3)	1.36
Optimum water content (%)	31.00
Unconfined compressive strength (kPa)	191.5

Table 1.2 Chemical properties BCS.

Minerals	Value (%)
Al_2O_3	10
Fe_2O_3	10
CaO and $MgCO_3$	8
K_2CO_3	< 0.5
Phosphate, Nitrogen, Humus	Low

Lime, iron, and magnesium are present in significant concentrations, while phosphorus, nitrogen, and organic materials are generally present in trace amounts. The way the BCS appears is depicted in Figure 1.1(a). The grain size distribution of BCS is displayed in Figure 1.2. Tables 1.1 and 1.2 display the chemical and physical characteristics of BCS, respectively.

1.2.2 Olivine

Olivine has high magnesium content. Before grinding was added, the olivine needed to have a large particle size in order to increase the reactivity with the BCS. With the chemical formula $(Mg^{2+}, Fe^{2+})^2SiO_4$, olivine is a magnesium-iron silicate. It belongs to the non-silicate or orthosilicate category. Figure 1.1(b) depicts how olivine looks. The chemical and physical characteristics of olivine are shown in Tables 1.3 and 1.4, respectively.

Table 1.3 Chemical characteristics of olivine.

Minerals	Concentration (%)
SiO_2	43.64
Al_2O_3	6.23
Fe_2O_3	8.46
CaO	0.36
MgO	43.86
LOI	1.96

Table 1.4 Physical characteristics of olivine.

Properties	Value
D50 (mm)	0.026
Specific gravity	2.86
pH	9.23
Electrical conductivity (mS/cm)	0.17
Color	Light Green

1.2.3 Leachate

The leachate utilized in this investigation was collected from multiple places in the densely populated city of Bhubaneswar, where municipal solid waste (MSW) is collected. Assuming that the occupants of the trash create a highly varied composition, the leachate was created and transported directly to the laboratory for characterization. The operating temperature range for the refrigerator used to store the leachate was 7 to 10°C. The leachate was maintained in opaque plastic bottles. Figure 1.1 (c) shows the appearance of leachate. Table 1.5, Table 1.6 and Table 1.7 represents the chemical, physical properties and environmental quality of olivine respectively. The grain size distribution of BCS is illustrated in Figure 1.2.

Table 1.5 Chemical properties of leachate.

Parameters	Quantity (mg/L)
Cr	0.42
Cd	0.03
Ni	0.31
Cu	0.41
Pb	0.07
Fe	358
Zn	6.33
Mn	5.34

Table 1.6 Physical properties of leachate.

Parameters	Quantity
pH	3.48
Electrical conductivity (mS/cm)	13.03
Density (gm/ml)	1.26
TDS (mg/L)	9583
Hardness (mg/L)	1895

Table 1.7 Environmental quality of leachate.

Parameters	Quantity (mg/L)
BOD5	24068
CODCr	61253
BOD5/CODcr	0.36
TSS	12046

1.2.4 Sample Preparation

After being oven dried at 105°C for three days, BCS reaches a consistent weight. The dry BCS was ground up before the test. Prior to testing, the required amount of olivine dry soil mass was combined with crushed soil samples that had been through a 2 mm sieve in line with the experimental plan. The acquired olivine was first cleaned of unwanted contaminants by handpicking. Olivine was then separated based on particle size; batches weighing roughly 1.5 kg were sieved via a number of sieves. To create typical feed olivine samples, each weighing roughly 3000 g, the various masses that were gathered and ready for testing were divided.

1.2.5 Test Procedures

Important engineering characteristics of the soil were ascertained using a variety of laboratory tests that were approved and conducted on the chosen earth sample in accordance with the IS Codes, as shown in Table 1.8.

Table 1.8 Codes recommended for various laboratory test.

Sl. no.	Test	Referred IS code
1	Analysis of Grain Size Distribution	IS: 2720 (PART-4): 1985 [28]
2	Specific Gravity Test	IS: 2720 (PART-3): 1980 [29]
3	Liquid Limit Test	IS: 2720 (PART-5): 1985 [30]
4	Plastic Limit Test	IS: 2720 (PART-5): 1985 [30]
5	Standard Proctor Test	IS: 2720 (PART-7): 1980 [31]
6	Modified Proctor Test	IS: 2720 (PART-8): 1983 [32]
7	Unconfined Compressive Strength Test	IS: 2720 (PART-10): 1991 [33]
8	Hydraulic Conductivity	IS: 2720 (PART-17): 1986 [34]

1.2.5.1 Wet Sieving for Grain Size Distribution

Wet sieving was performed to determine the percentage of sand, silt, and clay; that is, the proportion of coarse to fine grains that remain on a 75-micron screen sieve after the sample has been washed. After being cleaned via a 75 mm sieve, the soil sample that remains is gathered in a container and oven dried for a whole day at a temperature between 105 and 110°C. The soil sample is taken and weighed on a weighing scale following oven drying. After utilizing a series of sieves placed from top to bottom for a sieve analysis, the material retained on each sieve is weighed. The following sieve sizes were utilized for weight sieving: 4.75mm, 2.0mm, 425 micron, and 75 micron.

1.2.5.2 Dry Sieving for Grain Size Distribution

The sieve analysis method is employed to ascertain the granular material's grain size curve. In the sieve research, the grain size distribution is given as a percentage of the total dry weight. After going through a number of screens arranged from top to bottom, the material is weighed to assess its quality. A sieve analysis test, which determines the weight, particle size distribution, and amount of material suspended on each sieve as a percentage of the total weight, can be used to gradually reduce the size of the material.

The test results may have an impact since they will represent the state and characteristics of the aggregate from which the sample was collected. Fine particles are absent from soil that passes through the 4.75mm IS sieve and is caught on the 75μ IS sieve. Take a sample of the undisturbed soil, either 1 kg or 1000 g. Apply mesh analysis as 4.75 mm-2.36 mm-1.18 mm–600 μ-300 μ-150 μ-75 μ-Pan. Ten minutes can be spent sieving by hand or with a motorized sieve. Determine how much material each sieve can hold. Based on the overall weight of the sample obtained, the percentage retained on each sieve is computed. The percentage that passes through each sieve is calculated using this result.

1.2.5.3 *Liquid Limit Test*

It is defined as the lowest water content at which shear resistance to flow is starting to appear and the soil is still liquid. The amount of water present determines the differential in soil moisture between the liquid and plastic states. It is measured by Cassa grand's device, which has the W_L designation. The water content, a soil sample placed in a standard 10 mm cup and grooved or separated using a standard tool will close after 25 blows over the typical procedure is known as the liquid limit.

1.2.5.4 *Plastic Limit Test*

It is the quantity of water required for the soil to roll into three-millimeter-diameter threads. For the soil to transition from its plastic condition to its semi-solid state, it must contain a certain amount of water. A soil has a higher shear strength at the plastic limit than at the liquid limit. A soil paste that can be formed into a ball between the palms of hands is created by adding some distilled water to a sample of soil that weighs about 30 g and passes through an IS sieve with a 425-micron mesh size. After that, a little portion of the ball is rolled onto a smooth surface to form a thread with a diameter of 3 mm, and the thread is checked for cracks.

1.2.5.5 *Plasticity Index*

The plasticity index is one of the key elements of the fine-grained soil index. The plastic and liquid limits—the other two characteristics—can be used to compute it. This is the distinction between a plastic limit and a liquid limit. This indicator represents the lowest water content at which the soil becomes plastic. The plasticity index table is displayed in Table 1.9.

Table 1.9 Plasticity index.

Plasticity index (%)	Soil type	Degree of plasticity	Degree of cohesiveness
0	Sand	Non-Plastic	Non-cohesive
<7	Silt	Low Plastic	Partly cohesive
7-17	Silt Clay	Medium Plastic	Cohesive
>17	Clay	High Plastic	Cohesive

$$I_P = W_L - W_P \tag{1.1}$$

$$\text{Plasticity index } (I_P)(\text{From A-Line}) = (0.73(W_L\text{-}20)) \tag{1.2}$$

1.2.5.6 Specific Gravity

It is the ratio of the weights of a given volume of soil solids and an equal volume of distilled water at 27 degrees Celsius, both weights obtained in air. Calculations of the porosity ratio, saturation level, different unit weights, etc. are made using the soil-specific gravity. The specific gravity of soil is calculated using a 50 ml/100 ml density bottle. We find the bottle, M1, it's clean, dry substance. The bottle is filled with the required amount of the soil sample that has been oven dried. It is then chilled in a desiccator to determine the mass M2 of the soil-filled bottle. The soil within the bottle is then filled with distilled water, being careful to release all of the contained air. A bottle containing soil and water is found to have mass M3. After cleaning, the bottle's contents are taken out, and it is then filled exclusively with distilled water. Once the exterior of the bottle has been thoroughly cleaned, the water-filled bulk of M4 is revealed. The following is the soil's specific gravity:

$$\text{Specific Gravity }(G) = \frac{(M2 - M1)}{[(M2 - M2) - (M3 - M4)]} \tag{1.3}$$

1.2.5.7 Standard Proctor Test

The test employs the following instruments: (i) a collar with an effective height of 2", (ii) a ram weighing 2.5 kg with a drop height of 1', and (iii) a cylindrical metal mold with a removable base plate that has an internal diameter of 4", an internal height of 4.6", and an internal volume of 945 cm^3. Compaction of the soil at different water contents and determination of the proper dry densities are the main steps in the procedure. At every water content, the base plate mold is filled with three layers, each of which is given 25 blows from a standard rammer. In the density graph per dry, the compaction curve shows the relationship between dry density and water content. This test also contributes to the improvement of the foundation's bearing capacity, the structure's shear strength, the management of undesired volume change, and the decrease of hydraulic conductivity.

1.2.5.8 Modified Proctor Test

The modified proctor test was created to offer a better compaction standard in response to the introduction of larger vehicles and the need for higher compaction. Due to its establishment by the American Association of State Highway Officials, the test is often referred to as the modified AASHO test. The test method is comparable to the standard proctor test, which makes use of a proctor mold with a capacity of 1/30 Cu. Ft., or 945 ml. However, a 4.54 kg rammer with an 18" (45.72 cm) drop height is used to compact the soil in five stages, giving each layer 25 hits. The optimal moisture content decreases when the compaction effort is increased, but the maximum dry matter density increases.

1.2.5.9 Unconfined Compression Test

When no lateral pressure or confining pressure is applied, the unconfined compression test can be thought of as a specific example of the triaxial compression test. The soil sample is cylinder-shaped, with a length that is around two to five times its diameter. Unconfined compression tests can be performed in the lab using equipment that compresses the specimen at a constant rate of strain while monitoring the axial deformation and corresponding axial compressive force. The maximum compressive stress a specimen can bear before failing is known as its unconfined compressive strength. q_u denotes it, and the calculation is

$$q_u = \frac{F}{Ac} \tag{1.4}$$

Where F= axial compressive force at failure
A_c= the specimen's changed cross-sectional area at failure

The unconfined compression test is a quick test in which no drainage is allowed. The test is conducted on saturated clay and the volume change is assumed to be zero.

1.2.5.10 Hydraulic Conductivity

Permeability is a crucial consideration when choosing a waste containment liner or cover system since its performance is more closely related to it. In general, there is a connection between compaction and hydraulic conductivity (permeability). This might be accomplished by packing the material into the pores with fine material or compacting it to a higher dry unit weight. The coefficient of permeability value (k) of BCS is extremely low, measuring 0.000062 cm/s. This might be the case since more than 90% of the elements in BCS take the form of fine fraction (finer than 0.075mm). By filling up the pores, these tiny particles significantly reduce the hydraulic conductivity of BCS. The primary factor affecting permeability is the void ratio.

1.3 Results and Discussion

The effect of inclusion of various percentage of olivine in the geotechnical properties of BCS is examined. The different percentages of olivine inclusion in BCS are 10%, 20%, 30%, 35% and 40% of soil mass. The different geotechnical properties of treated BCS with olivine such as grain size distribution, consistency limit, dry density, UCS and hydraulic conductivity are investigated and explained.

1.3.1 Effect of Varying Olivine Percentages on Grain Size Distribution Curve on Olivine Treated BCS

The impact of adding different percentages of olivine to the soil mass of BCS is investigated in relation to the grain size distribution curve. Various percentages of olivine, ranging from 10% to 40% of the soil mass, are included in BCS and compared to BCS that has not been treated.

In Figure 1.3 below, the grading curves for BCS, both treated with olivine and untreated, are displayed. As can be observed from the curve, the amount of clay size often reduced as olivine content increased. For soil treated with 35 percent olivine, the clay-sized content decreased from 40.3

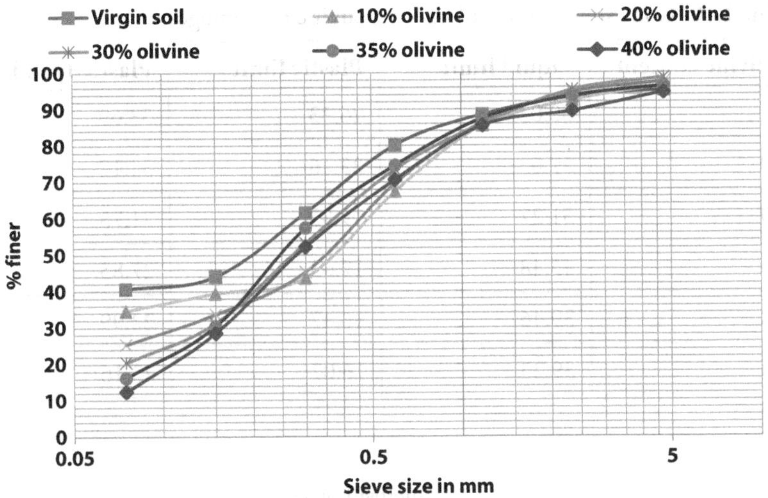

Figure 1.3 Distribution curve for grain size of olivine treated BCS.

percent to 16.7 percent. The amount of silt-sized content increased. For soil treated with 35 percent olivine, the silt size content rises from 41.6 percent to 61 percent. The variations in the concentration of silt and clay in the treated BCS are caused by both physical and chemical causes. The alterations are ascribed to flocculation on a chemical level. The gradation effect, which replaces the finer clay particles with olivine particles the size of sand and silt, is physically responsible for the alterations. As a result, the treated soil has more silt and sand than clay particles. Olivine particles with silt sand-sized particles react with clay and change the gradation of the soil.

1.3.2 Influence of Different Olivine Percentages on the BCS Consistency Limit

A study is conducted to examine the impact of varying olivine percentages on the BCS soil mass consistency limit. The different percentage of olivine such as 10%, 20%, 30%, 35% and 40% of soil mass is included in BCS and compared with untreated BCS. Table 1.10 shows the consistency limits of varying olivine percentages.

Figure 1.4 shows how the plasticity characteristics vary with olivine content. With an increase in olivine concentration of up to 40%, WL and Wp were generally reduced. The plasticity index (Ip), which normally decreased as olivine content increased, peaked at 24.3 percent at a 35 percent olivine level.

Table 1.10 Consistency limits of varying olivine percentages.

Olivine content	Liquid limit	Plastic limit	Plasticity index
0	60.02	26.33	33.69
10	57.63	23.46	34.17
20	51.27	19.52	31.75
30	45.48	17.63	27.85
35	39.46	16.47	22.99
40	39.23	16.07	23.16

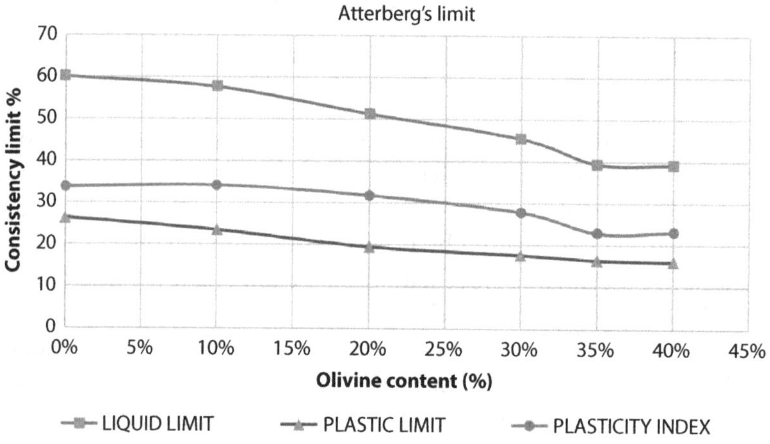

Figure 1.4 BCS consistency limit with different olivine (%).

1.3.3 Effect of Varying Olivine Percentages on Standard Proctor Test Result of Olivine Treated BCS

The effect of inclusion of various percentage of olivine on dry density of soil mass of BCS is examined by conducting standard proctor test. The different percentage of olivine such as 10%, 20%, 30%, 35% and 40% of soil mass is included in BCS and compared with untreated BCS. Tables 1.11-1.15 show the dry density BCS with different percentage of olivine mix (10%, 20%, 30%, 35% and 40% of soil mass).

Table 1.11 Standard proctor test result of BCS with 10% olivine mix.

Experiment no.	1	2	3	4	5
Mold Vol. (V) cm^3	1000	1000	1000	1000	1000
Mold Wt. (W1) g	3677	3677	3677	3677	3677
Mold Wt. + Compacted Soil (W2) g	5145	5255	5376	5485	5456
Weight of Compacted Soil ($W=W_2-W_1$) g	1468	1578	1699	1808	1779
Bulk density (Y=W/V) g/cm^3	1.4689	1.5781	1.6998	1.8084	1.779
Water Content (w)%	11.96	14.36	16.43	18.24	20.27
Dry Density ($Y_d=Y/1+w$)	1.312	1.380	1.467	1.531	1.485

Table 1.12 Standard proctor test result of BCS with 20% olivine mix.

Experiment no.	1	2	3	4	5
Mold Vol. (V) cm^3	1000	1000	1000	1000	1000
Mold Wt. (W1) g	3677	3677	3677	3677	3677
Mold Wt. + Compacted Soil (W_2) g	5155	5268	5391	5495	5448
Weight of Compacted Soil ($W=W_2-W_1$) g	1478	1591	1714	1818	1771
Bulk density (Y=W/V) g/cm^3	1.4785	1.5913	1.7143	1.8185	1.771
Water Content (w)%	8.64	11.83	13.76	15.1	17.32
Dry Density ($Y_d=Y/1+w$)	1.361	1.423	1.507	1.58	1.51

Table 1.13 Standard proctor test result of BCS with 30% olivine mix.

Experiment no.	**1**	**2**	**3**	**4**	**5**
Mold Vol. (V) cm^3	1000	1000	1000	1000	1000
Mold Wt. (W_1) g	3677	3677	3677	3677	3677
Mold Wt. + Compacted Soil (W_2) g	5395	5503	5604	5715	5686
Weight of Compacted Soil ($W=W_2-W_1$) g	1718	1826	1927	2038	2009
Bulk density (Y=W/V) g/cm^3	1.7184	1.8266	1.927	2.0385	2.009
Water Content (w)%	6.67	8.73	10.23	12.01	14.02
Dry Density (Y_d=Y/1+w)	1.611	1.680	1.749	1.82	1.762

Table 1.14 Standard proctor test result of BCS with 35% olivine mix.

Experiment no.	**1**	**2**	**3**	**4**	**5**
Mold Vol. (V) cm^3	1000	1000	1000	1000	1000
Mold Wt. (W_1) g	3677	3677	3677	3677	3677
Mold Wt. + Compacted Soil (W_2) g	5427	5560	5661	5790	5745
Weight of Compacted Soil ($W=W_2-W_1$) g	1750	1883	1984	2113	2068
Bulk density (Y=W/V) g/cm^3	1.7508	1.8836	1.9848	2.113	2.068
Water Content (w)%	6.376	8.449	9.66	11.25	13.026
Dry Density (Yd=Y/1+w)	1.646	1.737	1.81	1.90	1.83

Table 1.15 Standard proctor test result of BCS with 40% olivine mix.

Experiment no.	1	2	3	4	5
Mold Vol. (V) cm^3	1000	1000	1000	1000	1000
Mold Wt. (W_1) m	3677	3677	3677	3677	3677
Mold Wt.+ Compacted Soil (W_2) g	5421	5525	5641	5740	5721
Weight of Compacted Soil ($W=W_2-W_1$) g	1744	1848	1964	2063	2044
Bulk density (Y=W/V) g/cm^3	1.7448	1.8482	1.9643	2.063	2.0445
Water Content (w)%	6.462	8.349	10.112	12.12	14.027
Dry Density (Y_d=Y/1+w)	1.639	1.706	1.784	1.84	1.793

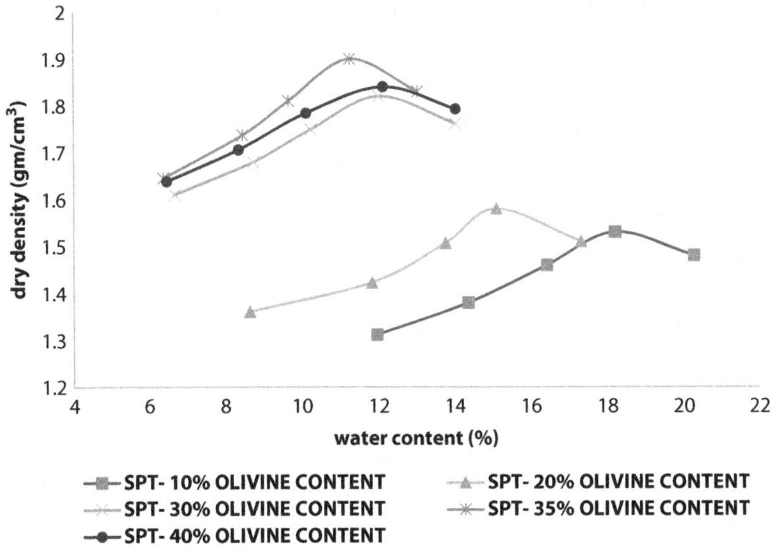

Figure 1.5 Dry density variation utilizing olivine-treated BCS for standard proctor compaction.

Using a typical compaction effort, Figure 1.5 illustrates how the olivine content affects the MDD and OMC of BCS. Up to a concentration of 35%, MDD typically increased as olivine content increased; after that, it considerably decreased. MDD values for traditional compaction efforts are 1.66–1.87 g/cm^3. On the other hand, when the olivine percentage reached 35 percent, the OMC usually climbed and then gradually reduced as the olivine level rose. From 20.27 to 11.96 percent, the OMC decreased from the lowest to the maximum compactive energy.

1.3.4 Effect of Varying Olivine Percentages on Modified Proctor Test Result of Olivine Treated BCS

The effect of inclusion of various percentage of olivine on dry density of soil mass of BCS is examined by conducting modified standard proctor test. The different percentage of olivine such as 10%, 20%, 30%, 35% and 40% of soil mass is included in BCS and compared with untreated BCS. Tables 1.16–1.20 shows the dry density BCS with different percentage of olivine mix (10%, 20%, 30%, 35% and 40% of soil mass).

Table 1.16 Modified proctor test result of BCS with 10% olivine mix.

Experiment no.	**1**	**2**	**3**	**4**	**5**
Mold Vol. (V) cm^3	1000	1000	1000	1000	1000
Mold Wt. (W_1) g	3677	3677	3677	3677	3677
Mold Wt. + Compacted Soil (W_2) g	5224	5331	5462	5543	5537
Weight of Compacted Soil ($W=W_2-W_1$) g	1547	1654	1785	1866	1860
Bulk density (Y=W/V) g/cm^3	1.5477	1.6549	1.7853	1.8667	1.8606
Water Content (w)%	10.87	12.43	14.37	16.02	18.36
Dry Density (Y_d=Y/1+w)	1.369	1.472	1.561	1.609	1.572

Table 1.17 Modified proctor test result of BCS with 20% olivine mix.

Experiment no.	**1**	**2**	**3**	**4**	**5**
Mold Vol. (V) cm^3	1000	1000	1000	1000	1000
Mold Wt. (W_1) gm	3677	3677	3677	3677	3677
Mold Wt. + Compacted Soil (W_2) g	5258	5362	5460	5539	5537
Weight of Compacted Soil ($W=W_2-W_1$) g	1581	1685	1783	1862	1860
Bulk density (Y=W/V) g/cm^3	1.5817	1.6853	1.7834	1.8627	1.8607
Water Content (w)%	7.53	9.72	11.26	13.12	15.43
Dry Density (Y_d=Y/1+w)	1.471	1.536	1.603	1.647	1.612

Table 1.18 Modified proctor test result of BCS with 30% olivine mix.

Experiment no.	**1**	**2**	**3**	**4**	**5**
Mold Vol. (V) cm^3	1000	1000	1000	1000	1000
Mold Wt. (W1) g	3677	3677	3677	3677	3677
Mold Wt. + Compacted Soil (W_2) g	5393	5486	5592	5725	5666
Weight of Compacted Soil ($W=W_2-W_1$) g	1716	1809	1915	2048	1989
Bulk density (Y=W/V) g/cm^3	1.7162	1.8094	1.9152	2.0488	1.9898
Water Content (w)%	5.16	6.252	8.76	10.81	12.36
Dry Density (Y_d=Y/1+w)	1.632	1.703	1.761	1.849	1.771

Table 1.19 Modified proctor test result of BCS with 35% olivine mix.

Experiment no.	**1**	**2**	**3**	**4**	**5**
Mold Vol. (V) cm^3	1000	1000	1000	1000	1000
Mold Wt. (W_1) g	3677	3677	3677	3677	3677
Mold Wt. + Compacted Soil (W_2) g	5428	5542	5676	5775	5764
Weight of Compacted Soil ($W=W_2-W_1$) g	1751	1865	1999	2098	2087
Bulk density (Y=W/V) g/cm^3	1.7510	1.8653	1.9991	2.0988	2.0876
Water Content (w%)	5.36	6.23	7.31	9.26	11.52
Dry Density (Y_d=Y/1+w)	1.662	1.756	1.863	1.921	1.872

Table 1.20 Modified proctor test result of BCS with 40% olivine mix.

Experiment no.	**1**	**2**	**3**	**4**	**5**
Mold Vol. (V) cm^3	1000	1000	1000	1000	1000
Mold Wt. (W_1) g	3677	3677	3677	3677	3677
Mold Wt. + Compacted Soil (W_2) g	5421	5508	5600	5708	5693
Weight of Compacted Soil ($W=W_2-W_1$) g	1744	1831	1923	2031	2016
Bulk density (Y=W/V) g/cm^3	1.7443	1.8318	1.9239	2.0316	2.0164
Water Content (w)%	5.21	6.123	7.361	9.823	11.037
Dry Density (Y_d=Y/1+w)	1.658	1.726	1.792	1.853	1.816

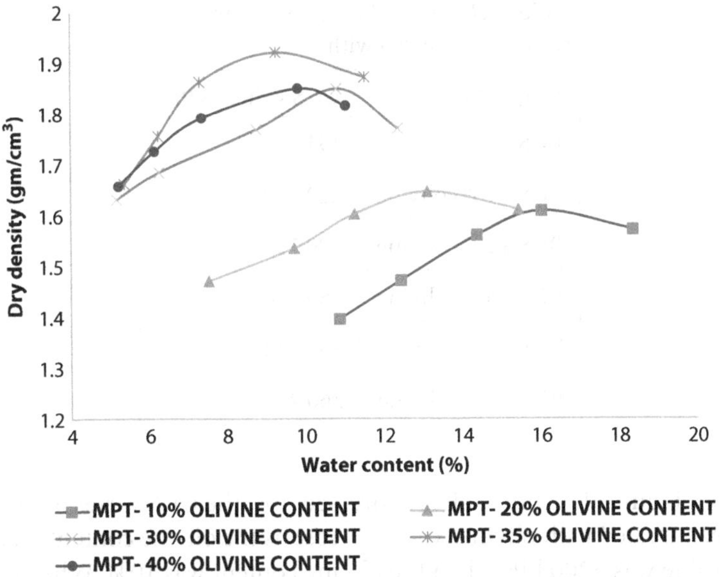

Figure 1.6 Modified proctor test for monitoring dry density change with olivine-fig.

Using changed compaction effort, Figure 1.6 illustrates how the olivine content affects the MDD and OMC of BCS. MDD generally rose with increasing olivine content up to 35% concentration, after which it significantly declined. The MDD values for the modified compaction effort vary from 1.64 to 1.83 g/cm^3. On the other hand, OMC generally rose when the olivine percentage hit 35% and then somewhat declined as olivine levels rose. From the lowest compaction energy to the greatest, the OMC decreased from 18.36 percent to 9.26 percent.

1.3.5 Effect of Varying Olivine Percentages on UCS on Olivine Treated BCS

The effect of the inclusion of various percentage of olivine on unconfined compressive strength soil mass of BCS is examined by conducting standard proctor test. The different percentage of olivine such as 10%, 20%, 30%, 35% and 40% of soil mass is included in BCS and compared with untreated BCS. Table 1.21 shows the UCS value according to varying olivine (%) with BCS.

Table 1.21 UCS value according to varying olivine (%) with BCS.

Proportion	UCS value (KPa)
BCS	191.5
BCS +10% olivine	262.532
BCS +20% olivine	290.732
BCS +30% olivine	359.236
BCS +35% olivine	362.843
BCS +40% olivine	360.234

From Figures 1.7 and 1.8, it is addressed how the UCS varies in BCS that has been treated with olivine. The UCS rose as the olivine content rose. The UCS value was >200 kPa when the olivine content was 10%. When olivine was added further, up to 35 percent, the UCS value climbed to a maximum of 362.843 kPa. The insoluble M-A-H and M-S-H phases produced by the hydration reaction were primarily responsible for the strength increase

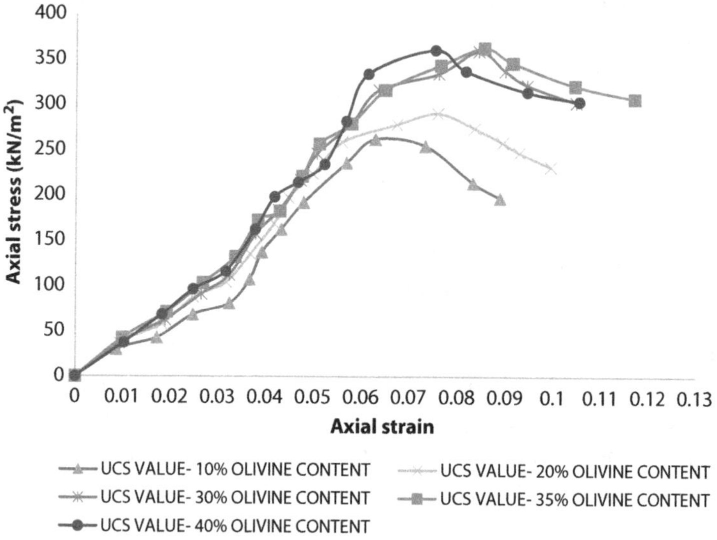

Figure 1.7 The behavior of treated BCS under stress and strain with different olivine (%).

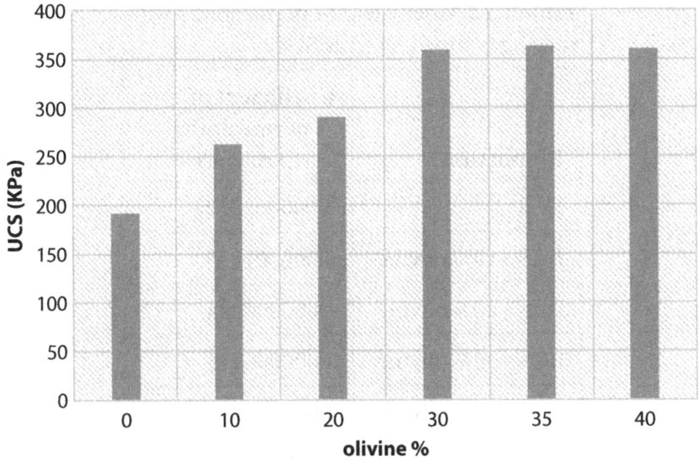

Figure 1.8 Change of UCS value of BCS with varying olivine (%).

with increased olivine content. The clay particles are joined together by these hydration products to form a new, stronger soil matrix, raising the treated sample's UCS. The UCS decreased when olivine was added in amounts greater than 35%. Hydration processes were produced and particle bonding took place for the BCS that had been treated and had an olivine content of up to 35%. The hydration reaction, however, is prevented for olivine contents more than 35%; moreover, the excess olivine particles serve as unbounded particles that degrade the material as a whole.

1.3.6 Effect of Varying Olivine Percentages on Hydraulic Conductivity on Olivine Treated BCS

The effect of the inclusion of various percentage of olivine on hydraulic conductivity of BCS is examined by conducting standard proctor test. The different percentage of olivine such as 10%, 20%, 30%, 35%, and 40% of soil mass is included in BCS and compared with the untreated BCS. Table 1.22 shows the coefficient of permeability with varying olivine (%) with BCS.

Since less holes resulted from increased compaction energy, the hydraulic conductivity generally decreased as the energy level increased. Additionally, the results demonstrate that, as illustrated in Figure 1.9, the hydraulic conductivity drops to a maximum of 35 percent of the olivine level before slightly increasing. As the amount of olivine in the mixture increased, M-S-A and M-A-H developed, blocking the pores and

Table 1.22 Coefficient of permeability with varying olivine (%) with BCS.

Proportion	Coefficient of permeability (k) in cm/s
BCS	.0000001532
BCS +10% olivine	.0000001135
BCS +20% olivine	.0000000824
BCS +30% olivine	.0000000213
BCS +35% olivine	.0000000078
BCS +40% olivine	.0000000096

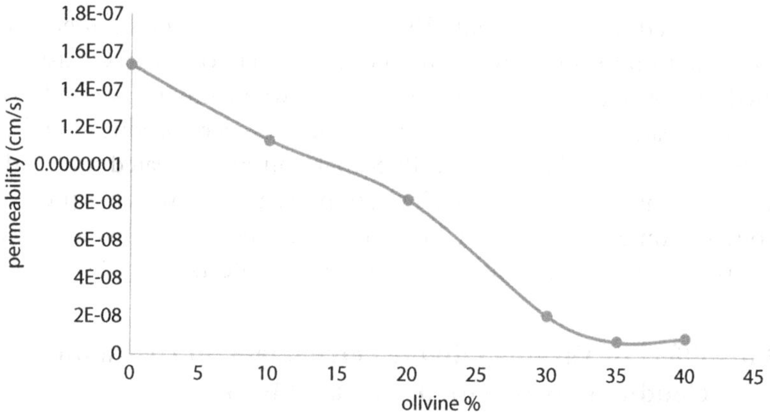

Figure 1.9 Variation of hydraulic conductivity of BCS with varying olivine (%).

decreasing hydraulic conductivity. However, extra olivine could alter the soil matrix and encourage flocculation, which would explain the rise in hydraulic conductivity at 35 percent olivine content.

1.4 Conclusion

In this work, a number of standard proctor test, modified proctor test, UCS, and hydraulic conductivity tests were administered, and the test results are shown in the form of tables and graphs. The conclusions reached in the light of these findings and the related discussions are outlined below:

1. After adding olivine at varied proportions of 0 percent to 40 percent, OMC value falls and MDD value rises. In standard proctor compaction test, the MDD value rises from 1.62 gm/cm^3 to 1.90 gm/cm^3, and the OMC falls from 19.1 percent to 11.25 percent. Additionally, the MDD value in modified proctor compaction test rises from 1.632 gm/cm^3 to 1.921 gm/cm^3, whereas the OMC value falls from 18.6% to 9.26%. The MDD value rises as the olivine content is mixed up to 35 percent, and then falls when more olivine is added.
2. After being treated with olivine up to 35 percent, virgin BCS with a UCS value of 191.5 kN/m^2 becomes 362.84 kN/m^2. Increasing the amount of olivine lowers the UCS value. The allowed UCS value > 200 kN/m^2 is reached after treatment with olivine; it works well in planned landfills as a bottom liner material.
3. The hydraulic conductivity of the treated BCS frequently decreased when the olivine content increased. The bottom liner materials meet the allowable hydraulic conductivity of 10^{-6} cm/sec at changing olivine content up to 35 percent.
4. As the leachate saturated the soil, the plasticity properties decreased, and the olivine concentration rose, indicating an improvement in the soil's index quality. The change is visible in clay properties for soil treated with 35 percent olivine, from highly plastic clay for untreated soil to intermediately plastic clay.

In summary, the results show that BCS treated with up to 40% olivine concentration can be employed as bottom cover material in landfills under consideration, with the best results coming from 35% olivine.

References

1. Vivek, Dutta, R.K., Parti, R., Effect of chemical treatment of the coir geotextiles on the interface properties of sand/clay-coir geotextile interface. *J. Inst. Eng. India Ser. A*, 100, 357–365, 2019.
2. Mohanta, N.R. and Samantaray, S., Study of Combined Effect of Metakaolin and Steel Fiber on Mechanical Properties of Concrete. *Pertanika J. Sci.*

Technol., 27, 3, 1381–1396, 2019, http://journals-jd.upm.edu.my/resources/files/ Pertanika%20PAPERS/JST%.
3. Vivek, and Dutta, R.K., Bearing ratio behavior of sand overlying clay with treated coir geotextiles at the interface. *J. Nat. Fibers*, 19, 14, 7534–7541, 2022.
4. Vivek, Effects of cyclic loading on sand overlaying clay model of unpaved roads reinforced with untreated/treated coir geotextiles. *J. Text. Inst.*, 155, 1694–1700, 2023.
5. Jaswal, P., Vivek, Sinha, S.K., Improvement in the performance of two layered model pavement with treated coir geotextile at the interface. *J. Ind. Text.*, 52, 15280837221114161, 2022.
6. Vivek, Shafi Mir, M., Sehgal, R., Studies of Modulus of Resilience on Unpaved Roads Reinforced with Untreated/Treated Coir Geotextiles. *J. Nat. Fibers*, 19, 16, 13563–13573, 2022.
7. Vivek, Shafi Mir, M., Sehgal, R., Study on bearing capacity of unpaved roads reinforced with coir geotextiles using finite element method (fem). *J. Nat. Fibers*, 19, 15, 11735–11748, 2022.
8. Jaswal, P. and Vivek, Experimental study on monotonic behaviour of two-layered unpaved road model reinforced with treated coir geotextiles. *Int. J. Pavement Res. Technol.*, 17, 1377–1388, 2023.
9. Singh, P., Dash, H.M., Samantaray, S., Effect of silica fume on engineering properties of expansive soil. *Mater. Today Proc.*, 33, 5035–5040, 2020, https://doi.org/10.1016/j.matpr.2020.02.839.
10. Kumar, M.H., Mohanta, N., Samantaray, S., Kumar, N.M., Combined effect of waste glass powder and recycled steel fibers on mechanical behavior of concrete. *SN Appl. Sci.*, 3, 1–18, 2021a, https://doi.org/10.1007/s42452-021-04422-w.
11. Kumar, M.H., Macharyulu, I.S., Ray, T., Mohanta, N.R., Jain, M., Samantaray, S., Sahoo, A., Effect of water absorption and curing period on strength and porosity of triple blended concrete. *Mater. Today Proc.*, 43, 2162–2169, 2021b, https://doi.org/10.1016/j.matpr.2020.12.092.
12. Kumar, M.H., Saikrishnamacharyulu, I., Kumar, U., Jena, S., Mohanta, N.R., Samantaray, S., Coupling effect of fly ash, metakaoline and different types of steel fibers on mechanical performance of concrete. *AIP Conf. Proc.*, 2417, 24–31, 2021c, https://doi.org/10.1063/5.0072573.
13. Ray, T., Mohanta, N.R., Kumar, M.H., Samantaray, S., Study of effect of temperature on behavior of alkali activated slag concrete. *Mater. Today Proc.*, 43, 1352–1357, 2021, https://doi.org/10.1016/j.matpr.2020.09.169.
14. Kumar, M.H., Saikrishnamacharyulu, I., Mohanta, N.R., Ashutosh, A., Mishra, P., Samantaray, S., Mechanical behaviour of high strength concrete modified with triple blend of fly ash, silica fume and steel fibers. *Mater. Today Proc.*, 65, 933–942, 2022a, https://doi.org/10.1016/j.matpr.2022.03.528.

15. Kumar, M.H., Mohanta, N.R., Patel, N., Samantaray, S., Reddy, S.V.B., Influence of Fly ash and Metakaoline on the crack resistance and shrinkage property of concrete. *Iran. J. Sci. Technol. Trans. Civ. Eng.*, 46, 2011–2026, 2022b, https://doi.org/10.1007/s40996-021-00748-y.
16. Emmanuel, E., Anggraini, V., Syamsir, A., Asadi, A., CO2 carbonation of olivine-admixed marine clay: suitability for bottom liner application. *Appl. Sci.*, 11, 7, 3164, 2021, https://doi.org/10.3390/app11073164.
17. Cokca, E., Use of class c fly ashes for the stabilizationof an expansive soil. *J. Geotech. Geoenviron. Eng.*, 127, 7, 568–573, 2001.
18. Phani Kumar, B.R. and Sharma, R.S., Effect of fly ash on engineering properties of expansive soils. *J. Geotech. Geoenviron. Eng.*, 130, 7, 764–767, 2004.
19. Satyanarayana, K.G., *Clay surfaces: fundamentals and applications*, Elsevier, Amsterdam, Netherlands, 2004.
20. Sivapullaiah, P.V., Sridharan, A., Bhaskar Raju, K.V., Role of amount and type of clay in the lime stabilization of soils. *Proc. Inst. Civ. Eng.-Ground Improv.*, 4, 1, 37–45, 2000.
21. Baytar, A.Ö., Effects of fly ash and desulphogypsum on the geotechnical properties of Çayırhan soil (Master's thesis), Middle East Technical University, 2005.
22. Brooks, R.M., Soil stabilization with flyash and rice husk ash, 2009.
23. Jakka, R.S., Datta, M., Ramana, G.V., Liquefaction behaviour of loose and compacted pond ash. *Soil Dyn. Earthquake Eng.*, 30, 7, 580–590, 2010.
24. Singh, S.P. and Sharan, A., Strength characteristics of compacted pond ash. *Geomech. Geoeng.*, 9, 1, 9–17, 2014.
25. Hervig, R.L. and Smith, J.V., Temperature-dependent distribution of Cr between olivine and pyroxenes in lherzolite xenoliths. *Contrib. Mineral. Petrol.*, 81, 184–189, 1982.
26. Richmond, N.C. and Brodholt, J.P., Incorporation of Fe3+ into forsterite and wadsleyite. *Am. Mineral.*, 85, 9, 1155–1158, 2000.
27. Papike, J.J., Karner, J.M., Shearer, C.K., Comparative planetary mineralogy: Valence state partitioning of Cr, Fe, Ti, and V among crystallographic sites in olivine, pyroxene, and spinel from planetary basalts. *Am. Mineral.*, 90, 2–3, 277–290, 2005.
28. IS 2720-4 (1985): Methods of test for soils, Part 4: Grain size analysis, Bureau of Indian Standards, 1985.
29. IS 2720-3-1 (1980): Methods of test for soils, Part 3: Determination of specific gravity, Section 1: Fine grained soils, Bureau of Indian Standards, 1980.
30. IS 2720-5 (1985): Methods of test for soils, Part 5: Determination of liquid and plastic limit, Bureau of Indian Standards, 1985.
31. IS 2720-7 (1980): Methods of test for soils, Part 7: Determination of water content-dry density relation using light compaction, Bureau of Indian Standards, 1980.

32. IS 2720-8 (1983): Methods of test for soils, Part 8: Determination of water content-dry density relation using heavy compaction, Bureau of Indian Standards, 1983.
33. IS 2720-10 (1991): Methods of test for soils, Part 10: Determination of unconfined compressive strength, Bureau of Indian Standards, 1991.
34. IS 2720-17 (1986): Methods of test for soils, Part 17: Laboratory determination of permeability, Bureau of Indian Standards, 1986.

2

Strength and Microstructural Characterization of Industrial Waste Amended Dispersive Soil

Samaptika Mohanty[1]*, Subhashree Samantasinghar[2], Amiya Ranjan Pandit[3] and Sandeep Samantaray[4]

[1]*Adharshila Architects and Engineers Private Limited, Rourkela, India*
[2]*Department of Civil Engineering, Odisha University of Technology and Research, Bhubaneswar, India*
[3]*University College Dublin, Dublin, Ireland*
[4]*Department of Civil Engineering, National Institute of Technology Srinagar, Hazratbal, Srinagar, India*

Abstract

Dispersive soil (DS) is a unique kind of clay that easily breaks down into smaller particles and floats in water. This makes it a great material for causing erosion and pipe damage to road slopes, channels, and dams. Because dispersive soils have a higher concentration of sodium content and are deflocculated when exposed to running water, they are more prone to erosion. The dispersive soil cannot be visually distinguished from other normal clay soils. Furthermore, visual classifications like Atterberg's limits, and particle size distribution don't deliver enough indication to distinguish between DS and other normal clayey soil. Dispersive soil was once considered to be non-erodible clayey soils. However, it was recently discovered that some clayey soils that contain montmorillonite and have medium-to-low plasticity (CL-CH and CL) also tend to erode. Chemistry and mineralogy of clays, and dissolved salts presence in the pore fluid, all affect soil dispersion leading to erosion. Various amounts of cement clinker (CC) (5 to 30%), blast-furnace slag in ground granulated form (GGBS) (5 to 15%), and various proportions of fly ash (FA) (5 to 20%) have been used in this investigation to stabilize the DS. To assess the strength of stabilized soils, specimens containing varying amounts of DS,

**Corresponding author*: mohantysamaptika4@gmail.com

Sandeep Samantaray, Subhashree Samantasinghar, Soumendra K. Mohanty, Vivek, Abinash Sahoo and Deba Prakash Satapathy (eds.) Waste Utilization in Geotechnical Practice, (31–72)

CC, FA and GGBS were collected. By mixing additives in different amounts, the results for unconfined compressive strength (UCS) can be substantially enhanced. The optimum mix proportion of 30% CC, 20% FA, and 15% GGBS in dispersive soils has been achieved from UCS test results. According to the study, a combined mix of CC, GGBS, and FA is more effective in improving strength than an alone mix. The formation of hydrated particles and changes in microstructure are shown by X-ray diffraction tests (XRD) and scanning electron microscopy (SEM). Ettringite, CASH, and CSH are important for enhancing strength due to the interaction between the additives and soil.

Keywords: DS, CC, GGBS, fly ash, UCS, microstructure

2.1 Introduction

1. Dispersion may occur in alluvial soil and clayey soils produced from mud deposits, when they are transported into the sea environment. These types of soils are typically found in beds of lake and submerged plain deposits. The electrochemical forces among particles in dispersive clays are unequal. This imbalance causes the small soil particles in dispersive soils (DS) to repel one another rather than clumping together. Farms with sinkholes and an upper layer of dispersive soils will lose their ability to use the land due to soil erosion. The ease with which particles are dispersed is influenced by plate shaped particles. The problematic soils have a lot of potential problems that need to be stabilized before application [1–4]. DS are not appropriate as fill/foundation material in the construction of the structure. Various methods can be used for reduction or elimination, depending on the dispersive mechanism. To assure engineering safety, dispersity should be eliminated. There are several places in the world where this kind of soil may be found, such as in Thailand, Greece, South Africa, Latin America, Australia, and India. Due to the problem of erosion, several earth dams and hydraulic structures along the roads had been damaged or collapsed. These soils were not utilized to build embankments in the past, which made the work more challenging and expensive [5]. Nevertheless, nowadays, the soils have been used in construction after being chemically treated [6, 7]. Over the last few decades, usage of lime and fly ash to stabilize dispersive soils has been investigated by a number of researchers [8–15]. Chemical stabilization of clay soil using lime, cement, or FA and other wastes are utilized in geotechnical engineering to enhance its workability and mechanical properties [16–22]. They did [23] double hydrometer test, often referred to as Soil Conservation Service Laboratory

Dispersion Test, one of the primary techniques for assessing dispersion in clay soils. A cylindrical dispersion test was developed [24] to investigate soil erosion and dispersion from cracked surfaces in water and this test is an extension of the crumb test. They [25] have observed that class C fly ash is more efficient alone to stabilize the dispersive soil than Class F fly ash. Chemical stabilizers such as GGBS, cement, and lime can greatly enhance the geotechnical properties of dispersive soils [26–28]. They [29, 30] have examined the strength and swell potential of dispersive soil stabilized with gypsum. The study revealed that addition of 5% gypsum considerably enhanced the strength of dispersive soil. To stabilize dispersive soil, [31] studied dispersive soil characterization with silica fume and hydrated lime. A substantial decrease in soil's dispersing characteristics has been achieved by adding 5% lime, 2% lime and 12% silica fume. They [32] evaluated the Portland cement as a geotechnical stabilizing agent for dispersive soils. A strength increases of 75% was observed in the specimen treated with 2% cement and cured for 90 days. After treatment for 7 to 14 days, the soil became non-dispersive. The soil became non-dispersive after treatment with 2% pozzolanic Portland cement (PPC) and cured for 7 to 14 days. When mixed with 2% cement and treated for 14 days, the plasticity index decreased from 17% to 11%. The most effective proportion for reducing mass loss, dispersion potential, and settling of dispersive soil was found to be 2% polyvinyl acetate (PVAc) [33]. The effect of GGBS on the inhibition of 'flower-like' ettringite formation in lime-stabilized kaolinite in response to a 4.2% Na_2SO_4 solution was examined [34]. They [35] have studied the increase of the strength of autoclaved PC quartz mix, where PC was partially substituted by alumina-silica-rich fired clay brick waste fines, confirmed supplementary cementing materials (SCM), and reactive magnesia. The improvement of strength was because of the formation of CSH including tobermorite. They [36] have studied gypsum-induced strength behavior of flash-lime. A compacted clay matrix is created by alumino-silicate dissolution and calcium silicate hydrate i.e., CSH, calcium aluminum silicate hydrate i.e., CASH, and calcium aluminum silicate hydroxide hydrate i.e., CASHH compound occurrence. Further compacted clay structure was revealed with the addition of 1% gypsum because ettringite needle filled in voids, thereby accelerating early strength and speeding up the early strength in soil-fly ash-lime samples. They [37] examined the impact of time-dependent mineralogical variations on UCS behavior of lime-GGBS treated naturally occurring acid sulfate soils (ASS) with varying pyrite amounts. For in the treatment of ASS, CSH which are cementitious compounds and ettringite were primary initiator of enhancement of mechanical properties. According to [38], adding 2% lime improves the strength characteristics. The cementitious compounds CSH showed in the lime stabilized GGBS-fly ash specimens. GGBS and lime mixture produced cementitious characteristics and a minor quantity of ettringite, according to research [39]. They [40] have investigated lithomargic clay stabilization using GGBS and cement. The improvement of strength was mainly an outcome of hydration reaction products formations. They [41] have analyzed, however, that the fly ash and lime-stabilized sediments improve the strength and modulus; in the case of cement-stabilized sediments, it weakens the mechanical characteristics. The increase in strength was caused by the pozzolanic reaction with 3% Plaster of Paris kiln dust (PKD) with dispersive soil, which occurred after a 28-day curing period. SEM and XRD were utilized to verify the development of cementitious products (CSH, CAH) [42]. Based on the current research, no one used CC alone and also combined a mixture of all, but rather a combination of CC, FA, and GGBS for stabilization of dispersive soil. The present study aims to assess the performance of soils with dispersive nature when treated with slag, fly ash, and cement clinker. Initially, dispersive soil is identified by performing a cylindrical dispersion test, double hydrometer test, and turbidity test. After that, the characterization of soil for geotechnical properties was examined. The blended impact of the industrial waste for mechanical, chemical and compaction characteristics has also been investigated in this study. The formation of reaction products from hydration products in stabilized dispersed soils is observed and confirmed by XRD analysis. The micro structural modification in the modified soil is studied by SEM analysis.

2.2 Materials and Methodology

2.2.1 Materials Used

Sodium bentonite has been used as a dispersive soil and FA, GGBS, and CC stabilizing agents are used during this investigation. From Cuttack, Odisha the dispersive soil was collected. From RSP Rourkela, GGBS and FA were collected. From Shiva Cement, Rourkela, Odisha CC was collected. After the CC had been received, the CC was further processed by crushing in a machine and then fine forms are attained through a ball mill after it had been received.

2.2.2 Test Procedure

Soil dispersiveness was evaluated through double hydrometer test following ASTM D4221-99 (2005) [43] and the cylindrical dispersion test. On the soil specimen, a parallel hydrometer test is conducted without inclusion of a chemical dispersant. If the dispersion value exceeds 50, then the soil has high dispersive character. Firstly, all the raw materials are oven-dried at 110°C. Secondly, the soil is mixed properly with GGBS, FA, and CC for mixing purposes until they achieve the optimum range To make the mixing and compaction process more efficient, water is added as needed. Compaction test shall be carried out directly after mixing. On untreated and stabilized soil, the light compaction tests are conducted in accordance with ASTM D698 (2007) [44]. These tests are done to measure the maximum dry density (MDD) and optimum moisture content (OMC). Unconfined compressive strength test in accordance with 2166 M-16 (2016) [45] is performed in this investigation to measure the strength characteristics of soil. Required amount of soil must have been oven-dried for 24 hours before the samples can be prepared. According to field requirements, the MDD and OMC measurements were utilized for preparing soil specimens, which may define *in-situ* moisture content. A 45 mm diameter steel rod was used for compaction to eliminate air pockets and enhance homogeneity of sample. To prepare the soil specimens, various quantities of GGBS, CC and FA are mixed with the dispersive soil. When establishing an optimum number of additives in the process of developing soils, the strength of soil should be taken into consideration as an essential parameter. The specimen was wrapped in polyethylene film and stored at room temperature for 0, 7, 14, 28 and 90 days. The specimens are removed from the polyethylene wrap after completion of curing period, then the test is carried out. The sample shall be orientated between the lower and the platen at the upper part of a mechanical loading frame without confinement or any lateral support after completion of an interaction period. A 1% per minute axial strain is produced by the loading when it is applied in this manner. Each composition was examined using three samples. Furthermore, to avoid the effect of any probable errors, the results have been averaged. Cylindrical dispersion test methodologies were adopted with the recommendations made [24]. The water quality in which the soil samples were submerged is monitored in this experiment, and the results show a colloidal suspension that may be categorized as either dispersive or non-dispersive. The most important component in determining each pavement layer's thickness when designing a pavement is CBR testing. Samples that had cured for 7, 28, and 90 days were removed and a powder having particle size minus 75 μm were obtained by crushing using mortar and pestle before the micro-analysis, i.e., SEM and XRD testing. The mineralogical composition of stabilized and untreated dispersive soil has been determined through the XRD test. Using a Ni filter and a copper target, the Rich-Siefert X-ray diffractometer was used to perform XRD test. With 0.05°/sec step size, the scan rate was measured at a 2θ angle that ranged from 5° to 60° degree/sec. The reaction products have been observed after the peaks have been analyzed using Philips X' Pert High Score. Untreated and stabilized soil samples were subjected to SEM examination using a model JEOL-JSM-6480LV attached with an energy-propagation X-ray spectroscopy (EDX) analysis system. Chemical characteristics of the chemistry in the surfaces of the materials were examined using EDX.

2.3 Results and Discussion

2.3.1 Details of Tests Conducted for Identification of Soil

2.3.1.1 Double Hydrometer Test (DHT)

Basically, two hydrometer tests are carried out using this method: first is the conventional hydrometer test; and the second one is the hydrometer test conducted without the use of a chemical dispersant (sodium hexameta phosphate). As shown in Table 2.1, when the value exceeds 50, it is considered to be highly dispersive. The following equation gives the % dispersion (ASTM D4221-99, 2005) [43]:

$$\%\,Dispersion = \frac{\%\,clay\left(0.005\,mm\right) without\ dispersing\ agent}{\%\,clay\left(0.005mm\right) with\ a\ dispersing\ agent} \times 100 \tag{2.1}$$

Table 2.1 Percentage of dispersion from DHT (Volk 1938).

Percent of dispersion	Degree of dispersion
<30%	Non-dispersion
30% - 50%	Intermediate dispersion
>50%	Highly dispersion

Figure 2.1 Equipment used for determination DHT in laboratory.

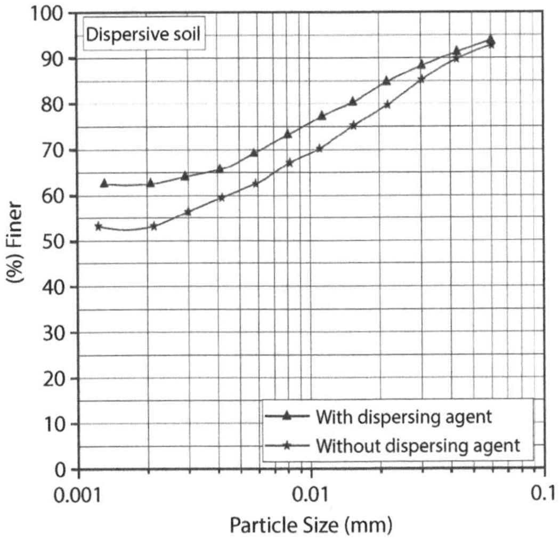

Figure 2.2 Curves obtained from double hydrometer test.

Figure 2.1 depicts the setup of a DHT. As shown in Figure 2.2, a dispersion ratio of soil is found to be 90.66%, detected from the double hydrometer test and is considered to be highly dispersive soil. Similar observations have been found [16, 45, 46].

2.3.1.2 Crumb Test

When exposed to water, soils are prone to deflocculating and can be determined by performing a simple physical test known as the crumb test. The potential erosion properties of clayey soils can be determined successfully using qualitative results provided by this test. This test was done on the fresh mix according to ASTM D6572-13e2 (2013) [47] code. A white porcelain vessel filled with 250 ml of distilled water shall be used to place a cubical sample with a 15 mm equal volume. Colloidal particles segregate and form a suspension as an outcome of reaction amid soil and water. Observations shall be recorded at specified intervals for classification. The dispersive soil may be graded into 4 grades based on the cloudiness in turbidity measurement. Grade 1: Soil may crumble, but the water is not cloudy and it is classified as non-dispersive. Within the first hour, each particle settles. Grade 2: A barely detectable colloidal suspension turbid water has developed in the nearby area of the soil, and it is categorized as an intermediate dispersive. Grade 3: A suspension with cloud is clearly discernible surrounding the surface of soil and it is classified as dispersive. Grade 4: Over the entire surface, a thick cloud of colloidal suspension is visible, which is categorized as highly dispersive. After immersion, it is clearly shown in Figure 2.3

Figure 2.3 Crumb test results of dispersive soil.

that the soil is getting dispersed within 5 minutes, and cloud suspension is detected on the complete surface. The corresponding soil is highly dispersive, as divulged by the test results (Grade-4).

2.3.1.3 Cylindrical Dispersion Test

By using a cylindrical dispersion test, the resistance to erosion can be measured in terms of true coherence, and the impact of the pore water of dispersion properties can also be evaluated. The cylinder dispersion test submerges a saturated sample in water to investigate soil properties at zero effective stress. Cylindrical specimens have been prepared according to the length, to diameter ratio considered as 2, the same as required for the UCS test. The prepared specimen was first wrapped in a polythene film and cured at ambient room temperature and after achievement of the curing period, the test was performed. Figure 2.4 shows the cylindrical

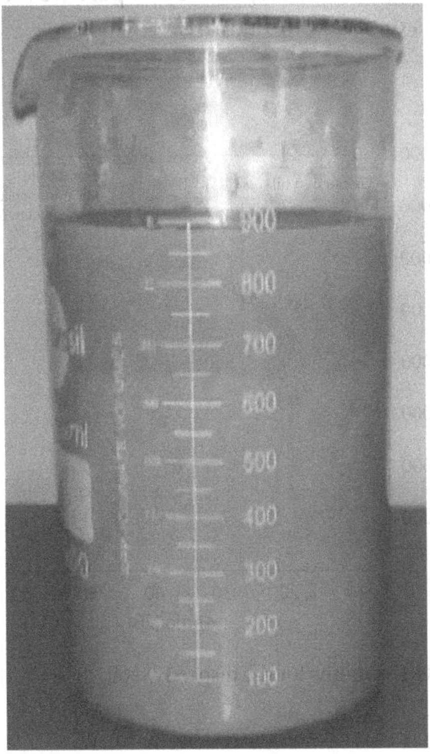

Figure 2.4 Cylindrical dispersion test of dispersive soil.

dispersion test of the dispersive soil. It has been noted that soil collapses in the presence of water with no resistance, and emphasizes substantial highly dispersion behavior.

2.3.2 Raw Material Characterization

Details of the material chosen for the present investigation are explained below. Both the GGBS and CC used in this experimental study were processed in the form of powder in a ball mill to make it finer than 150μm; then, it was utilized. To eliminate the moisture, the materials were dried in an oven at 110°C.

2.3.2.1 Dispersive Soil

The dispersive soil XRD analysis, as shown in Figure 2.5, reveals that the crystalline phase consists of quartz, montmorillonite, and kaolinite. The flaky and angular shape of the dispersive soil particle may be detected from the SEM images, as depicted in Figure 2.6.

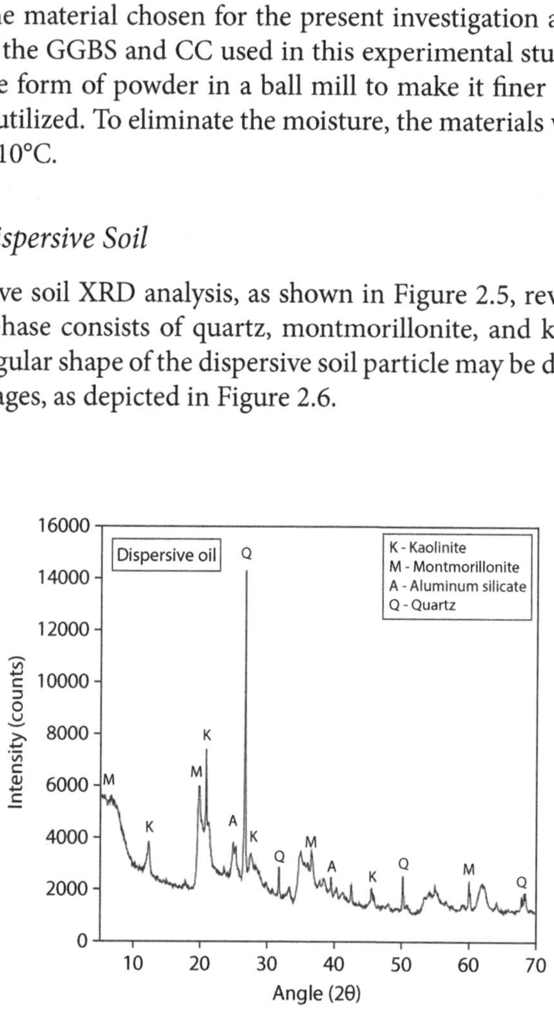

Figure 2.5 Result of XRD patterns for the dispersive soil.

Figure 2.6 SEM images of the dispersive soil.

2.3.2.2 *Fly Ash*

Fly ash's cementitious properties and chemical reaction to form cementitious material in the presence of water can increase the soil's incompressibility and strength. The gases that are produced by coal-fired furnaces, mostly from a thermal power plant, are the source of fly ash as a waste material. According to the XRD analysis reported in Figure 2.7, the fly ash mainly includes glassy in addition to hematite's' crystalline phase, mullite and quartz. The SEM images (Figure 2.8) show that fly ash particles are spherical in shape. The fine-grained fly ash contains 73% finer particles

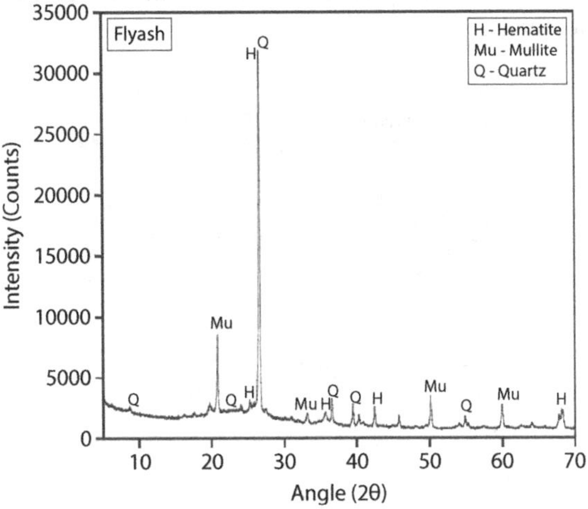

Figure 2.7 XRD patterns of the fly ash.

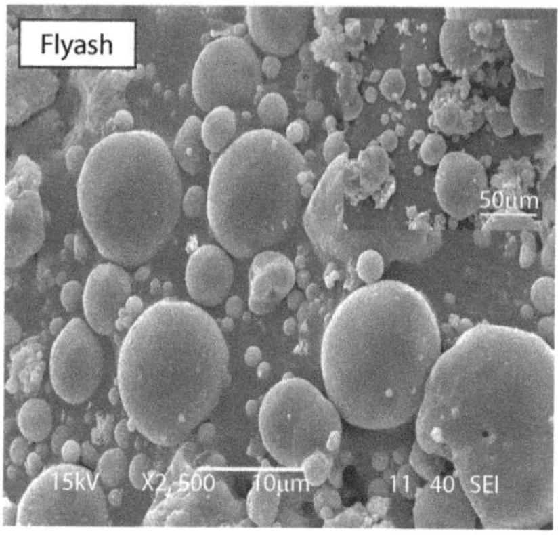

Figure 2.8 SEM micrograph of the fly ash.

that are less than 75 μm. The particle size analysis curve of the raw materials is presented in Figure 2.13.

2.3.2.3 Cement Clinker

CC is flaky in shape and comprises tricalcium aluminates, tetra calcium alumina ferrite and calcium silicates, which can be observed to be in an amorphous phase, according to the SEM images (Figure 2.9) and XRD (Figure 2.10). According to the results, the CC used in this study is poorly graded and coarse-grained. It has been found that the CC used in this study is coarse grained soil with poorly graded gradation as the values of the co-efficient of uniformity (Cu) and curvature co-efficient (Cc) are 2 and 1.12, respectively.

Figure 2.9 SEM micrograph of the cement clinker.

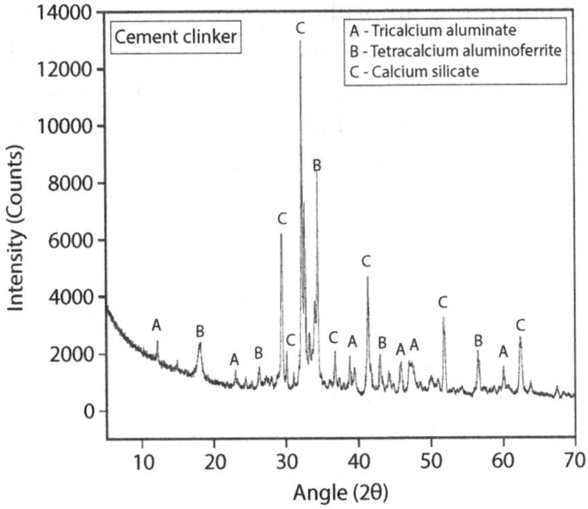

Figure 2.10 Results of XRD patterns forcemeat clinker.

2.3.2.4 *Ground Granulated Blast Furnace Slag*

The GGBS is a by-product material produced from steel manufacturing. There are substantial amounts of CaO present in GGBS, which increase in strength. As it has a cementitious property, so it acts as binding material for the soil. The minerals coesite, gehlenite, melilite and akermanite are noticed in GGBS, and at 30°, large humps are spotted indicating the

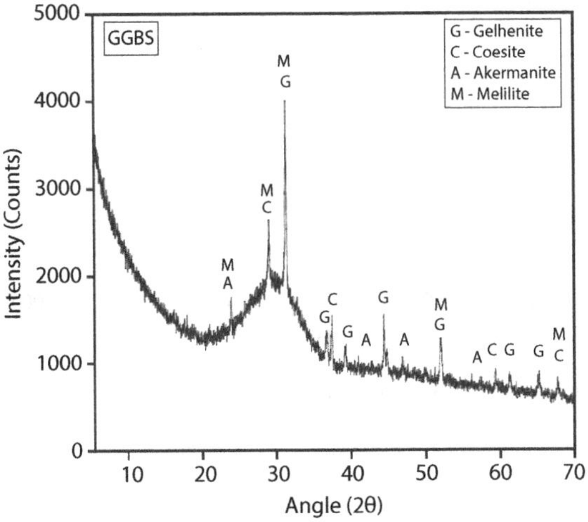

Figure 2.11 Results of XRD patterns for GGBS.

presence of an enormous amorphous phase, according to the XRD data depicted in Figure 2.11. High pozzolanic activity is enhanced by this amorphous phase [48]. The SEM images for GGBS, as revealed in Figure 2.12 reveals that the particle is angular in shape. The results show that GGBS is poorly graded, with Cu and Cc values of 4.5 and 1.38, respectively.

Figure 2.12 SEM micrograph of the GGBS.

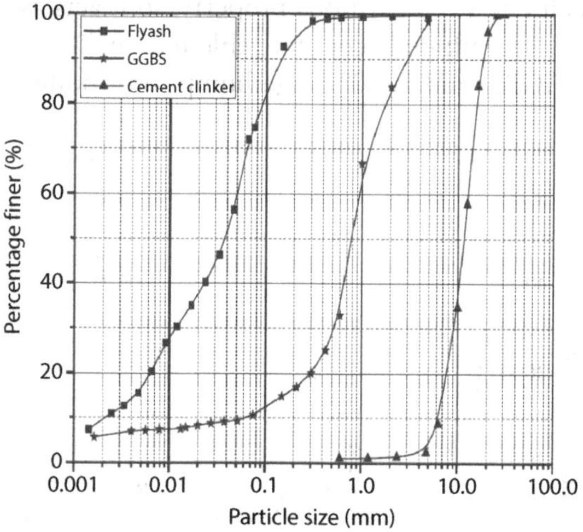

Figure 2.13 Particle size analysis curve of raw materials.

2.3.2.5 *Physical Characteristics*

The physical characteristics of raw materials are shown in Table 2.2. The plastic limit of CC, GGBS and FA could not be determined, and therefore, these samples are reported as non-plastic (NP). On the other hand, when these specimens are compacted to their MDD and OMC, they develop pseudo-cohesive strength in addition to particle attractive force. These impart strength to the specimens and are measured through the UCS test (ASTM D2166M, 2016 code) [49] at 0 days. The specific surface area has been done by the Brunauer-Emmett-Teller instrument of AUTOSORB-1, Quanta chrome instrument. The specific surface area of FA, CC and GGBS are 29.18, 22.08, and 21.84 respectively.

Table 2.2 Index and engineering characteristics of soil and industrial by-products.

Properties	DS	GGBS	FA	CC
Liquid limit (%)	428	32.8	31.5	33.02
Plastic limit (%)	62	NP	NP	NP
Plasticity index (%)	366	-	-	-
Specific gravity	2.74	2.84	2.45	3.24
Specific surface area (m^2/g)	-	21.84	29.18	22.08
Color	Brown	White	Grey	Light brown
Shape	Flaky and angular	Angular	Spherical	Flaky
pH	8.94	10.85	9.68	13.18
OMC (%)	34.12	14.17	37.8	14.36
MDD (kN/m^3)	13.98	16.60	11.48	18.01
UCS, q_u (kPa)	175.47	176.19	125.87	110.38

2.3.2.6 *Leachate Analysis*

To establish the leachable content of raw materials such as fly ash or GGBS, water sorption specimens have been prepared at ASTM D4793-09, 2009 [50] and results are presented in Table 2.3. Initially, the samples were dried for 24hrs and mixed with distilled water at a mass ratio of 1:1. The mixture was agitated continuously for 18hrs using a magnetic stirrer, and the supernatant liquid was filtered through 0.45μm filter paper, and analysis was made by using PerkinElmer AAnalyst200 atomic absorption spectroscopy. Prior to analyzing an element's content, the instrument was calibrated using 1, 2, and 3 ppm standard solutions of that element, which had been prepared utilizing distilled water. The metal concentrations of calcium (Ca), magnesium (Mg), sodium (Na), iron (Fe), copper (Cu), lead (Pb), nickel (Ni), arsenic (As), chromium (Cr), manganese (Mn), mercury (Hg),

Table 2.3 Concentration of metals in leachate specimens of fly ash and GGBS.

Elements	Metal concentration (mg/L)		WHO's acceptable limit for drinking water
	Fly ash	GGBS	
Na	4.5	-	-
Ca	9.7	52	75
Fe	0.043	0.210	3.0
Cu	0.067	0.074	2.0
Mn	0.048	0.062	0.1
Zn	0.026	0.041	0.05
Pb	0.279	0.052	0.01
Ni	0.140	0.024	0.07
Cr	0.217	0.360	5.00
Mg	2.012	3.044	30
As	0.186	0.682	5.00
Hg	0.038	0.044	0.20

and zinc (Zn) were found in the leachate specimens. They were subjected to measurement and comparisons with the acceptable limits of various elements in drinking water, as specified by WHO.

2.3.2.7 Designation of Test Specimens

Since there were no guidelines available for the percentage of these components, which are shown in Table 2.4, the soil was mixed with FA (5 to 30%), GGBS (5 to 25%), and CC (5 to 40%) of dry weight for the preparation of the mixtures. The designations used in this investigation are S –soil of dispersive nature, SS - soil-slag mix, SF - soil-fly ash mix, SC –soil-cement clinker mix, S2M10 - Soil-FA-GGBS mix, S3M10 - Soil-FA-CC mix, S4M10 - Soil-GGBS-CC mix, S5M10 - Soil-GGBS-FA-CC mix, S6M10 - Soil-FA-GGBS-CC mix, S7M12-Soil-CC-GGBS-flyash mix respectively. Nine combinations were selected for additional testing based on the findings of UCS test, out of a total of 82 combinations which were examined. These combinations of materials gave a higher strength than other combinations after 90 days curing time.

Table 2.4 Mixture proportion of stabilized soil specimens.

Specimen code	**DS (%)**	**FA (%)**	**CC (%)**	**GGBS (%)**
S	100	-	-	-
SC7030	70	-	30	-
SS8515	80	15	-	-
SF8020	80	20	-	-
S2M10	65	20	-	15
S3M10	50	20	30	-
S4M10	55	-	30	15
S5M10	50	5	30	15
S6M10	55	10	30	5
S7M12	50	10	30	10

2.3.2.8 Compaction Characteristics

The OMC and MDD variations with additive content are depicted in Figures 2.14 and 2.15, respectively. The compaction curves of various mix proportions, as shown in Figure 2.16. The MDD is between 13.86 and 14.01 kN/m^3, with an OMC value reduction varying between 34.12% - 28.58%. Furthermore, decreases in OMC and increases in MDD have been observed when GGBS mixed with dispersive soil. A similar observation has reported [51]. It has been observed that when CC percentage increases, MDD decreases and OMC increases. This development may be caused by CC absorbing water, which would decrease dry unit weight by increasing water content. MDD increases and OMC decreases when fly ash is added to the dispersive soil. The reason for this addition of fly ash has contributed to the development in gradation of clay-predominated soil. They [52] have found a similar study. A decrease in OMC because of soil particle flocculation indicates that mixes may be compacted with lesser water content. At the same time, as a result of reduced levels of repulsion between soil particles, these particles become closer and subsequently, an increase in MDD. Soil-FA-GGBS mix (S2M10) the OMC value is 26.70% and MDD value is 14.46 kN/m^3. Soil-FA-CC mix (S3M10) the OMC value is 25.93% and MDD value is 13.40 kN/m^3. Soil-GGBS-CC mix (S4M10) the OMC value is 28.08% and MDD value is 14.21 kN/m^3. Soil-GGBS-FA-CC mix (S5M10) the OMC value is 28.58% and MDD value is 14.01 kN/m^3. Soil-FA-GGBS-CC mix (S6M10) the OMC value is 28.05% and MDD value is

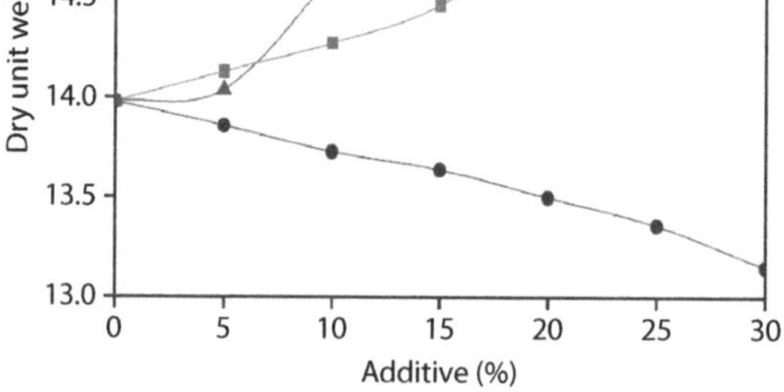

Figure 2.14 Variation of MDD with additive content.

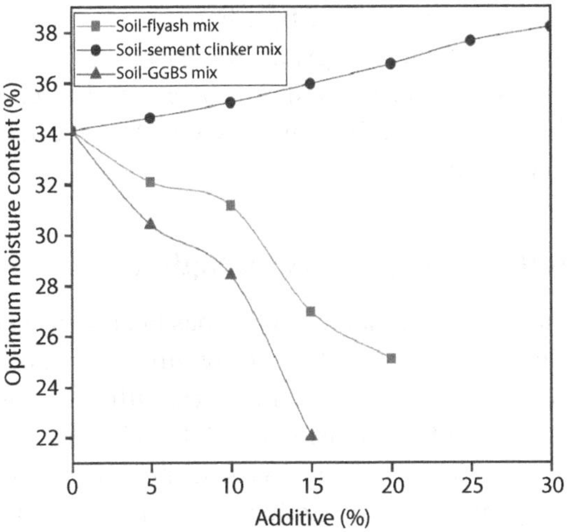

Figure 2.15 OMC variation with additive content.

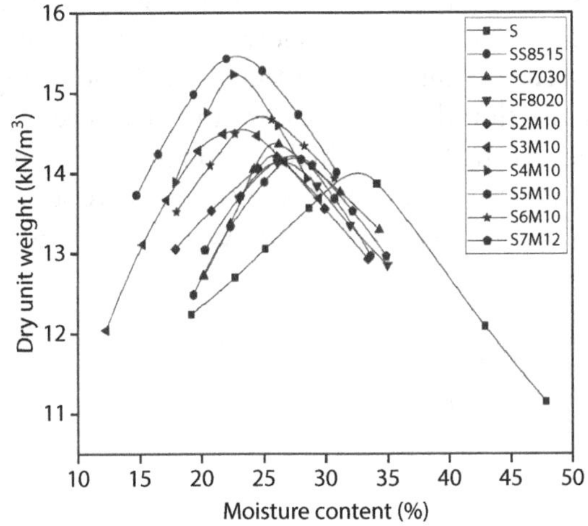

Figure 2.16 Compaction curves of various mix proportion.

14.16 kN/m^3. Soil-CC-GGBS-FA mix (S7M12) the OMC value is 26.90% and MDD value is 14.45 kN/m^3. When GGBS, FA and CC are added to the soil, because of the non-plastic behavior, the resistance is reduced for a similar compaction. For improving the soil properties, CC is considered as an effective stabilizing agent than fly ash and GGBS. Since more calcium ions are accessible from CC, compared to GGBS and FA, the CC particles are finer. This leads to reduction in MDD and enhancement in OMC, as a strong flocculated soil matrix develops. For use as construction material, for a particular compactive energy, an increased dry weight density will be necessary because it indicates soil improvement. Moreover, it improves the quality of the mixture and increases its MDD by adding percentages of stabilizing agents to soil.

2.3.3 Unconfined Compressive Strength (UCS)

To determine the soil strength, a UCS test has been carried out, according to ASTM D2166M (2016) [49]. This study aims to evaluate the blended impacts for fly ash, CC and GGBS on the strength of soil with dispersion character. The impact of the inclusion of GGBS, CC and FA on the unconfined compressive strength of dispersive soil specimens at different curing periods 0 to 90 days at room temperature is conducted as shown in Figures 2.17–2.20. The addition to fly ash to the soil shows a minimal effect on the UCS of untreated and stabilized soil specimens of up to 90 days. The change between UCS and curing periods for a range of FA incorporated into dispersive soils is shown in Figure 2.17. When fly ash is added to the soil with an increase of up to 20%, it can be shown that the strength has increased and then decreased. A similar trend is found for all the curing periods. This is a pozzolanic reaction that depends on time. However, occasionally the sample preparations and testing which is enough for response to switch the particle size, influence and increased the compressive strength. After adding 15% slag in the soil, the strength increases but thereafter, it decreases

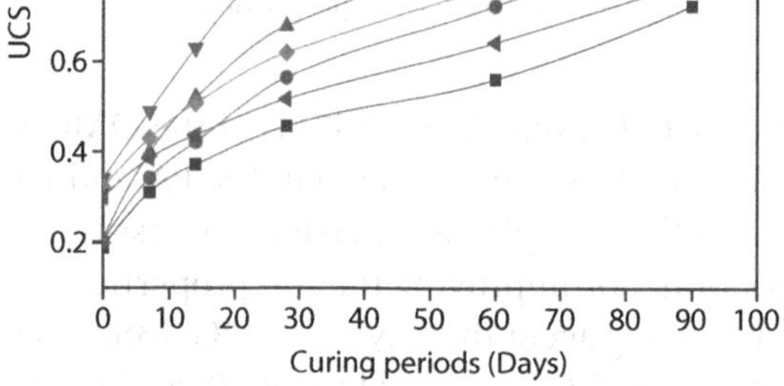

Figure 2.17 Variation in UCS for soil-FA mix with curing periods.

with further increasing GGBS percentages as represented in Figure 2.18. The increased compressive strength is attributed to a higher percentage of GGBS, as well as the curing time. They [28] have made a similar observation. Figure 2.19 depicts the UCS vs the curing periods for the soil-CC mix. Various ratios of CC are mixed with dispersive soil, and the mixture is allowed to cure at room temperature. The results indicate that the development of cementitious compounds may have contributed to the increase in strength with the inclusion of 30% of the CC in the soil. Studies have

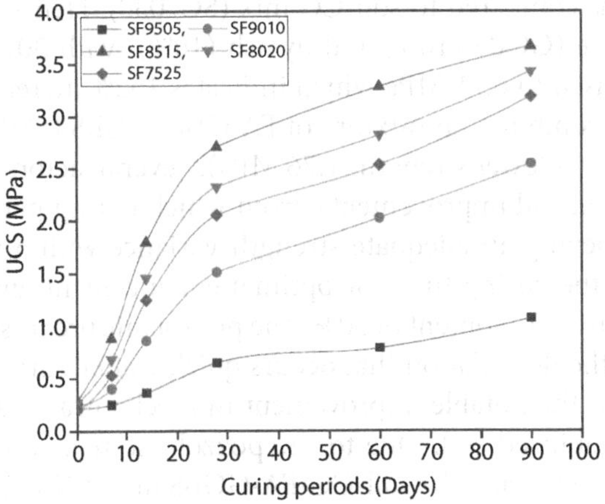

Figure 2.18 Variation in UCS for soil-GGBS mix with curing periods.

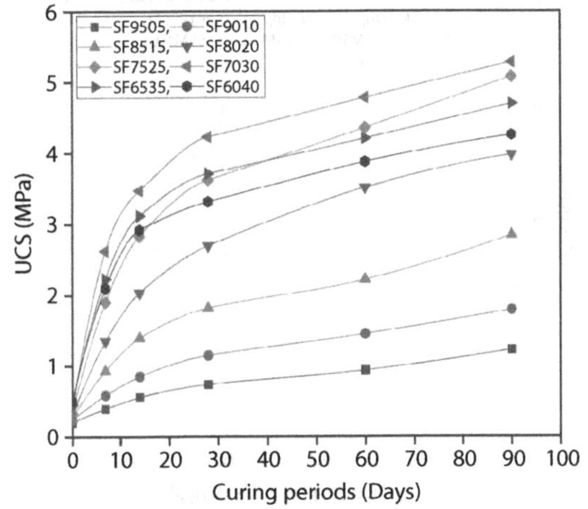

Figure 2.19 Variation in UCS for soil-CC mix with curing periods.

also revealed that the pozzolanic reaction may be prolonged by additional curing time. A minor percentage of CC is not sufficient to enhance the hydration reaction. The difference between the UCS vs curing time of different mixture is shown in Figure 2.20. As compared to different mixes with the soil, the addition of 30% CC the UCS value varies from 0.18 MPa to 5 MPa. The UCS value of soil containing GGBS (15%) and fly ash (20%) after 90 days curing is found to be 2.66 MPa compared to that of 0.18 MPa of virgin soil. The strength gain value is very slow due to the lack of cementitious material. The soil-CC-FA mix (S3M10) and soil-CC–GGBS mix (S4M10) specimen gives higher UCS values i.e., 5.76 MPa and 5.93 MPa respectively, as compared to soil-CC mix (SC7030). Hence, an additional 20% of additive (GGBS (10%) and fly ash (10%)) with 30% CC the UCS value is increased to 6.71 MPa which indicates 152% increase in strength. Whereas, the combination mixture of FA (5%), GGBS (5%) and CC (5%) with the soil gives lower strength (1.76 MPa). Several factors determine the effects of CC on soil improvements, from which curing period is important [53]. Achieving an adequate strength variance with a 90-day curing period yields the curing time for optimal enhancement, given a mixture with a maximum CC content of 30%. The pozzolanic process, which can be enhanced by the flocculation that occurs quickly after curing, is primarily responsible for the notable improvement in mechanical characteristics of CC and FA stabilized soil. Due to the pozzolanic reaction that develops during the curing phase, the soil-CC-FA-GGBS mix (S7M12) combination achieves maximum compressive strength compared to other proportions.

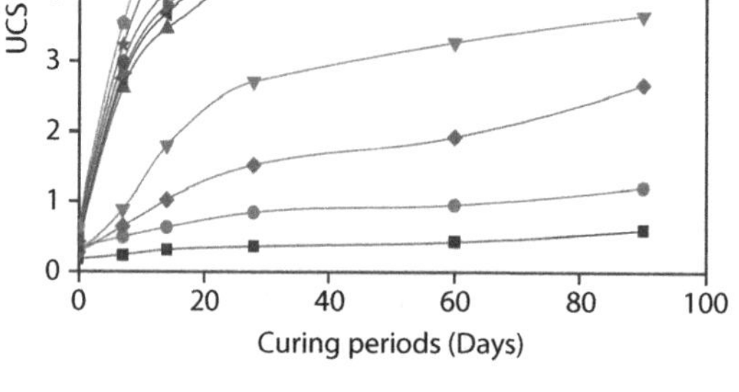

Figure 2.20 UCS variation with different mix proportions and curing periods.

The change of strength during pozzolanic reaction depends on the availability of free Al^{+3}, Si^{+4}, Ca^{+2} ions in the reaction medium and their ratios. By changing the mix proportion, an optimum ratio of the above elements is achieved, which is observed for the mix S7M12. The chemical reactions that occur when GGBS, FA and CC are mixed with soil include cation exchange, pozzolanic reaction, and cementation. With the addition of 20% FA with soil, the UCS is increased up to 188% after 7 days, 264% at 14 days, 394% on 28 days, 458% after 60 days and 605% after the curing period of 90 days. The addition of GGBS with soil increased the strength to 417% for a 7-day curing period, 952% for a 14-day curing period, 1494% for a 28-day curing period, 1823% for a 60-day curing period, and 2052% for a 90-day curing period. The specimen treated with 30% CC and cured for 7, 14, 28, 60 and 90 days showed a strength increment of 1441%, 2018%, 2388%, 2711% and 3005% respectively. An increase up to 1982% after 7 days of curing, 2823% after 14 days, 3105% after 28 days, 3470% after 60 days and 3847% after 90 days of curing is recorded for the S7M12 mix specimen. For proper stabilization of the soil, 30%, 15% and 20% are considered optimal CC, GGBS and FA content. Numerous studies [54, 55] have found a decrease in soil strength when the amount of additive exceeds its optimal value. Comparing untreated specimen with treated specimen, there is a noticeable improvement in compressive strength. The failure pattern ranges between brittle and ductile, depending on the different combination. The binding action of CC during the hydration process is responsible for the brittle behavior caused by the stabilized soil samples. The void ratio has decreased due to these binding actions, and the stabilized soil has become harder. As the stabilized soil matrix goes through its cementitious phase, hydration and pozzolanic reactions strengthen and reinforce it [56]. From the above-presented results, it can be portrayed that lower percentages of additive percentage give lower strength. Therefore, in this study higher percentages of the mix are selected.

2.3.4 California Bearing Ratio (CBR) Test

To measure the bearing capacity of the water attack, the CBR value of stabilized dispersive soil in both unsoaked and soaked conditions shall be used. Subsequently, before the CBR test (ASTM D2166 -2016) [57], the samples were submerged in water for 96 hours. In pavement design, CBR is the main component that establishes the thickness of each pavement layer. This is the ratio expressed as a percentage of the force per unit area, which is essential to penetrate the mass of the soil. At penetrations of 2.5 mm and 5 mm, the CBR value was determined. In general, a 2.5 mm penetration should have a higher CBR value than a 5 mm penetration. Despite the dispersive soil's low CBR value, the soil can be modified before being used as a subgrade material. To achieve this, the specimens were prepared by static compression of mixtures into an appropriate OMC and MDD according to a standard proctor test in CBR molds. For each mixture, a number of soaked CBR tests were conducted. For 7 days, the cured soil specimens, were covered in plastic bags and before testing, the specimens were allowed to drain for fifteen minutes after being kept at room temperature and immersed in water for four days. Three similar specimens were made and examined for a specific combination. The compressive strength and bearing capacity of compacted subgrade are closely related to CBR values. Therefore, the suitability of all compaction grades should be assessed in this test [52]. The California bearing ratios of all mixtures, for both the conditions i.e., unsoaked and soaked mixtures, are depicted in Figures 2.21 and 2.22. Figure 2.22 illustrates variations in CBR with varying mix proportions. With the combination of SF8020, SS8515, and S2M10, for the soaked state, it is revealed that CBR values are lesser than those in the unsoaked state. The California bearing ratio is higher for soaked state than for unsoaked state for the mixtures of S3M10, SC7030, S4M10, S6M10, S5M10, and S7M12.The CBR for soaked condition shall be increased than that when unsoaked, when stabilizing agents such as GGBS, CC and FA are used in the soil. The frictional resistance between the stabilizing materials is the reason for an increase in CBR. When compared to the FA, which has

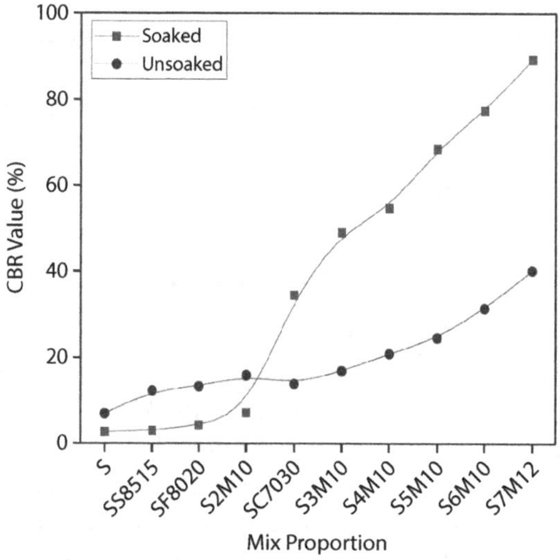

Figure 2.21 CBR value versus mix proportion.

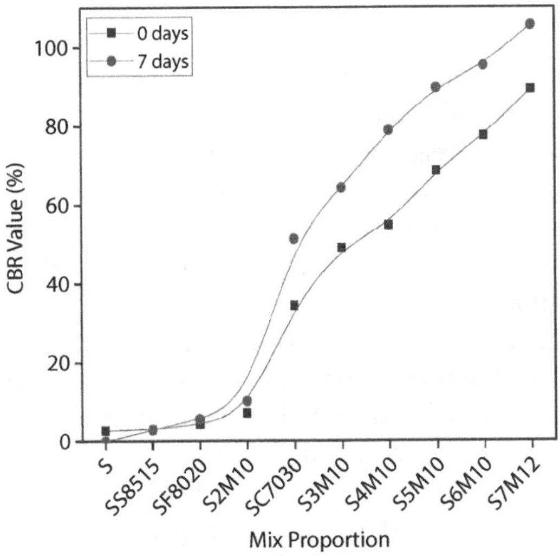

Figure 2.22 CBR value in a soaked condition versus mix proportion.

a spherical shape, the CC and GGBS, due to their irregular shapes, exhibit higher CBR values caused by their increased resistance to penetration. In soaked CBR conditions, the CBR value reduces with an increment in the percentages of fly ash and GGBS. The CBR value increased with the 11 days of curing i.e., 7 days curing with 4 days soaking, illustrated in Figure 2.22. CBR value improved by 2.9%-105.5% (SS8515 to S7M12). The soil with dispersion when modified by CC of 30%, fly ash of 10%, and slag of 10%, a maximum California bearing ratio of 105.5% was observed for S7M12. CBR values have been increased because of the CC hydration process, in which soil particles are bound and developed strength. A significant increase of CBR value is achieved by soaking the specimens in water for 4 days. The increased strength and CBR values, which justify this trend, is due to the hydration reactions and chemical bonds that are occurring between CC and soil with increasing dosages of by-products.

2.3.5 Micro Structural Characterization

After 7, 28, and 90 days of curing, the broken portion of the samples were accumulated from fractured parts of the failed UCS test. The hydration process was stopped by immersing them in anhydrous ethanol, and they were oven-dried for 24 hours at 110°C before being tested for XRD and SEM.

2.3.5.1 X-Ray Diffraction

To determine the variations in the mineralogical phase due to pozzolanic reactions, stabilized soil XRD analysis is essential. This test was carried out to identify the several reaction compounds; those that have cementitious property formed in the soil when modified by 15% of GGBS, 20% of FA, and 30% of CC; that is, on specimens that indicated higher strength. The figure depicts the amorphous structures at 10 to 60, 2θ angle that represent silica-aluminum reaction products. The XRD patterns of stabilized soil following a seven-day curing period are revealed in Figures 2.23–2.25. SC7030 mix shows the presence of CSH, CASH, calcite, and ettringite. The main reaction products for the SS8515 mix include CASH, CH, CSH, and calcite. SF8020 mix of the reaction products present CAH and CSH, but calcium sulfo-aluminate (ettringite) is not found. For S2M10 mix hydration products are CASH, CSHH, and calcite peaks are noticed. The results of XRD analysis for S3M10 to S7M12 mix reveal that a decrease of montmorillonite peaks influenced changing the mineralogy of the treated soil consequently the reaction products, calcium alumino-silicate hydrates i.e., CASH, CSH, calcium aluminum oxide chloride hydrate (CAOCH), ettringite, and calcite are formed.

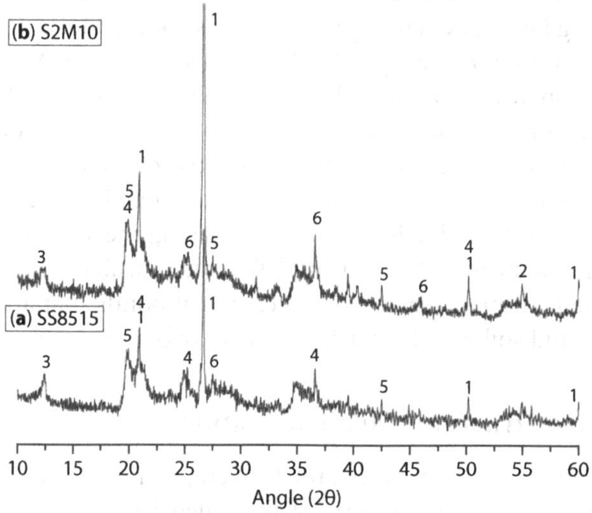

Figure 2.23 XRD analysis of SS8515, S2M10 at 7 days curing (1 – Quartz (Q); 2 – Calcite (C); 3 – Montmorillonite (M); 4 – CSH; 5 – CASH; 6 – CSHH).

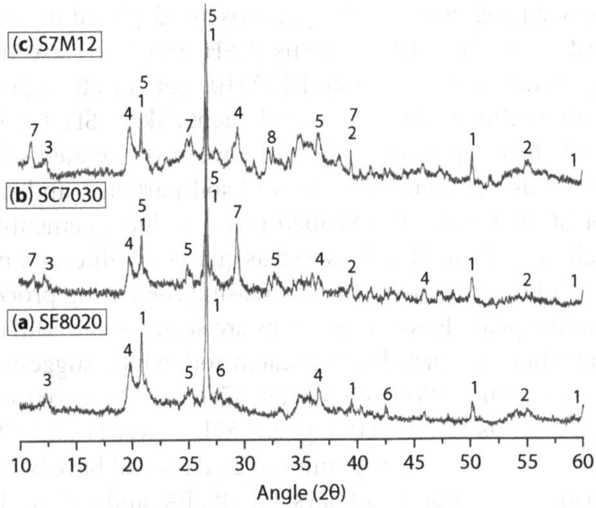

Figure 2.24 XRD analysis of SF8020, SC7030 and S7M12 at 7 days curing (1 – Q; 2 – C; 3 – M; 4 – CSH; 5 – CASH; 6 – Calcium aluminate hydrate (CAH); 7 – Calcium sulfo-aluminate; 8 – CAOCH).

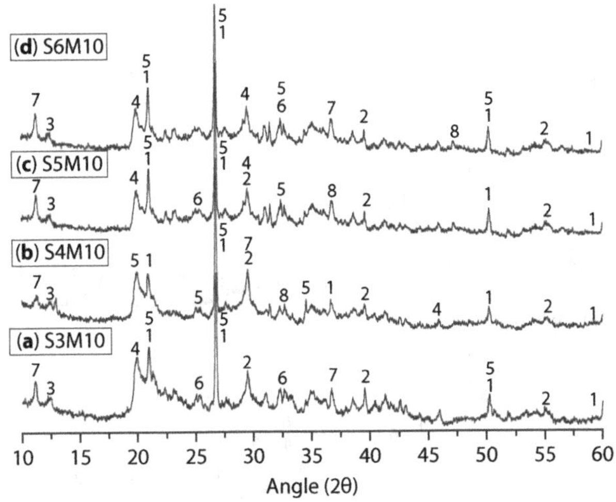

Figure 2.25 XRD analysis of S3M10, S4M10, S5M10 and S6M10 at 7 days curing (1 – Q; 2 – C; 3 – M; 4 – CSH; 5 – CASH; 6 – CSHH; 7 – Ettringite; 8 – CAOCH).

The 28-day curing period XRD patterns are depicted in Figures 2.26–2.28. According to the XRD patterns, CSH and CAH are the primary hydration products for SS8515 and S2M10, but no ettringite peaks are found in both mixtures. The existence elements like, CSH, CASH, ettringite, and calcite forming a strong bond between the treated soil particles, make great bonds between the stabilized soil particles, and improve the strength for SC7030 mix. The SF8020 mix, produces cementitious compounds, such as CAH and CSH, as well as quartz, mullite, and montmorillonite peaks. It has been noted that increasing the curing process reduces montmorillonite peak. Fewer new peaks are seen for the S3M10 - S7M12 combination when compared to untreated soil, which suggests the development of new cementitious chemicals. Cementitious compounds with larger peaks, such as CASH, ettringite, CSH, CAOCH and CSHH, have been produced at shorter curing times of 28 days and have been shown to increase curing periods up to 90 days. GGBS, FA, and CC produce CASH and CSH when they substantially counteract with the ions of calcium generated by calcium oxide hydrolysis. These compounds also depend on the presence of silica and hydrated alumina from soil. Numerous researchers have reported that the main reaction products are CASH and CSH, which contribute to enhancing strength [38, 58].

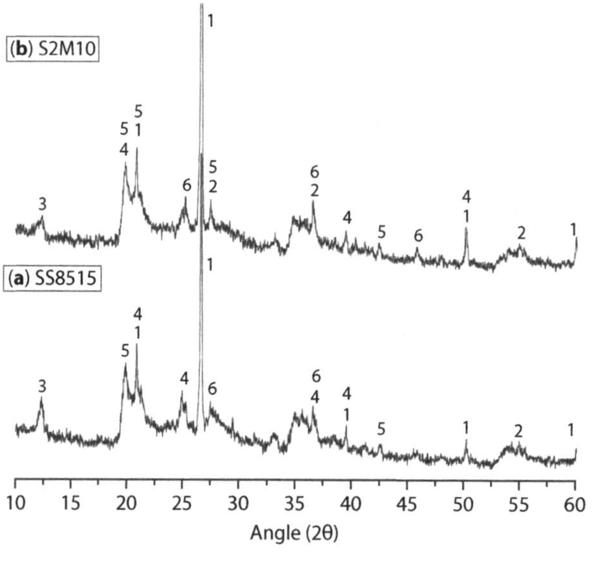

Figure 2.26 XRD analysis of SS8515 and S2M10 at 28 days curing (1 – Q; 2 – C; 3 – M; 4 - CSH; 5 - CASH; 6 - CSHH).

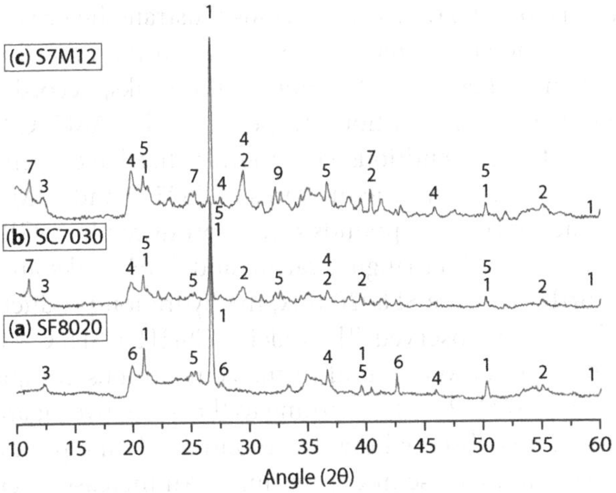

Figure 2.27 XRD analysis of SC7030, SF8020 and S7M12 at 28 days curing (1 – Q; 2 – C; 3 – M; 4 – CSH; 5 – CASH; 6 – CAH; 7 – Ettringite; 8 – CSHH; 9 – CAOCH).

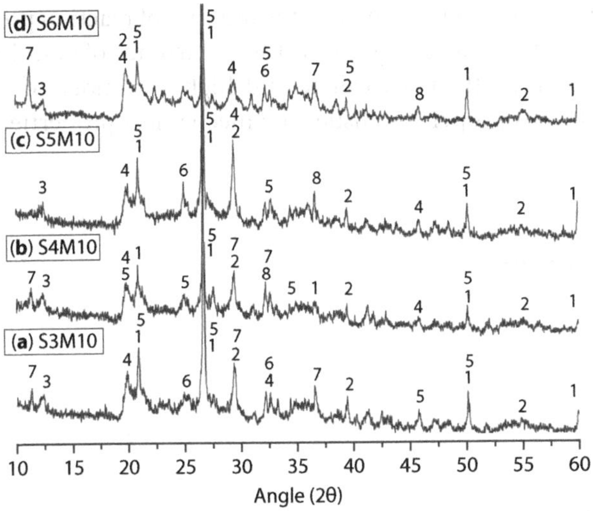

Figure 2.28 XRD analysis of S3M10, S4M10, S5M10 and S6M10 at 28 days curing (1 – Q; 2 – C; 3 – M; 4 – CSH; 5 – CASH; 6 – CSHH, 7 – Ettringite; 8 – CAOCH).

XRD patterns for 90 days curing periods illustrate the broader peaks of cementitious, as shown in Figures 2.29–2.31. On the other hand, longer curing periods have been seen to improve the peaks, according to reaction compounds with cementitious property. CSH, CASH, CAH, CSHH, and $CaCO_3$ are the cementitious compounds that have been identified. Ettringite is not detected for the mixtures of SS8515 and S2M10. A small number of cementitious compounds can be produced by GGBS alone. In the SC7030 mix, CASH, ettringite, calcite and CSH peaks are noticeable reflections. Further, for the SF8020 mix, the hydration products CSH and CAH peaks have been observed. The Calcite, CSHH, CSH, CASH, ettringite and CAOCH peaks were found in the cementitious compounds produced by S3M10 - S7M12 mix, according to the XRD investigation. Due to the stabilization, it is observed that the montmorillonite peak and quartz are reduced with an increase in curing times. An increase in curing periods appears to enhance the intensity of peaks associated with cementitious compounds. The sharper intense peaks of cementitious products were considered for CC inclusions. It can be concluded that CC, GGBS, and FA have an important influence on the development of strength and CC played a significant role among these by-products. It is believed that the benefits of the CC have a continuous increase in the number of reaction products generated for the enduring curing times. The occurrence of additional hydration products, i.e., ettringite, calcite and higher number of compounds with cementitious properties ensues the mechanical properties. When an

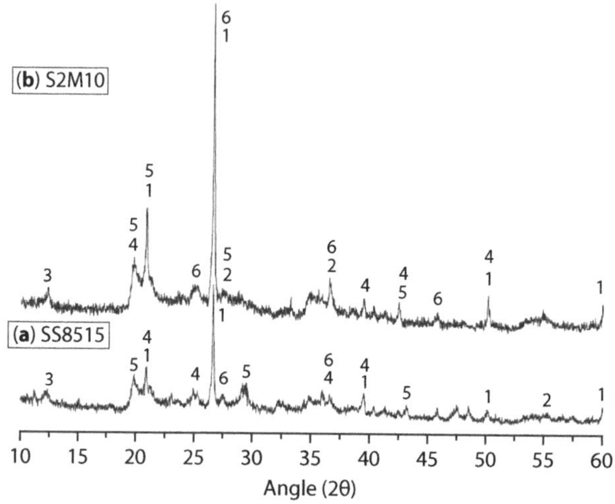

Figure 2.29 XRD analysis of SS8515 and S2M10 at 90 days curing (1 – Q; 2 – C; 3 – M; 4 - CSH; 5 - CASH; 6 - CSHH).

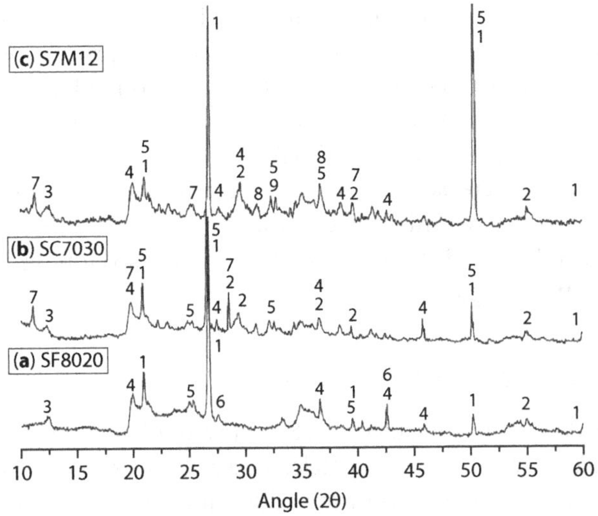

Figure 2.30 XRD analysis of SC7030, SF8020 and S7M12 at 90 days curing (1 – Q; 2 – C; 3 – M; 4 – CSH; 5 – CASH; 6 – CAH; 7 – Ettringite; 8 – CSHH; 9 – CAOCH).

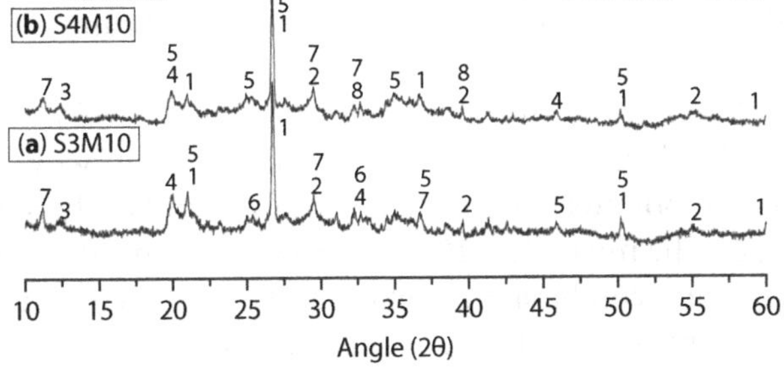

Figure 2.31 XRD analysis of S3M10, S4M10, S5M10 and S6M10 at 90 days curing (1 – Q; 2 – C; 3 – M; 4 – CSH; 5 – CASH; 6 – CSHH; 7 – Ettringite; 8 – CAOCH).

additive is replaced, it results in hydration and pozzolanic reactions, which occur at all proportion. In the pozzolanic reaction process, which takes time to complete, there is an increase in calcium content accompanied by a depletion in silica and alumina in the soil-admixtures after a curing period of 90 days. After such time, the soil FA-GGBS-CC registers a decrease in Si and Al and an increase in Ca due to the pozzolanic reaction, which is a long-term procedure. This suggests that additional cementitious chemicals, such CAH, CASH, CAOCH and CSH, are developing and are the cause of the strength gain.

2.3.5.2 Scanning Electron Microscopy

The stabilized soils are subjected to a SEM test to identify any variations in the microstructure in the soil particle. These changes in soil micro structural development resulted from the addition of stabilizers, which have a substantial influence on the geotechnical characteristics of stabilized soils. In comparison to the untreated soil, the treated soil has a notably changed microstructure.

The stabilized soil SEM micrographs of the seven-day-cured specimens are depicted in Figure 2.32. XRD analysis confirmed that the presence of certain reaction products subsequently developed the strength. For SF8020 mix, it seems to be smooth as that of the unreacted fly ash particles, due to the slow reaction of alumina silicates. Moreover, the compacted matrixes within the soil matrix and hydrated gel have been observed for SS8515 and S2M10 mixture. Whereas, for cement-treated specimens SC7030, S3M10, S4M10, S5M10, S6M10, and S7M12, appear to have very less formation of hydrated gel with some needle-shaped crystals; possibly, ettringite are seen. Initially, the compressive strength enhancement can be attributed to ettringite formation.

Figure 2.33 shows the major hydration reaction products at 28-days curing time. The development of a needle-like crystal between soil and cementitious compounds is significant because it makes the soil structure filled and stiffer. The mixes S2M10 and SS8515 showed a dense matrix and hydration gel. For a combination SS8515 and S2M10, a hydration gel and dense matrix were observed. The inclusion of GGBS in the micrograph demonstrates the development of cementitious product patches and provides noticeable evidence of a flocculated structure. Additionally, a few unreacted fly ash particles were found, and a hydrated gel is formed by a mixture of specimens SF8020 around the surface of the soil particles. For the SC7030 mix specimen, a thick matrix and ettringite crystal is formed. A dense matrix, compacted matrix, flocculated structure, a substantial

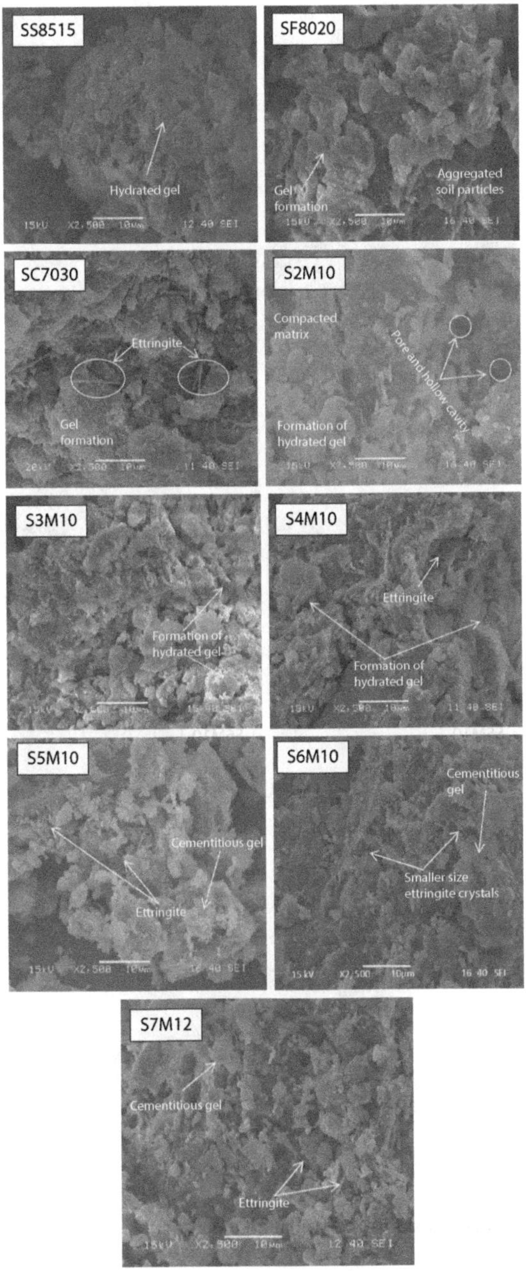

Figure 2.32 SEM images of stabilized soil at 7 days curing.

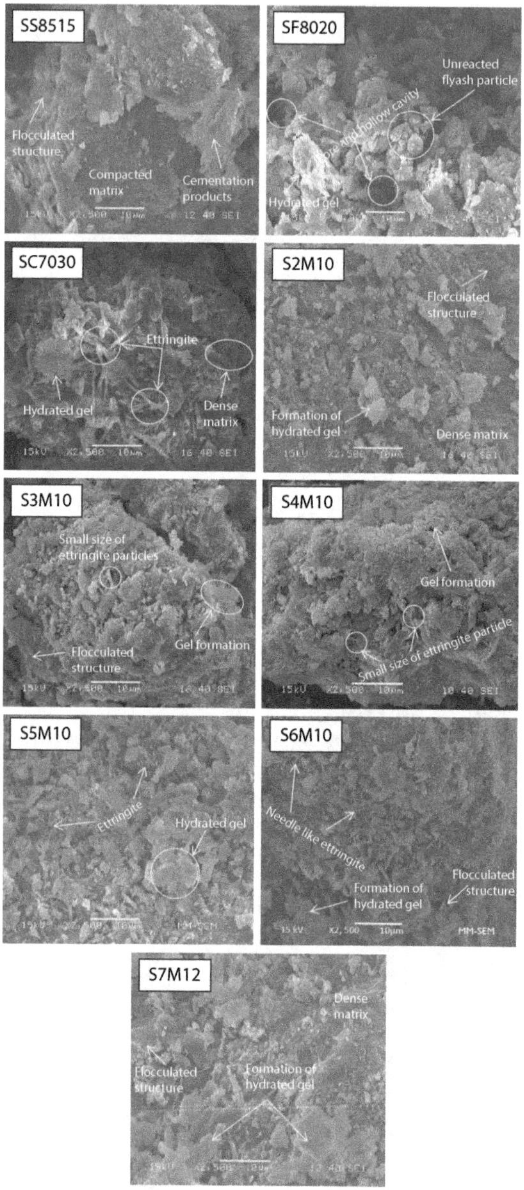

Figure 2.33 SEM images of stabilized soil at 28 days curing.

amount of crystalline hydrate products is detected for S3M10, S4M10, S5M10, S6M10, and S7M12 mix specimens. However, compared to the GGBS and FA mixture, the CC-treated mixture is revealed to have higher hydration products, which may improve the strength of the treated soil. The development of ettringite crystals within the soil pores results in a decrease of void space due to increment of curing periods. The usage of CC is effective in increasing the strength of the soil, as the void spaces are filled with the cementitious substances, strengthening the dispersive soil and making it denser because of the 28-day curing period.

The SEM images of the treated specimens after a 90-day curing period are presented in Figure 2.34. At the 90-day curing period, more reaction products are seen, which leads to enhanced strength. With the treatment of CC, FA, and GGBS, it has been observed that the increase in the duration of the curing period has as significant effect on the soil microstructure. Consequently, the binding of soil-GGBS particles to the SS8515 mix has increased the development of a compacted matrix. Aggregated soil particles, honeycomb structures, and flocculated structures are observed for the SF8020 and S2M10 mix. For the SC7030 mix, ettringite is wrapped with other cementitious materials. Ettringite can be formed in the system due to the abundance of aluminum and sulfate ions. The presence of calcium ions in solution plays a vital role in ettringite formation. However, the mixed combination (S3M10 to S7M12) shows a gel-like mass, and more amorphous ettringite crystals in various forms, like needles and flowers cover completely the composite particles and fill the inter particle spaces. A cementitious matrix is formed with binding particles of a cementitious gel due to the increased content of FA, CC and GGBS. The relative dominance of ettringite crystals, reticulated structure, and cemented and compacted matrix, together with the production of cementitious patch-like compounds, are observed with curing periods in treated soil, resulting in associated changes in the microstructure. The compacted structure makes additional connection, and better cohesion in the soil matrix contributing to developed rigidity and better strength [59]. A substantial amount of CSH gel is formed as a result of stabilization, as clearly illustrated by the formation of hydration products, which results in a denser and more stable structure for the specimen. The size and shape of the ettringite crystals determine the strength behavior. The larger ettringite crystals are less expansive and have an increased strength [60]. A compacted matrix and a flocculated structure are formed, therefore, out of the aggregated particles' more pronounced association with the cementitious gel formation as the curing time rises.

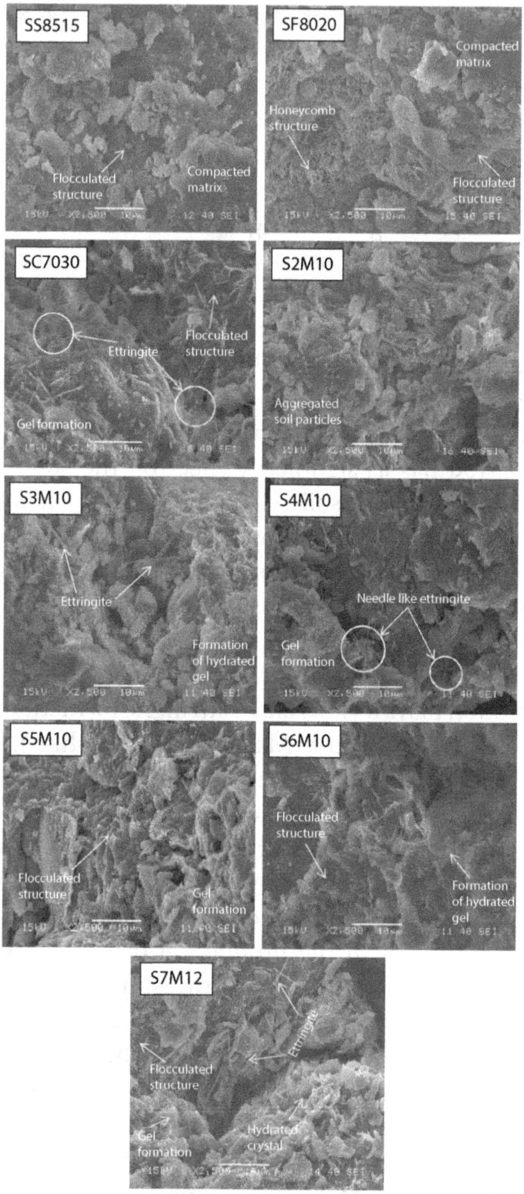

Figure 2.34 SEM images of stabilized soil at 90 days curing.

2.4 Conclusion

The resulting conclusions concerning the stabilization of soil with dispersive nature using GGBS, FA, and CC have been derived from this experimental study presented here:

1. The dispersion % of soil is determined to be 90.66% based on the double hydrometer test. After being treated with GGBS and FA the soil specimen is dispersed in water based on crumb test, cylindrical dispersion test, and turbidity test. However, with the addition of CC, and great stability under water the soil tends to be non-dispersive.
2. The OMC of dispersive soil is found to be 34.12% at an MDD of 13.98 kN/m^3. The maximum dry unit weight value reduced as CC content increased from 0 to 30% with the soil. However, as the amount of GGBS (0 to 15%) in the soil increases, OMC reduces to 22.04% and the maximum dry weight density value enhanced to 13.98 kN/m^3 from 15.42 kN/m^3. Similarly, for flash (0 to 20%) with the soil, increased optimum dry weight density value (14.86 kN/m^3) and decreased moisture content (25.11%) occurs.
3. The strength of dispersive soil significantly increases due to the treatment with additives and an increase in the curing period. The water content of the treated soil samples will decrease with increasing cure time, making them brittle. The strength of the untreated soil is found to be 0.18 MPa. After treatment, the UCS values obtained are 3.54 MPa, 4.97 MPa, 5.45 MPa, 6.07 MPa and 6.71 MPa after 7, 14, 28, 60 and 90 days respectively. Hence, it is concluded that the specimen treated with 30% CC, 10% FA, and 10% GGBS is capable of enhancing the UCS by 1982% after 7 days of curing, 2823% after 14 days of curing, 3105% after 28 days of curing, 3470% after 60 days of curing and 3847% after 90 days of curing.
4. The unsoaked CBR value of untreated soil is 6.9% and the soaked CBR value is 2.7%. Whereas, with the inclusion of CC 30% with the soil, the soaked CBR value is found to be 147% higher than the unsoaked condition. However, the opposite trend is shown in the case of 15% GGBS and 20% fly ash with the soil, the soaking condition reduces the CBR value by 120% that of unsoaked condition. Except for

the SS8515, SF8020, and S2M10 mixtures, CBR values are observed to be higher in a soaked condition of each mixture after being cured for 7 days. The soaked California bearing ratio increases because of the flaky and angular shaped structure of CC particles.

5. The development of a newly created compound is an observation of the morphological modification in the stabilized peat that is further examined using XRD According to the XRD test results, calcium alumina silicate-based hydration reaction products were developed because of the hydration of reactive minerals into a glassy phase. These cementitious gels bind and coat the individual soil particles resulting in a rigid structure, which advances in contraction of voids.
6. he SEM image of the untreated soil indicates that there are voids and excessive sodium. As a result of the water and sodium ions deflocculating processes, soil particles are loose enough so they can easily be dispersed. The microstructure of the soil particles is more compact and resistant to dispersibility due to the addition of CC, FA, and GGBS. This change in the microstructure correspond to the variations in the geotechnical properties of the soil.
7. The overall results provided in this investigation are a significant contribution towards the effective stabilization of soils with dispersive nature with industrial by products, i.e., blast furnace slag, fly ash, and cement clinker. FA and GGBS are insufficient to stabilize the dispersive soil due to the low calcium content; whereas, the combination of all the industrial by products proved to be the utmost effective stabilizers of the dispersive soil.

References

1. Vivek, Dutta, R.K., Parti, R., Effect of chemical treatment of the coir geotextiles on the interface properties of sand/clay-coir geotextile interface. *J. Inst. Eng. India Ser. A*, 100, 357–365, 2019.
2. Mohanta, N.R. and Samantaray, S., Study of Combined Effect of Metakaolin and Steel Fiber on Mechanical Properties of Concrete. *Pertanika J. Sci. Technol.*, 27, 3, 1381–1396, 2019, http://journals-jd.upm.edu.my/resources/files/Pertanika%20PAPERS/JST%.

3. Vivek, and Dutta, R.K., Bearing ratio behavior of sand overlying clay with treated coir geotextiles at the interface. *J. Nat. Fibers*, 19, 14, 7534–7541, 2022.
4. Vivek, Effects of cyclic loading on sand overlaying clay model of unpaved roads reinforced with untreated/treated coir geotextiles. *J. Text. Inst.*, 15, 9, 1694–1700, 2023.
5. Abbaslou, H., Hadifard, H., Poorgohardi, A., Characterization of dispersive problematic soils and engineering improvements: a review. *Comput. Mater. Civ. Eng.*, *1*, 2, 65–83, 2016.
6. Penner, D. and Lagaly, G., Influence of anions on the rheological properties of clay mineral dispersions. *Appl. Clay Sci.*, *19*, 1-6, 131–142, 2001.
7. Abbasi, N. and Nazifi, M.H., Assessment and modification of Sherard chemical method for evaluation of dispersion potential of soils. *Geotech. Geol. Eng.*, *31*, 1, 337–346, 2013.
8. Choquette, M., Berube, M.A., Locat, J., Mineralogical and microtextural changes associated with lime stabilization of marine clays from Eastern Canada. *Appl. Clay Sci.*, *2*, 3, 215–232, 1987.
9. Locat, J., Bérubé, M.A., Choquette, M., Laboratory investigations on the lime stabilization of sensitive clays: shear strength development. *Can. Geotech. J.*, *27*, 3, 294–304, 1990.
10. Brown, T.H., Brown, M.A., Sorini, S.S., Huntington, G., The use of coal fly ash for soil stabilization (No. DOE/MC/11076-3620), University of Wyoming Research Corp., Laramie, WY (United States), Western Research Inst, 1991.
11. Bell, F., *Engineering treatment of soils*, CRC Press, USA, 1993.
12. Macphee, D.E., Black, C.J., Taylor, A.H., Cements incorporating brown coal fly ash from the Latrobe Valley region of Victoria, Australia. *Cem. Concr. Res.*, *23*, 3, 507–517, 1993.
13. Kolias, S., Kasselouri-Rigopoulou, V., Karahalios, A., Stabilisation of clayey soils with high calcium fly ash and cement. *Cem. Concr. Compos.*, *27*, 2, 301–313, 2005.
14. Sharma, N.K., Swain, S.K., Sahoo, U.C., Stabilization of a clayey soil with fly ash and lime: a micro level investigation. *Geotech. Geol. Eng.*, *30*, 1197–1205, 2012.
15. Ogundipe, O.M., An investigation into the use of lime-stabilized clay as subgrade material. *Int. J. Sci. Technol. Res.*, *2*, 10, 82–86, 2013.
16. Stavridakis, E.I., Al-Rawas, A., Goosen, Z., Assessment of anisotropic behavior of swelling soils on ground and construction work, in: *Expansive soils: recent advances in characterization and treatment*, pp. 371–384, Taylor and Francis, England, 2006.
17. Singh, P., Dash, H.M., Samantaray, S., Effect of silica fume on engineering properties of expansive soil. *Mater. Today Proc.*, 33, 5035–5040, 2020, https://doi.org/10.1016/j.matpr.2020.02.839.

18. Kumar, M.H., Mohanta, N., Samantaray, S., Kumar, N.M., Combined effect of waste glass powder and recycled steel fibers on mechanical behavior of concrete. *SN Appl. Sci.*, 3, 1–18, 2021a, https://doi.org/10.1007/s42452-021-04422-w.
19. Kumar, M.H., Macharyulu, I.S., Ray, T., Mohanta, N.R., Jain, M., Samantaray, S., Sahoo, A., Effect of water absorption and curing period on strength and porosity of triple blended concrete. *Mater. Today Proc.*, *43*, 2162–2169, 2021b, https://doi.org/10.1016/j.matpr.2020.12.092.
20. Ray, T., Mohanta, N.R., Kumar, M.H., Samantaray, S., Study of effect of temperature on behavior of alkali activated slag concrete. *Mater. Today Proc.*, *43*, 1352–1357, 2021, https://doi.org/10.1016/j.matpr.2020.09.169.
21. Kumar, M.H., Saikrishnamacharyulu, I., Mohanta, N.R., Ashutosh, A., Mishra, P., Samantaray, S., Mechanical behaviour of high strength concrete modified with triple blend of fly ash, silica fume and steel fibers. *Mater. Today Proc.*, *65*, 933–942, 2022a, https://doi.org/10.1016/j.matpr.2022.03.528.
22. Kumar, M.H., Mohanta, N.R., Patel, N., Samantaray, S., Reddy, S.V.B., Influence of Fly ash and Metakaoline on the crack resistance and shrinkage property of concrete. *Iran. J. Sci. Technol. Trans. Civ. Eng.*, 46, 2011–2026, 2022b, https://doi.org/10.1007/s40996-021-00748-y.
23. Volk, G.M., The method of determination of degree of dispersion of the clay fraction of soils as used in investigation of abnormal characteristics of soils in region eight of the soil conservation service. *Soil Sci. Soc. Am. J.*, *2*, C, 561–565, 1938.
24. Atkinson, J.H., Charles, J.A., Mhach, H.K., Examination of erosion resistance of clays in embankment dams. *Q. J. Eng. Geol. Hydrogeol.*, *23*, 2, 103–108, 1990.
25. Savaş, H., Türköz, M., Seyrek, E., Ünver, E., Comparison of the effect of using class C and F fly ash on the stabilization of dispersive soils. *Arabian J. Geosci.*, *11*, 1–13, 2018.
26. Vakili, A., Selamat, M.R., Moayedi, H., An assessment of physical and mechanical properties of dispersive clay treated with lime. *Casp. J. Appl. Sci. Res.*, *2*, 197–204, 2013a.
27. Vakili, A.H., Selamat, M.R., Moayedi, H., Effects of using pozzolan and portland cement in the treatment of dispersive clay. *Sci. World J.*, *2013*, 1, 547615, 2013.
28. Goodarzi, A.R. and Salimi, M., Stabilization treatment of a dispersive clayey soil using granulated blast furnace slag and basic oxygen furnace slag. *Appl. Clay Sci.*, 108, 61–69, 2015.
29. Yilmaz, I. and Civelekoglu, B., Gypsum: an additive for stabilization of swelling clay soils. *Appl. Clay Sci.*, *44*, 1-2, 166–172, 2009.
30. Rengasamy, P., Tavakkoli, E., McDonald, G.K., Exchangeable cations and clay dispersion: net dispersive charge, a new concept for dispersive soil. *Eur. J. Soil Sci.*, *67*, 5, 659–665, 2016.

31. Amina, S.M. and Rani, V., Control of dispersivity of soil using lime and silica fume. *Int. J. Res. Appl. Sci. Eng. Technol.*, 5, 5, 386–391, 2017.
32. Vakili, A.H., Kaedi, M., Mokhberi, M., bin Selamat, M.R., Salimi, M., Treatment of highly dispersive clay by lignosulfonate addition and electroosmosis application. *Appl. Clay Sci.*, *152*, 1–8, 2018.
33. Vakili, A.H., Salimi, M., Keskin, I., Abujazar, M.S.S., Shamsi, M., Effects of polyvinyl acetate content on contact erosion parameters of pavement embankment constructed by dispersive soils. *Bull. Eng. Geol. Environ.*, *82*, 10, 398, 2023.
34. Tasong, W.A., Wild, S., Tilley, R.J., Mechanisms by which ground granulated blastfurnace slag prevents sulphate attack of lime-stabilised kaolinite. *Cem. Concr. Res.*, 29, 7, 975–982, 1999.
35. Liu, B., Ray, A.S., Thomas, P.S., Strength development in autoclaved aluminostilicate rich industrial waste cement systems containing reactive magnesia. *J. Australas. Ceram. Soc.*, 43, 1, 82–87, 2007.
36. Sivapullaiah, P.V. and Jha, A.K., Gypsum induced strength behaviour of fly ash-lime stabilized expansive soil. *Geotech. Geol. Eng.*, *32*, 1261–1273, 2014.
37. Islam, S., Haque, A., Wilson, S., Ranjith, P.G., Time-dependent strength and mineralogy of lime-GGBS treated naturally occurring acid sulfate soils. *J. Mater. Civ. Eng.*, 28, 1, 04015077, 2016.
38. Sharma, A.K. and Sivapullaiah, P.V., Ground granulated blast furnace slag amended fly ash as an expansive soil stabilizer. *Soils Found.*, 56, 2, 205–212, 2016.
39. Keramatikerman, M., Chegenizadeh, A., Nikraz, H., Effect of GGBFS and lime binders on the engineering properties of clay. *Appl. Clay Sci.*, 132, 722–730, 2016.
40. Sekhar, D.C. and Nayak, S., SEM and XRD investigations on lithomargic clay stabilized using granulated blast furnace slag and cement. *Int. J. Geotech. Eng.*, 13, 6, 615–629, 2017.
41. Wang, D., Zentar, R., Abriak, N.E., Di, S., Long-term mechanical performance of marine sediments solidified with cement, lime, and fly ash. *Mar. Georesour. Geotechnol.*, *36*, 1, 123–130, 2018.
42. Fatima, B., Alshameri, B., Hassan, W., Maqsood, Z., Jamil, S.M., Madun, A., Sustainable incorporation of Plaster of Paris kiln dust for stabilization of dispersive soil: A potential solution for construction industry. *Constr. Build. Mater.*, *397*, 132459, 2023.
43. Standard test method for dispersive characteristics of clay soil by double hydrometer, ASTM D 4221-99, 2005.
44. Standard test methods for laboratory compaction characteristics of soil using standard effort, ASTM D698, 2007.

45. Bhuvaneshwari, S., Robinson, R.G., Gandhi, S.R., Stabilization of expansive soils using fly ash. *Fly Ash India*, *8*, 5, 1–10, 2005.
46. Umesh, T., Dinesh, S., Sivapullaiah, P.V., Characterization of dispersive soils. *Mater. Sci. Appl.*, 2, 6, 629–633, 2011.
47. Standard test methods for determining dispersive characteristics of clayey soils by the crumb test, ASTM D6572-13e2, 2013.
48. Quina, M.J., Bordado, J.C., Quinta-Ferreira, R.M., The influence of pH on the leaching behaviour of inorganic components from municipal solid waste APC residues. *Waste Manage. (Oxford)*, 29, 9, 2483–2493, 2009.
49. Standard test method for unconfined compressive strength of cohesive soil, ASTM D2166/D 2166 M-16, 2016.
50. Standard Test Method for Sequential Batch Extraction of Waste with Water, ASTM D4793-09, 2009.
51. Yadu, L. and Tripathi, R.K., Stabilization of soft soil with granulated blast furnace slag and fly ash. *Int. J. Res. Eng. Technol.*, *2*, 2, 115–119, 2013.
52. Indraratna, B., Nutalaya, P., Kuganenthira, N., Stabilization of a dispersive soil by blending with fly ash. *Q. J. Eng. Geol.*, *24*, 3, 275–290, 1991.
53. Estabragh, A.R., Beytolahpour, I., Javadi, A.A., Effect of resin on the strength of soil-cement mixture. *J. Mater. Civ. Eng.*, 23, 7, 969–976, 2011.
54. Sridharan, A., Prashanth, J.P., Sivapullaiah, P.V., Effect of fly ash on the unconfined compressive strength of black cotton soil. *Proc. Inst. Civ. Eng.-Ground Improv.*, *1*, 3, 169–175, 1997.
55. Basha, E.A., Hashim, R., Mahmud, H.B., Muntohar, A.S., Stabilization of residual soil with rice husk ash and cement. *Constr. Build. Mater.*, *19*, 6, 448–453, 2005.
56. Wong, L.S., Hashim, R., Ali, F., Utilization of sodium bentonite to maximize the filler and pozzolanic effects of stabilized peat. *Eng. Geol.*, *152*, 1, 56–66, 2013.
57. Standard test method for California bearing ratio (CBR) of laboratory compacted soils, ASTM D2166-16, 2016.
58. Horpibulsuk, S., Rachan, R., Chinkulkijniwat, A., Raksachon, Y., Suddeepong, A., Analysis of strength development in cement-stabilized silty clay from microstructural considerations. *Constr. Build. Mater.*, 24, 10, 2011-2021, 2010.
59. Mishra, M.K. and Karanam, U.R., Geotechnical characterization of fly ash composites for backfilling mine voids. *Geotech. Geol. Eng.*, *24*, 1749–1765, 2006.
60. Mehta, P.K., Pozzolanic and cementitious byproducts as mineral admixtures for concrete-a critical review. *Spec. Publ.*, *79*, 1–46, 1983.

3

Effect of Geotechnical-Properties of Soil by Varying the Proportions of Plastic Waste and Ceramic Tile Waste

Pradeep Kumar Sahoo[1], Chhabirani Tudu[1], Manaswinee Patnaik[2] and Siprarani Pradhan[1*]

[1]School of Infrastructure and Planning, Odisha University of Technology and Research, Bhubaneswar, Odisha, India
[2]Civil Engineering Department, Government College of Engineering, Kalahandi, Odisha, India

Abstract

In India, a sizable section of the land is covered in red earth. These earths are prevalent in regions with little precipitation, and they don't seem to be able to retain moisture. These types of earths need to be stabilized to enhance their shear strength and dilatancy. This research focuses on waste utilization in the process of earth stabilization. For this purpose, low-density polyethylene (LDPE) in the form of milk pouches and ceramic tile waste are employed as stabilizers in this investigation. An experimental program was undertaken to study the individuals and admixed effect of ceramic tile waste and plastic strips (LDPE) on the geotechnical characteristics of the collected red earth with varying ceramic tile dust and plastic content. Ceramic tile dust which passed through a 4.75mm IS sieve and LDPE plastic strips, 10cm long were randomly mixed with virgin earth. Ceramic tile dust was mixed with the earth sample by diverging proportions of 5%, 10%, 15%, 20%, 25%, and 30% of the dry wt. of the earth. Divergent tests, like standard penetration test (SPT), UCS test, CBR test, and direct shear test (DST) were carried out in this kind of earth. It was found that red earth with 25% tile dust gives the maximum CBR and UCS value. Furthermore, keeping 25% of marble dust constant, the earth was mixed with LDPE with varying proportions of 0.25%, 0.50%, 0.75%, 1.00%, and 1.25% of the dry wt. of the earth. A series of SPT, unsoaked CBR tests, and

**Corresponding author*: srpradhance@outr.ac.in

Sandeep Samantaray, Subhashree Samantasinghar, Soumendra K. Mohanty, Vivek, Abinash Sahoo and Deba Prakash Satapathy (eds.) Waste Utilization in Geotechnical Practice, (73–98)

soaked CBR tests were conducted for these earth samples. The results showed that taking 25% tile dust with 0.75% LDPE gives the maximum values for maximum dry density (MDD), CBR, and UCS value. Based on the outcome, waste ceramic tile dust and waste LDPE plastic can be used, sooner or later, as very good additives to upgrade the geotechnical properties of red earth.

Keywords: MDD, CBR, UCS, LDPE, ceramic tile dust

3.1 Introduction

The biological, chemical, or mechanical moderation of earth engineering possessions is known as earth stabilization. It is also used in civil engineering to refine and improve the engineering properties of earths [1–4]. Certain natural earth properties can be created to be useful for a planned engineering purpose. Stabilization was developed to create something in a stable condition, which may be a challenging task to attempt, as well as the development of any earth properties by increasing the bearing ratio of earth with the help of waste materials like plastic wastes or other solid wastes that is being used up for the earth [5–10]. We need to stabilize the earth to make it easier to estimate the earth's load-bearing capacity and maybe even become greater than capacity because earth qualities vary greatly and the construction of structures heavily relies on that capacity [11–14]. While working with earths, it's crucial to keep the gradation of the earth in mind. The earths could be uniformly graded, which seems stable but have more voids, compared to well-graded earths, which are preferable because they have fewer voids. Therefore, it is preferable to combine diverse earth types to improve the earth's strength qualities. Because it would be highly expensive to replace the poor earth totally, earth stabilization is what you want in. these situations. Earth stabilization upgrades the strength of the earth; thus, upgrading the earth bearing capacity. It is occasionally used to stop earth erosion or dust from forming, which is quite helpful, especially in arid conditions. Earth stabilization is also done to water-proof the earth, which helps the earth retain its strength by preventing water from penetrating the earth. It aids in minimizing earth volume changes brought on by temperature or moisture changes. Stabilization also rises the earth's durability and usability.

3.2 Background of the Study

3.2.1 Previous Research

They [15] implemented three distinct proportions of stone dust in the compositions of 4%, 8%, and 12%. After determining the MDD and UCS, the optimal proportion of stone dust was used to generate a blend with the three LDPE proportions (i.e., 1%, 2%, and 3%), and the properties engineering properties were checked. The fibers have a length of 6 mm each. After testing it was observed that this material is more cost-effective for reinforcing sub-grade earths in flexible pavements than methodically reinforced earth. They [16] investigated the effectiveness of using fly ash and waste ceramic in combination with sand for stabilizing clayey earth. The sand was added to clayey earth, which resulted in a drop in OMC while the CBR improved. Later on, fly ash was also added to clay earth, which enhanced OMC value but lowered the MDD value and upgraded CBR value. According to their findings, adding ceramic tile waste lowers the MDD value while the OMC and the CBR value rise. In collation with clayey earth or fly ash, composite material's drainage properties performed better. They [17] investigated the silty earth's behavior while combining with sugarcane bagasse ash and marble dust. The parent earth was mixed with diverging proportion of marble dust i.e., 5%, 10%, 15% and 20%. The maximum UCS occurred at 15% of marble dust. Further, with 15% marble dust the sugarcane bagasse ash was added with 3% and 5%. With 15% marble dust and 5% sugarcane bagasse ash mixed with silty earth resulted in maximum UCS value. Again, the results of SPT test indicated that there was higher OMC and low MDD with the rise in the stabilizer content. The UCS test result showed that there was enhancement in the strength value with the rise in the stabilizer content. They [18] utilized the waste materials, stone dust and plastic bottles fiber for the stabilization of clayey earth, as well as to evaluate the geotechnical properties of the stabilized earth. The optimum content of stone dust was found to be 7%, and further tests such as SPT and CBR were performed on this optimum value with 0.4%, 0.8%, 1.2%, 1.6%, and 2.0% of the plastic fibers with the earth. They observed that the CBR value goes on increasing up to 7% of stone dust and 2.0% of plastic strips content.

They [19] investigated the geotechnical properties of red earth after the incorporation of polypropylene (P) in it. They observed that the addition of 3% of polypropylene strips to the earth, improves its shear parameters such as cohesion and friction. The CBR value of red earth rose with 1%, 2%, and

3% of PP addition. Their findings indicated that reinforcing red earth with PP fiber could improve its properties. They [20] recommended the use of PVC wrappers for earth stabilization. According to their experimental results the density and the CBR values were improved with the addition of PVC wrappers. They [21] utilized LDPE plastic waste as strips with a width of about 5 mm to stabilize expansive earth. The use of 0.8% plastic as an additive has stabilized the earth by having greater dry density and reducing the moisture content. The CBR and UCS value was initially enhanced by the application of optimum proportion (0.8%) of plastic strips. They [22] utilized fibers made of polyethylene (PE) and polypropylene (PP) in their investigation. According to their test results, the plastic pieces lowered the MDD and OMC of the stabilized earths, which are necessary for the building of lightweight embankments. The CBR outcomes showed that adding plastic fibers to clayey earths enhances the earth's strength and deformation behavior, particularly when the fiber content is 4% for both lengths of 1cm and 2cm. They [23], found that 30% of ceramic waste may be used to replace pavement sub grades, significantly lowering construction costs. The liquid limit, plastic limit, and plasticity index of earth decreases with the addition of ceramic waste, but the CBR of clay earth became greater. By adding 30% more ceramic dust to the treated earth than to the untreated earth, the performance of the soaking CBR is raised and swelling pressure is lowered if more ceramic dust was utilized.

They [24], studied the stabilization of dune sand using tile waste fiber as an additive. Different proportions like 10%, 20%, and 30% of ceramic tile waste size (passing through 4.75mm, 2.0mm, and 1.18mm sieves) were mixed with dune sand, and the permeability qualities and performance of dune sand were assessed. They [25], utilized ceramic tiles waste for stabilization of fine sand. Tests were carried out on mixtures containing 2%, 4%, 8%, and 12% ceramic tile waste and fine sand with a dry density of 1.58 gm/cc. Based on the results of the trials, it has been established that stabilizing fine sand using ceramic tile waste as an additive rises the strength qualities of the fine sand, making it suitable for embankment building. The significance of the research is as follows:

- To provide a substitute solution for the disposal of plastic waste and ceramic tile dust.
- To provide cost-effective materials like plastic waste and ceramic tile dust for earth stabilization.

- To achieve the above, red earth that is locally available was utilized; to be stabilized by LDPE waste and ceramic waste dust.
- Geotechnical properties of the red earth stabilized with plastic waste in the form of strips of length 10cm and ceramic tile dust are to be examined.
- To determine the optimum plastic and ceramic tile dust content for red earth stabilization.

3.3 Material

3.3.1 Red Earth Sample

Earth is a vital element in all civil engineering construction as most of them are made on the earth strata. For the current research work, the earth sample was gathered from the nearby area of Institute of Gastro and Kidney Care Hospital, Kalinga Nagar, Bhubaneswar, Odisha, India, 751003 (20°16' 24.4"N 85°45' 54.7"E). The earth was excavated by machinery tools at a depth of 2 to 2.5m from the ground surface (Figure 3.1). The site of the sample was 1.5 km from our institute (OUTR, Bhubaneswar). The earth sample collected from this campus is red in color. Red earth is the 3rd largest earth group of India covering an area of about 3.5 lakh sq. km (10.6% of India's Area). This red color of earth is mainly due to ferric oxide's occurrence as thin coating on the earth particles while the iron oxide occurs as hematite or as hydrous ferric oxide. But if it is formed as limonite, the earth gets yellow color.

(a) (b)

Figure 3.1 Red earth (a) at the site, and (b) earth sample.

3.3.2 Plastic Waste

Plastic is a plentiful waste material on the planet's surface. These are human-made things that are damaging the environment as well as the surface of the earth, and they are not biodegradable. Plastic waste management has arisen as a challenge for the modern-day population. For the current research, waste household plastic bags are utilized, and these are also a large source of plastic waste. They are made of LDPE. This is a semi-rigid and translucent polymer. It is a thermoplastic made from the monomer ethylene. LDPE is produced in either a stirred auto clave or a tube reactor (Figure 3.2). In general, it is made by compressing ethylene gas, polymerizing, using an initiator, and gas separation. It has a density of 0.910 to 0.940 gm/cm^3 and thickness of 0.12 to 0.14mm. In this research LDPE has been used as stabilizer to enhance the earth's loading capacity. The waste plastic bags were cut 10cm long and were mixed with earth in varying proportion of 0.25%, 0.50%, 0.75%, 1%, and 1.25% of the dry wt. of the earth sample.

3.3.3 Ceramic Tile Waste

Ceramic tiles are largely used nowadays during building construction. These are inorganic materials. Ceramic materials can be crystalline, partially crystalline, or amorphous (Figure 3.3). Majority of common ceramics are crystalline. Heat and subsequent cooling are done to form this nonmetallic solid. The potters utilized clay to manufacture glazed tiles, thus the name ceramic tiles. A significant amount of ceramic tile waste is created during the production, shipping, and placement of ceramic tiles. This waste substance is inorganic and dangerous. As a result, its disposal is

(a) (b)

Figure 3.2 LDPE plastic (a) in pouch form, and (b) in strip form.

Figure 3.3 Ceramic tile (a) waste pieces, and (b) waste dust.

an issue that may be solved by considering it as a combination of stabilized dune sand, such that the mix is highly affordable and can be utilized as sub grade in low traffic highways or village roads. In this study, experimental investigation is done to evaluate the functionality of ceramic tile waste as a stabilizing agent for earth. The earth is mixed with the ceramic tile waste of diversifying proportion i.e., 5%, 10%, 15%, 20%, 25%, and 30% of the dry wt. of earth. Ceramic tile waste passing through 4.75mm IS sieve and retained in the pan were utilized. The density and water absorption of tile waste was found to be 2.82 gm/cm^3 and 4% respectively.

3.4 Methodology

The red earth that was collected from the site was stabilized by mixing it with waste LDPE plastic pouch strips and waste ceramic tile dust. The waste milk pouches (LDPE) were collected from milk shops and tea shops. They are made of low-density polyethylene (LDPE) which is confirmed by taking the help of IS: 11805: 2007 [26]. Ceramic tile waste was collected from the nearby under-construction buildings and tile shops that were in broken form or waste that was not used in tile work. An experimental investigation was done to calculate the different properties of the collected red earth sample mixed with the milk pouch strips and ceramic tile waste with varying proportions of the dry wt. of earth.

1. Initially the ceramic tile dust and red earth sample were mixed at different proportions, i.e., 5%, 10%, 15%, 20%, 25%, and 30%. The optimum proportion of tile waste would

be fixed concerning the optimum MDD and OMC value and increasing the CBR, and UCS value.

2. Later on, the ceramic tile dust with the optimum dosage of ceramic tiles will be mixed with varying proportions of plastic milk pouch strips to get the optimum value. Plastic milk pouches strips (LDPE) are mixed with red earth and tile waste at 0.25%, 0.5%, 0.75%, 1% and 1.25%.

The basic properties of virgin earth have been found out through laboratory tests. Tests have been conducted with the inclusion of variant proportions of LDPE plastic strips and ceramic tile waste to red earth to analyze the earth behavior, and find out the relationship between them. The laboratory test includes specific gravity, liquid limit, plastic limit, and grain size distribution analysis, SPT, UCS, DST and CBR. The methodology of current study is illustrated in Figure 3.4.

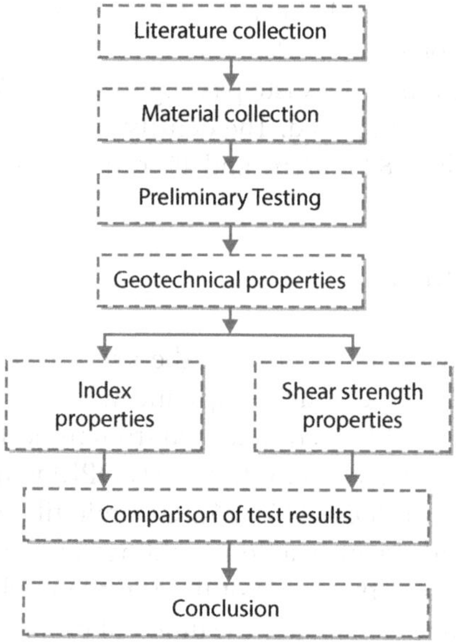

Figure 3.4 Flowchart of methodology.

3.5 Geotechnical Investigation

Important engineering properties that are useful for pavement construction were determined by various laboratory tests which are approved on the selected earth sample, as given in respective IS Codes.

3.5.1 Grain Size Distribution Test

Sieve analysis test was conducted on red earth and TD according to specifications IS: 2720 (Part-4) - 1985 [27]. In the semi-logarithmic graph, we can draw the grain size distribution curve for the provided sample to find out D10, D30, and D60; and from this value, we can find out Cu and Cc.

$$\text{Coefficient of uniformity (Cu)} = D_{60}/D_{10} \tag{3.1}$$

$$\text{Coefficient of curvature (Cc)} = D_{30}^{2}/\,(D_{60}{}^{*}D_{10}) \tag{3.2}$$

3.5.2 Specific Gravity Test

This test was conducted on red earth by density bottle method, i.e., pycnometer according to IS: 2720 (Part-3) - 1980 [28].

3.5.3 Liquid Limit Test

This test was conducted on red earth by using Casagrande apparatus according to specifications IS: 2720 (Part-5) - 1985 [29].

3.5.4 Plastic Limit Test

This test was conducted on red earth according to specifications of IS: 2720 (Part-5) - 1985 [29].

3.5.5 Plasticity Index

It is one of the important index properties of fine-grained earth. It is the difference between the liquid limit and the plastic limit. This index is the minimum water content at which earth displays its plasticity.

$$\text{Plasticity index } (I_P) \text{ (From A- Line)} = (0.73\ (W_L - 20)) \quad (3.3)$$

3.5.6 Standard Protector Test

This test was conducted on red earth and mixed with varying proportion of TD and plastic strips according to specifications of IS: 2720 (Part-7) – 1980 [30]. It is a laboratory method of experimentally calculating the OMC at which a given earth type will become most dense and achieve its MDD.

3.5.7 California Bearing Ratio (CBR) Test

This test was conducted on red earth and mixed with varying proportion of TD and plastic strips according to specifications IS: 2720 (Part-16) – 1965 [31].

$$CBR = (P_T/P_S)*100 \quad (3.4)$$

P_T = Corrected unit load corresponding to the chosen penetration
P_S = Unit standard load for the same depth of penetration as for PT

3.5.8 Unconfined Compressive Strength (UCS) Test

This test was conducted on red earth and mixed with varying proportion of TD and plastic strips according to specifications IS: 2720 (Part-10) – 1991 [32].

$$S_{u=} Q_u/2 \quad (3.5)$$

Where, S_u = shear strength of earth
Q_u = compressive strength of earth

3.5.9 Direct Shear Test

DST test was conducted on red earth and mixed with varying proportion of TD and a plastic strip according to specifications of IS: 2720 (Part-13) 1986 [33].

The formula is as follows:

$$(\tau = c + \varsigma * \tan \varphi) \tag{3.6}$$

3.6 Results and Discussion

3.6.1 Index Properties of Virgin Earth

The properties and various characteristics of red earth collected locally are given in Table 3.1.

Table 3.1 Index properties of virgin earth.

Properties	Results
Specific gravity (G)	2.628
Liquid limit (LI)	25.605%
Plastic limit (PI)	14.190%
Plastic index (I_p)	11.415 By A-Line I_p = 4.092
Coefficient of uniformity (Cu)	4.832
Coefficient of curvature (Cc)	0.948
OMC	15.820%
MDD	1.747 gm/cm^3
CBR (Un-Soaked)	7%
CBR (Soaked)	6.38%
UCS	119.507 kN/m^2
Cohesion (C)	0.2567 kg/cm^2
Internal angle of friction (Ø)	29.3^o

3.6.2 Standard Proctor Test/Light Compaction Test

3.6.2.1 OMC and MDD of Ceramic Tile Dust Mixed Earth

From the below Table 3.2 and Figure 3.5, it has been observed that the value of lowered and MDD raised as the proportion of tile dust content rises from 5% to 30%. The MDD value increased from 1.747 gm/cm^3 to 1.843 gm/cm^3 when ceramic dust incorporation is from 0% to 30%. The reason for such behavior could be due to the replacement of ceramic dust particle having high specific gravity (2.82) with the earth particles having less specific gravity (2.62). However, with a rise in dosages of tile dust from

Table 3.2 OMC and MDD of stabilized earth with ceramic tile dust.

Tile dust (TD) mixed soil (%)	OMC (%)	MDD (gm/cm^3)
Virgin soil	15.820	1.747
Soil + 5% T.D	14.800	1.771
Soil + 10% T.D	14.500	1.784
Soil + 15% T.D	13.850	1.793
Soil + 20% T.D	13.560	1.808
Soil + 25% T.D	13.400	1.822
Soil + 30% T.D	13.290	1.843

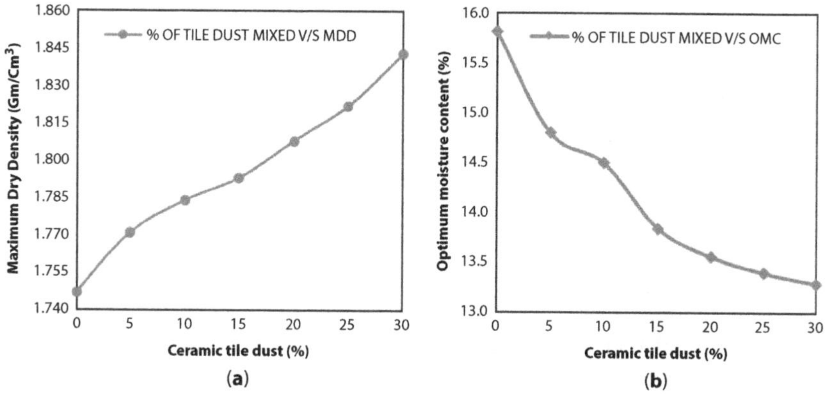

Figure 3.5 Content of ceramic waste dust V/S (a) MDD, and (b) OMC.

0% to 30%, the OMC of earth lowered from 15.82% to 13.29%. The reason for this reduction could be due to the lesser capacity of ceramic dust particles to attract the water molecules compared to the earth particles.

3.6.2.2 OMC and MDD of Earth With Optimum Dosage of Tile Dust Mixed with Plastic (LDPE)

From the below Table 3.3 and Figure 3.6, it has been observed that OMC increased from 13.15% to 13.75% and MDD also increased from 1.853

Table 3.3 OMC and MDD of stabilized earth with optimum tile dust mixed with plastic.

Optimum tile dust with plastic mixed soil	OMC (%)	MDD (gm/cm³)
Virgin soil + 25% TD	13.400	1.822
Soil + 0.25% LDPE + 25% TD	13.150	1.853
Soil + 0.50% LDPE + 25% TD	13.410	1.877
Soil + 0.75% LDPE + 25% TD	13.750	1.902
Soil + 1.00% LDPE + 25% TD	14.300	1.882
Soil + 1.25% LDPE + 25% TD	14.700	1.870

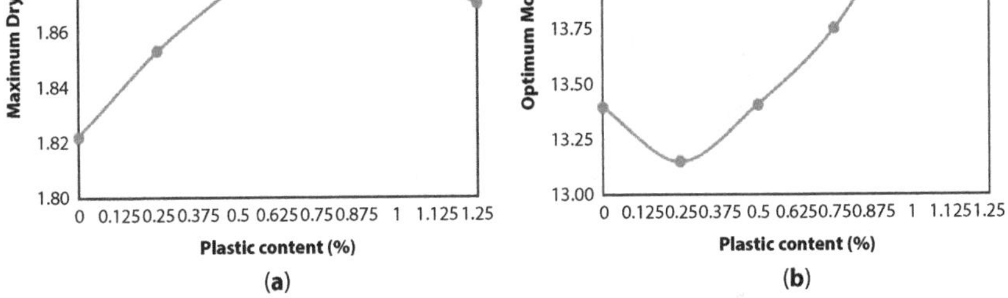

Figure 3.6 Ceramic tile dust (25%) mixed with different percentages of plastic V/S (a) MDD, and (b) OMC.

gm/cm^3 to 1.902 gm/cm^3 as the dosages of plastic strip content raised from 0.25% to 0.75%. However, a further rise in dosages of plastic strip content from 0.75% to 1.25% resulted in a rise of OMC and lowered MDD. The conceivable reason for this rise in MDD of the earth by the addition of tile dust and LDPE is due to the proper rearrangement of earth particles and the addition of non-plastic material which improves the binding capacity.

3.6.3 Unsoaked California Bearing Ratio (CBR)

3.6.3.1 CBR Value (Unsoaked) for Earth Stabilized with Ceramic Tile Dust

From the below Table 3.4 and Figure 3.7, it has been observed that the value of unsoaked CBR increased with a rise in the proportion of tile dust content up to a certain limit; after that, it lowered. The maximum CBR value observed with 25% of ceramic tile dust by dry wt. of earth. The unsoaked CBR raised from 7% to 8.88% as the ceramic dust was raised from 0% to 25%. The unsoaked CBR value depends upon the OMC, MDD, and the amount of tile dust present in the earth. The CBR value of the earth, by the addition of tile dust in comparison with the original earth is due to the greater density of modified earth mass having more strength. As MDD raised with a rise in the proportion of ceramic dust, it resulted in the increase in both the UCS and unsoaked CBR values of the earth.

Table 3.4 CBR (unsoaked) results of tile dust stabilized earth.

Tile dust (TD) mixed in soil (%)	CBR value (%)
Virgin Soil	7.00
Soil + 5% TD	7.11
Soil + 10% TD	7.42
Soil + 15% TD	7.84
Soil + 20% TD	8.36
Soil + 25% TD	8.88
Soil + 30% TD	8.57

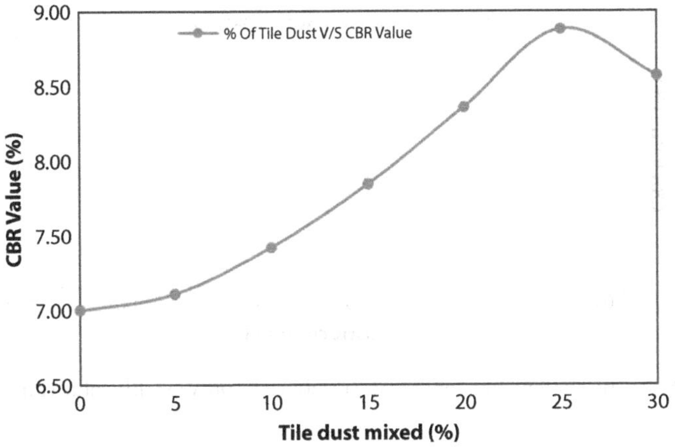

Figure 3.7 Ceramic tile dust stabilized earth V/S CBR value (un-soaked).

3.6.3.2 CBR Value (Unsoaked) for Earth Stabilized with Tile Dust and LDPE

From Table 3.5 and Figure 3.8, it has been observed that the value of unsoaked CBR improved for earth with 25% tile dust as the proportion of plastic strip content increased, up to a certain limit; later on, it lowered. The maximum CBR value occurred at 0.75% of LDPE content. Any further addition of plastic strips content, i.e., from 0.75% to 1.25%, the CBR value lowered. Unsoaked CBR value depends upon the OMC, MDD, and the amount of plastic strips present in the earth. The rise in CBR value may be attributed to the reduction in the void ratio of the earth with the addition

Table 3.5 CBR (unsoaked) results of earth stabilized with tile dust and LDPE.

LDPE plastic and 25% tile dust mixed in soil	CBR value (%)
Virgin Soil + 25% TD	8.88
0.25% LDPE + 25% TD	9.62
0.50% LDPE + 25% TD	10.03
0.75% LDPE + 25% TD	12.44
1.00% LDPE + 25% TD	11.92
1.25% LDPE + 25% TD	11.50

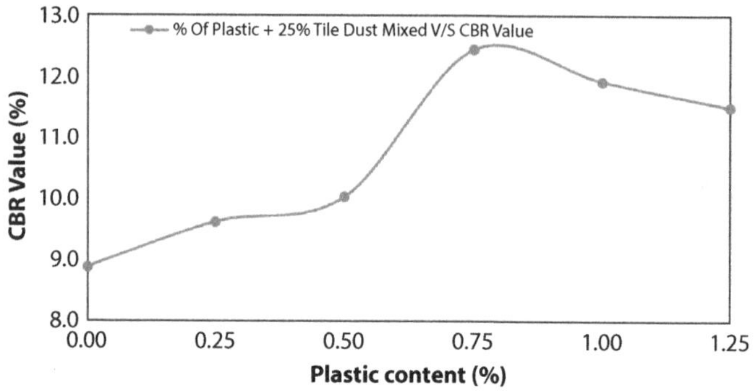

Figure 3.8 Plastic content and 25% tile dust stabilized soil V/S CBR value (un-soaked).

of tile dust, and also due to the high specific gravity of this additive. So, the binding between earth particles is not properly done or the binding capacity of earth becomes less. Therefore, the unsoaked CBR value as well as the strength of earth specimen with LDPE plastic strips also becomes less.

3.6.4 Soaked California Bearing Ratio (CBR)

3.6.4.1 CBR Value (Soaked) for Earth Stabilized with Ceramic Tile Dust

From the below Table 3.6 and Figure 3.9, it has been observed that the value of soaked CBR raised up to 25% of tile dust of the dry wt. of earth.

Table 3.6 CBR (soaked) results of tile dust stabilized earth.

Tile dust (TD) mixed in soil (%)	CBR value (%)
Virgin Soil	6.38
Soil + 5% TD	6.59
Soil + 10% TD	6.90
Soil + 15% TD	7.11
Soil + 20% TD	7.42
Soil + 25% TD	8.05
Soil + 30% TD	7.84

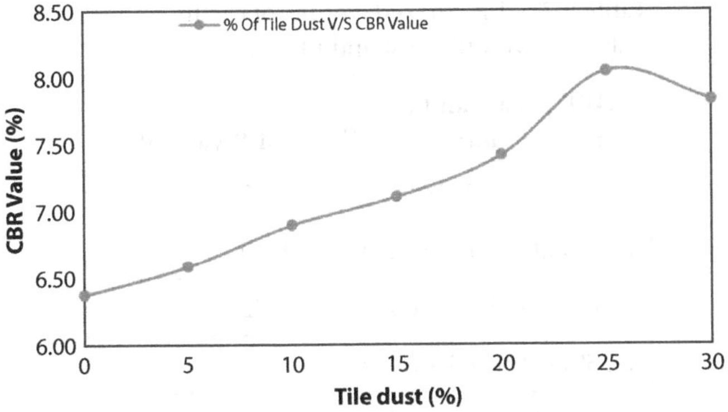

Figure 3.9 Ceramic tile dust stabilized earth V/S CBR value (soaked).

The soaked CBR increased from 6.38% to 8.05% with the ceramic dust content from 0% to 25%. Soaked CBR value depends upon the OMC, MDD, and the amount of tile dust present in the earth. The greater CBR value of the earth is due to the addition of tile dust in comparison with the original earth due to the larger density of modified earth mass having more strength. As MDD enhanced with a rise in the proportion of ceramic dust, it raised both the UCS and soaked CBR values of the earth. The CBR values of the earth in a soaked condition are less than that in an unsoaked condition. This is because soaking the sample specimen in water for four days results in a higher water absorption by the earth, which in turn lowers the penetration resistance.

3.6.4.2 CBR Value (Soaked) for Earth Stabilized with Tile Dust and LDPE

From the below Table 3.7 and Figure 3.10, it has been observed that the value of soaked CBR increased with a rise in the proportion of plastic strip content up to a certain limit, in stabilized earth with 25% tile dust by dry wt. of earth. Maximum soaked CBR value was observed with 0.75% of plastic content. Further addition of plastic strips i.e., from 0.75% to 1.25%, the CBR value lowered. Soaked CBR value depends upon the OMC, MDD, and the amount of plastic strips present in the earth. The rise in value may be attributed to the reduction in void ratio of the earth with the addition of tile dust, and also due to the high specific gravity of this additive.

Table 3.7 CBR (soaked) results of earth stabilized with tile dust and LDPE.

LDPE plastic and 25% tile dust mixed in soil	CBR value (%)
Virgin Soil + 25% TD	8.05
0.25% LDPE + 25% TD	9.20
0.50% LDPE + 25% TD	9.72
0.75% LDPE + 25% TD	12.02
1.00% LDPE + 25% TD	11.71
1.25% LDPE + 25% TD	11.18

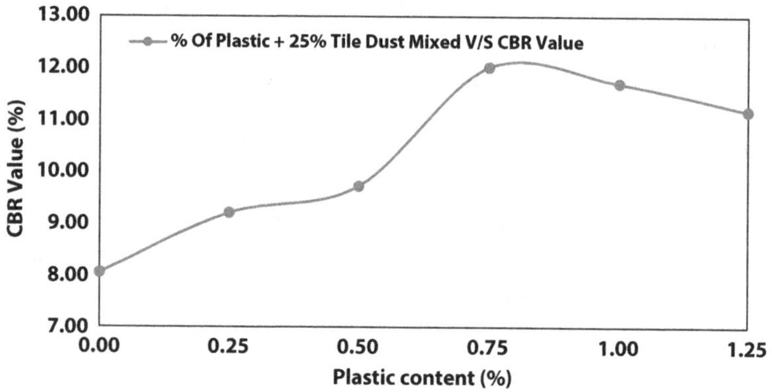

Figure 3.10 Plastic content and 25% tile dust stabilized soil V/S CBR value (soaked).

3.6.5 Unconfined Compressive Strength Test (UCS)

3.6.5.1 UCS Value of Ceramic Tile Dust Stabilized Earth

From the below Table 3.8 and Figure 3.11, it has been observed that the value of UCS raised with the rise of tile dust content up to a certain limit. The maximum UCS value was at 25% of tile dust by dry wt. of earth. The UCS increased from 119.507 kN/m^2 to 201.489 kN/m^2, when ceramic dust content was raised from 0% to 25%. After that, the strength lowered to 193.368 kN/m^2 at 30% tile dust content. As MDD raised with the rise of ceramic dust; both the UCS and CBR values of the stabilized earth also raised.

Table 3.8 UCS results of tile dust stabilized earth.

Tile dust (TD) mixed in soil (%)	UCS (kN/m²)
Virgin Soil	119.507
Soil + 5% TD	132.564
Soil + 10% TD	153.645
Soil + 15% TD	161.123
Soil + 20% TD	176.154
Soil + 25% TD	201.489
Soil + 30% TD	193.368

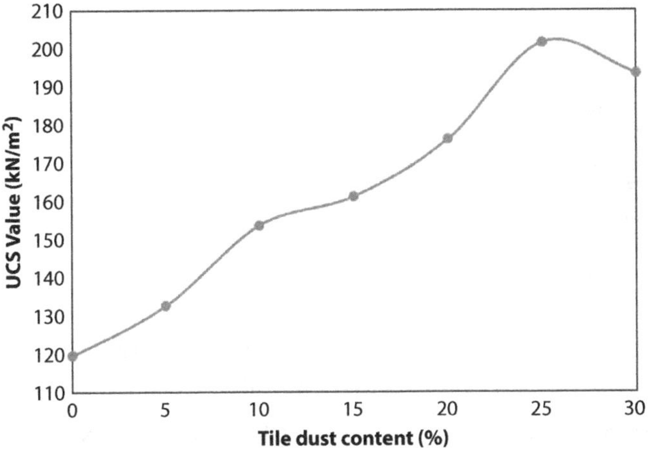

Figure 3.11 Ceramic tile dust content V/S UCS value.

3.6.5.2 UCS Value for Earth Stabilized with Tile Dust and LDPE

From the below Table 3.9 and Figure 3.12, it has been observed that the value of UCS raised with the rise of plastic strip content up to a certain limit in 25% tile dust-added earth. The maximum UCS value was obtained at 0.75% of plastic content. Further addition of the proportion of plastic strips, i.e., from 0.75% to 1.25%, the CBR value lowered. The UCS value rises from 201.489 kN/m² to 282.241 kN/m², when plastic content was raised from 0.25% to 0.75%. After that, the strength decreases to 252.282 kN/m² at 1.25% plastic content. As MDD raised with a rise in ceramic dust content, the UCS and CBR values of the stabilized earth also raised.

Table 3.9 UCS results of earth stabilized with tile dust and LDPE.

LDPE plastic and 25% tile dust mixed in soil	UCS (kN/m²)
Virgin Soil + 25% TD	201.489
0.25% LDPE + 25% TD	217.861
0.50% LDPE + 25% TD	245.491
0.75% LDPE + 25% TD	282.241
1.00% LDPE + 25% TD	266.705
1.25% LDPE + 25% TD	252.282

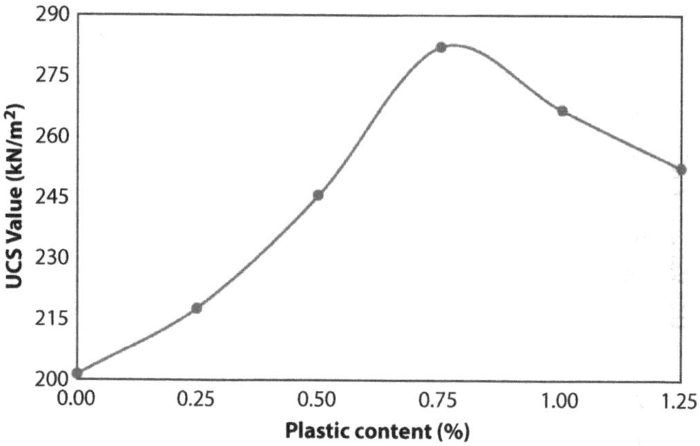

Figure 3.12 Plastic content and 25% tile dust stabilized earth V/S UCS value.

3.6.6 Direct Shear Test (DST)

3.6.6.1 Cohesion (C) and Angle of Internal Friction (Ø) of Ceramic Tile Dust Stabilized Earth

From the below Table 3.10 and Figure 3.13, it has been observed that cohesion (C) lowered but the internal angle of friction (Ø) increases as the proportion of tile dust content rises from 5% to 30%. As the content of ceramic dust rises from 0% to 30%, Ø of earth goes on increasing from 29.3° to 38.92°. However, with the rise in proportion of ceramic dust from 0% to 30%, the C of earth goes on decreasing, i.e., 0.2567 kg/cm² to 0.1435 kg/cm².

Table 3.10 Direct shear test results of tile dust stabilized earth.

Tile dust (TD) mixed in soil (%)	Cohesion (C) (kg/cm²)	Angle of internal friction (Ø) (Degree)
Virgin Soil	0.2567	29.3
Soil + 5% TD	0.2219	31.17
Soil + 10% TD	0.2134	32.28
Soil + 15% TD	0.1942	34.21
Soil + 20% TD	0.1871	35.72
Soil + 25% TD	0.1608	37.82
Soil + 30% TD	0.1435	38.92

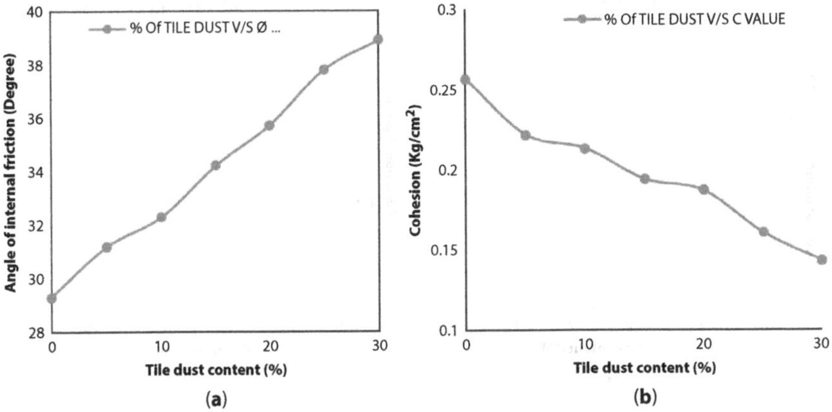

Figure 3.13 Ceramic tile dust content V/S (a) angle of internal friction (Ø), and (b) cohesion.

3.6.6.2 Cohesion (C) and Angle of Internal Friction (Ø) for Earth Stabilized with Tile Dust and LDPE

From the below Table 3.11 and Figure 3.14, it has been observed that both cohesion (C) and internal angle of friction (Ø) raised with a rise in the proportion of plastic strip content from 0.25% to 0.75%. Further, a rise in the proportion of plastic strip content from 0.75% to 1.25% resulted in a

Table 3.11 Direct shear test results of tile dust stabilized soil stabilized with tile dust and LDPE.

LDPE plastic and 25% tile dust mixed in soil	Cohesion (C) (kg/cm^2)	Angle of internal friction (Ø) (Degree)
Virgin Soil + 25% TD	0.1608	37.82
0.25% LDPE + 25% TD	0.1812	40.59
0.50% LDPE + 25% TD	0.2069	42.26
0.75% LDPE + 25% TD	0.2371	44.12
1.00% LDPE + 25% TD	0.2605	45.23
1.25% LDPE + 25% TD	0.2532	43.59

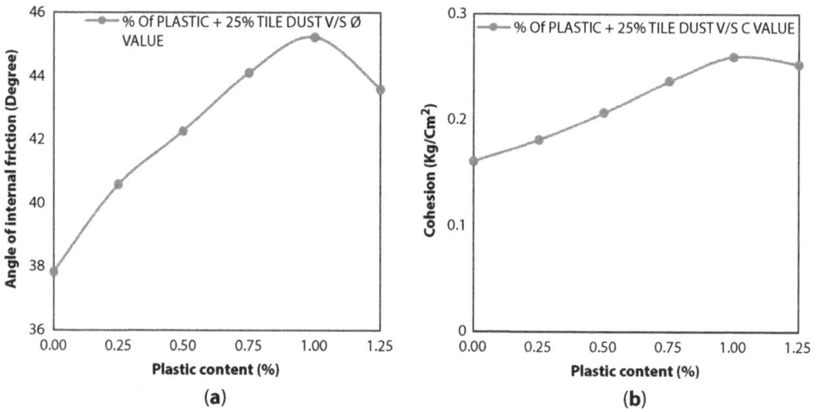

Figure 3.14 Plastic content and 25% tile dust stabilized soil V/S (a) angle of internal friction (Ø), and (b) cohesion.

reduction of C and Ø. Initially, the Ø value raised from 37.82° to 45.23° and then lowered to 43.59°. In the same way, C value raised from 0.1608 kg/cm^2 to 0.2605 kg/cm^2; after that, it lowered to 0.2532 kg/cm^2.

3.7 Conclusion

In the current research work, a series of SPT, CBR, UCS, and DST were conducted and the results of the tests have been presented and discussed in the above sections. The conclusion drawn based on these results and discussion was summarized below:

1. OMC value lowered but MDD value increased after adding tile dust at varying proportions of 0% to 30%. MDD value increased from 1.747 gm/cm^3 to 1.843 gm/cm^3 and OMC lowered from 15.82% to 13.29%. Such performance is due to the substitution of tile dust particles with earth particles. After mixing the optimum tile dust content (25%) with varying proportions of LDPE plastic, OMC value increased and MDD also raised up to a certain limit, then lowered.
2. UN-Soaked CBR value increased from 7% to 8.57% with the variations of tile dust content from 0% to 25%. In the same way, soaked CBR value also raised. After adding plastic waste with optimum content of tile dust (25%), the unsoaked CBR value rises to 12.44% and the soaked CBR value rises to 12.02%.
3. Virgin earth UCS value was found to be 119.507 kN/m^2. After mixing tile dust up to 30%, UCS raised to 155.115 kN/m^2. Then, with the addition of LDPE plastic waste with optimum content of tile dust, the UCS value raised up to 243.919 kN/m^2.
4. Cohesion (C) value decreased with the addition of tile dust; however, the angle internal friction (Ø) increased. The maximum Ø value was observed to be 38.92°. Cohesion is lowered up to 0.1435 kg/cm^2. After adding LDPE plastic waste with varying proportions, C and Ø value raised by up to 1% of plastic content, and after that, it started to lower.
5. From the current study, it can be inferred that an addition of 25% of ceramic tile dust with 0.75% LDPE plastic strips is the optimum amount of admixture that can be efficiently used for red earth.
6. The combination of ceramic tile waste and waste LDPE plastic strips can be used as very good stabilizing agents to enhance the geotechnical properties of earth.

References

1. Vivek, Dutta, R.K., Parti, R., Effect of chemical treatment of the coir geotextiles on the interface properties of sand/clay-coir geotextile interface. *J. Inst. Eng. India Ser. A*, 100, 357–365, 2019.
2. Mohanta, N.R. and Samantaray, S., Study of Combined Effect of Metakaolin and Steel Fiber on Mechanical Properties of Concrete. *Pertanika J. Sci. Technol.*, 27, 3, 1381–1396, 2019, http://journals-jd.upm.edu.my/resources/files/Pertanika%20PAPERS /JST%.
3. Vivek, and Dutta, R.K., Bearing ratio behavior of sand overlying clay with treated coir geotextiles at the interface. *J. Nat. Fibers*, 19, 14, 7534–7541, 2022.
4. Vivek, Effects of cyclic loading on sand overlaying clay model of unpaved roads reinforced with untreated/treated coir geotextiles. *J. Text. Inst.*, 115, 9, 1694–1700, 2024.
5. Singh, P., Dash, H.M., Samantaray, S., Effect of silica fume on engineering properties of expansive soil. *Mater. Today Proc.*, 33, 5035–5040, 2020, https://doi.org/10.1016/j.matpr.2020.02.839.
6. Kumar, M.H., Mohanta, N., Samantaray, S., Kumar, N.M., Combined effect of waste glass powder and recycled steel fibers on mechanical behavior of concrete. *SN Appl. Sci.*, 3, 1–18, 2021a, https://doi.org/10.1007/s42452-021-04422-w.
7. Kumar, M.H., Macharyulu, I.S., Ray, T., Mohanta, N.R., Jain, M., Samantaray, S., Sahoo, A., Effect of water absorption and curing period on strength and porosity of triple blended concrete. *Mater. Today Proc., 43*, 2162–2169, 2021b, https://doi.org/10.1016/j.matpr.2020.12.092.
8. Ray, T., Mohanta, N.R., Kumar, M.H., Samantaray, S., Study of effect of temperature on behavior of alkali activated slag concrete. *Mater. Today Proc., 43*, 1352–1357, 2021, https://doi.org/10.1016/j.matpr.2020.09.169.
9. Kumar, M.H., Saikrishnamacharyulu, I., Mohanta, N.R., Ashutosh, A., Mishra, P., Samantaray, S., Mechanical behaviour of high strength concrete modified with triple blend of fly ash, silica fume and steel fibers. *Mater. Today Proc., 65*, 933–942, 2022a, https://doi.org/10.1016/j.matpr.2022.03.528.
10. Kumar, M.H., Mohanta, N.R., Patel, N., Samantaray, S., Reddy, S.V.B., Influence of Fly ash and Metakaoline on the crack resistance and shrinkage property of concrete. *Iran. J. Sci. Technol. Trans. Civ. Eng.*, 46, 2011–2026, 2022b, https://doi.org/10.1007/s40996-021-00748-y.
11. Jaswal, P., Vivek, Sinha, S.K., Improvement in the performance of two layered model pavement with treated coir geotextile at the interface. *J. Ind. Text., 52*, 15280837221114161, 2022.
12. Vivek, Shafi Mir, M., Sehgal, R., Studies of Modulus of Resilience on Unpaved Roads Reinforced with Untreated/Treated Coir Geotextiles. *J. Nat. Fibers, 19*, 16, 13563–13573, 2022.

13. Vivek, Shafi Mir, M., Sehgal, R., Study on bearing capacity of unpaved roads reinforced with coir geotextiles using finite element method (fem). *J. Nat. Fibers*, *19*, 15, 11735–11748, 2022.
14. Jaswal, P. and Vivek, Experimental study on monotonic behaviour of two-layered unpaved road model reinforced with treated coir geotextiles. *Int. J. Pavement Res. Technol.*, 17, 1377–1388, 2023.
15. Singh, J., Singh, B., Singla, S., Enhancement of expensive soil by addition of stone dust and LDPE Fibre. *Int. J. Res. Appl. Sci. Eng. Technol.*, *9*, 1, 910–916, 2021.
16. Sharma, R.K., Utilization of fly ash and waste ceramic in improving characteristics of clayey soil: a laboratory study. *Geotech. Geol. Eng.*, *38*, 5, 5327–5340, 2020.
17. Mishra, P. and Arora, V.K., Stabilization of silty soil with marble dust and sugarcane bagasse ash, in: *Recycled Waste Materials: Proceedings of EGRWSE 2018*, pp. 51–58, 2019.
18. Phonsa, R., Stabilization of clayey soil by using stone dust and plastic bottle strips in subgrades, *IJITEE*, 8, 7, 1627–1629, 2019.
19. Suriya, P.A., Sangeetha, S.P., Abirami, R., Subathra, P., Stabilization of red soil using polypropylene. *Mater. Today Proc.*, *46*, 5881–5884, 2021.
20. Mali, S., Kadam, S., Mane, S., Panchal, K., Kale, S., Navkar, Y., Soil stabilization by using plastic waste. *Int. Res. J. Eng. Technol. (IRJET)*, *6*, 4056–4060, 2019.
21. Harini, V., Harsha, R., Durga Dhanalakshmi, S., Effect on Properties of Expansive Soil by using LDPE Plastic Wastes. *SSRG Int. J. Civ. Eng. (SSRG - IJCE)*, 5, 7, 13–16, 2018.
22. Hassan, H.J.A., Rasul, J., Samin, M., Effects of plastic waste materials on geotechnical properties of clayey soil. *Transp. Infrastruct. Geotechnol.*, 8, 3, 390–413, 2021.
23. Iravanian, A. and Saber, S.A., Using ceramic wastes in stabilization and improving soil structures: A review study. *IOP Conf. Ser.: Earth Environ. Sci.*, 614, 1, 012081, December 2020.
24. Hiranandani, P., Ameta, K., Purohit, D.G.M., Performance of ceramic tiles wastage on permeability of dune sand. *JETIR*, 4, 11, 265–269, 2017.
25. Panwar, K. and Ameta, N.K., Stabilization of Fine Sand with Ceramic Tiles Waste as Admixture for Construction of Embankment. *Am. J. Eng. Res.*, 5, 8, 206–212, 2016.
26. IS: 11805: 2007, Polyethylene Pouches for Packaging Liquid Milk- Specification, 2007.
27. IS: 2720 (PART-4): 1985, Methods of test for soils: Grain size analysis, Bureau of Indian Standards, 1985.
28. IS: 2720 (PART-3): 1980, Methods of test for soils: Determination of specific gravity, Section 1: Fine grained soils, Bureau of Indian Standards, 1980.
29. IS: 2720 (PART-5): 1985, Methods of test for soils: Determination of liquid and plastic limit, Bureau of Indian Standards, 1985.

30. IS: 2720 (PART-7): 1980, Methods of test for soils: Determination of water content-dry density relation using light compaction, Bureau of Indian Standards, 1980.
31. IS: 2720 (PART-16): 1965, Methods of test for soils, Determination of California Bearing Ratio, Bureau of Indian Standards, 1965.
32. IS: 2720 (PART-10): 1991, Methods of test for soils, Determination of unconfined compressive strength, Bureau of Indian Standards, 1991.
33. IS: 2720 (PART-13): 1986, Methods of test for soils, Direct Shear Test, 1986.

4

Experimental Investigation of the Effect of Bentonite on the Cyclic Resistance of Pond Ash

Anuj Kumar Patel, Vamsi Alla, G. Suneel Kumar and Rabi Narayan Behera*

Department of Civil Engineering, NIT Rourkela, Odisha, India

Abstract

This study evaluates the viability of pond ash and bentonite mixture as a dynamically stable lining material. In 5% increments, bentonite content in pond ash-bentonite (PA+B) mixture is increased from 5% to 25% by weight. Various geotechnical properties of PA+B mixes were examined in the laboratory. The behavior of waste liner systems under dynamic loading, with a focus on how adding bentonite affects cyclic resistance of pond ash is investigated. Stress-controlled, undrained cyclic triaxial tests were done at 1 Hz to check the influence of confining pressure on resistance to liquefaction. For this, three test sets were done at 100, 200, and 300 kPa confining pressure. To find out how the confining pressure affects liquefaction resistance, the combination of 80% PA and 20% B was given reversible shear stress in a planned way by changing the Cyclic Stress Ratio (CSR) from 0.1 to 0.9 at 0.2 intervals. A compacted mixture of pond ash-bentonite (PA+B) with 20% bentonite possessed the strength and hydraulic conductivity needed for the liner. The pore water pressure in PA+B mixtures with 10% bentonite is high, but it moves down as the quantity of bentonite goes up to 20% in the pond ash mixture. This causes the cyclic resistance to go up.

***Keywords*:** Pond ash, bentonite, hydraulic conductivity, seismic response, landfill liner

**Corresponding author*: rnbehera82@gmail.com

Sandeep Samantaray, Subhashree Samantasinghar, Soumendra K. Mohanty, Vivek, Abinash Sahoo and Deba Prakash Satapathy (eds.) Waste Utilization in Geotechnical Practice, (99–116) © 2025 Scrivener Publishing LLC

4.1 Introduction

The growing municipal waste due to industrialization and urbanization presents disposal challenges [1–4]. Land filling, favored for lower costs, demands a robust hydraulic barrier to contain waste effectively. Clay soils and other earth materials are used for various geotechnical constructions [5–9]. Seismic activities pose risks like ground deformations and liner damage, urging geotechnical assessments. Large earthquake-induced ground deformations have been observed to cause severe damage to liners. Under significant amplitude cyclic loading, additional pore water pressure is generated, leading to reduction in soil rigidity and strength. During an earthquake, increase in pore pressure causes soil to lose strength and rigidity.

When the water content and dry density of the coal ash-bentonite mix are at their optimum levels, the bonding between the mix particles is enhanced, increasing the strength and longevity of the liner [10]. They [11], verified local lining materials: pros and cons of red earth, - 20% bentonite mix, illite - 20% bentonite mix, fly ash - 20% bentonite, and illite - 1% lime. Wetting-drying cycles affected sand-bentonite swelling pressures with and without lime. Lime stabilized sand-bentonite combinations had lower swelling pressures than those without lime. Wetting-drying cycles reduced the lime stabilization's swelling pressure-reducing impact [12]. Bentonite, being a less permeable clay helps in stopping leachate from reaching groundwater. But bentonite exhibits high swelling and shrinkage characteristics because of its high montmorillonite content that poses the risk of crack development. The compacted fly ash-bentonite mixture's permeability reduced with bentonite content, 20% of bentonite in fly ash met landfill liner permeability requirements at 1×10^{-7} cm/sec. For pond ash, 12% bentonite was used [13]. They [14], examined the effect of saturation degree and compaction energy on properties pertaining to the mechanical strength of compacted pond ash. Shear parameters were obtained from normal stress vs shear stress plots. They [15], found that specific gravity, swell index, maximum dry unit weight, and plasticity index increase with the increase in bentonite content. They found that 25% bentonite and 75% fly ash mixture as optimum mix, satisfying the criteria of permeability and UCS to be utilized as landfill liner material. EPA soil liners must have 1×10^{-7} cm/s hydraulic conductivity. They [16], found no correlation amid the clay content in sand and its cyclic shear strength, and they concluded that an increased value of plasticity index leads to an increase in cyclic strength. They [17], found pond ash and bentonite as landfill liners in a laboratory setting. Bentonite increased the liner's hydraulic conductivity, whereas pond ash increased its shear strength. The study suggested using pond ash and bentonite as cheaper landfill liners. They [18], studied the mechanical properties of clay-bentonite mixes and determined that adding bentonite increases the total number of loading cycles required by a clay specimen before undergoing liquefaction at a particular cyclic stress ratio under undrained cyclic triaxial conditions. They [19], simulated a geo synthetic clay liner's cycle. Cyclic loading decreased GCL shear strength, causing liner deformation and failure. The researchers suggested placing a drainage layer between the GCL and compacted clay liner to lower pore pressure and improve cycle efficiency. They [20], conducted landfill simulations and found that cyclic loading decreased liner system's shear strength, causing cracks and failure. To optimize liner system cycling, researchers suggested adding a drainage layer or thickening of the geomembrane.

4.2 Materials and Methodology

4.2.1 Materials

Cohesion less material includes pond ash, collected from Rourkela Steel Plant, Odisha. Due to its primary mineral composition, the montmorillonite group, bentonite shows very significant swelling and shrinkage properties with changes in water content. The bentonite employed in this investigation is a sodium-based kind that may be purchased in nearby markets.

4.2.2 Methodology

Geotechnical properties of pond ash, bentonite and their mixes were determined through conducting various laboratory tests such as test on specific gravity following IS 2720 Part-3/Sec 2 (1980) [21], Atterberg limits from IS 2720 Part-5 (1985) [22], particle size distribution obtained from IS 2720 Part-4 (1985) [23], free swell index test (DFS) as specified in IS 2720 Part-40 (1977) [24], linear shrinkage in accordance with IS 2720 Part-6 (1972) [25], relationship between water content and dry density by light compaction in accordance with IS 2720 Part-7 (1980) [26] and by heavy compaction following IS 2720 Part-8 (1980) [27], hydraulic conductivity following IS 2720 Part-17 (1986) [28], shear strength as specified in IS 2720-11 (1993) [29], unconfined compressive strength (UCS) test in accordance with IS 2720-10 (1991) [30]. Based on the CPCB guidelines for liner material, PA+B mixes satisfying the required criteria will be selected for dynamic analysis. The unconsolidated undrained tests on the soil specimens were conducted using triaxial testing apparatus in accordance with

ASTM D 5311 [31] to determine cohesion intercept, angle of internal friction, and Mohr-Coulomb shear failure envelopes.

4.3 Experimental Outcomes and Discussion

Basic geotechnical properties of bentonite, pond ash, and their mixes with bentonite content ranging from 0% to 25% with 5% intervals were determined and their variation with respect to bentonite content is discussed.

After having around 12% of bentonite added to it, the non-plastic pond ash mixture began to show signs of plasticity after this additive was put into it. The variation of plasticity index (Ip) can be seen in Figure 4.1 (d). It was determined that 5% was the maximum allowable shrinkage for bentonite. The amount of shrinkage that could be tolerated by the mixture of bentonite and pond ash was significantly elevated.

4.3.1 Specific Gravity

Specific gravity (G) of both materials was measured through four consecutive repetitions, and the average values of specific gravity (G) of both pond ash and bentonite were determined as 2.12 and 2.64 respectively.

4.3.2 Atterberg Limits

Pond ash's (PA) liquid limit was found to be 40.23%, which is incredibly low when compared to bentonite's (B). The liquid limit of bentonite used for the study was 264%. The PA+B mixture's liquid limit was substantially increased by the bentonite addition to the pond ash. Figure 4.1(a) shows the variation.

4.3.3 Swelling and Shrinkage Characteristics

Differential Free Swell (DFS) for bentonite was 55.6%. Because of the significant cation exchange capability it possesses, bentonite exhibits extremely high swelling behavior. Variation of DFS concerning bentonite content in PA+B mixes is shown in Figure 4.2a. The bentonite that was utilized in this investigation had a high linear shrinkage index of 44.65% and displayed clear signs of desiccation cracking. From Figure 4.2b, it is clear that the addition of more bentonite in PA+B mixes increased the linear shrinkage index (LSI). For bentonite content less than 20%, the LSI was within 10%.

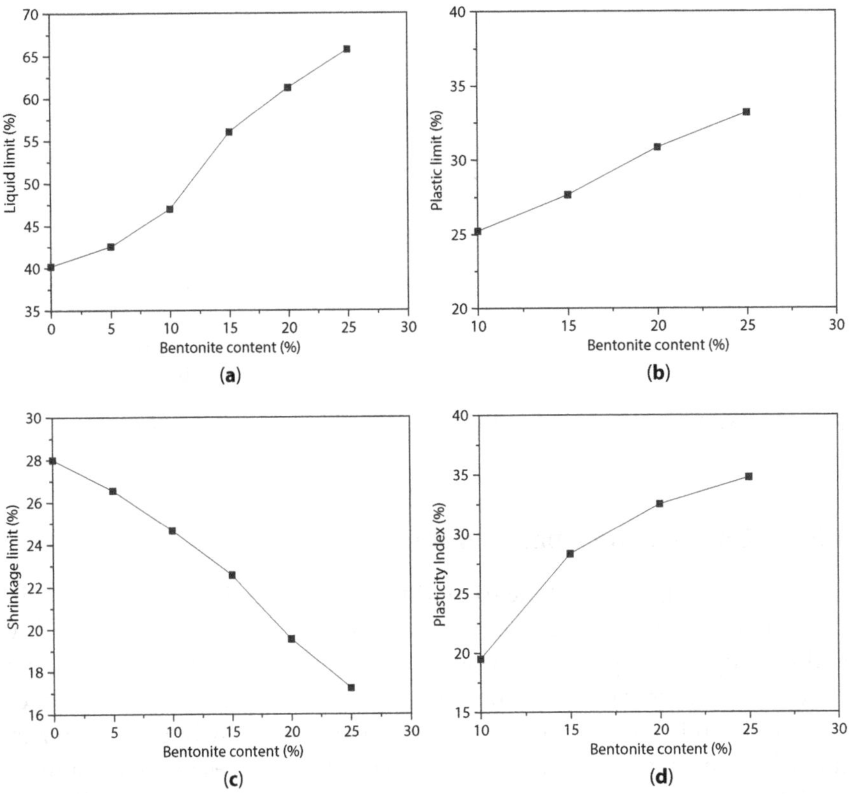

Figure 4.1 Variation of Atterberg limits of PA+B mixes with increase in bentonite content.

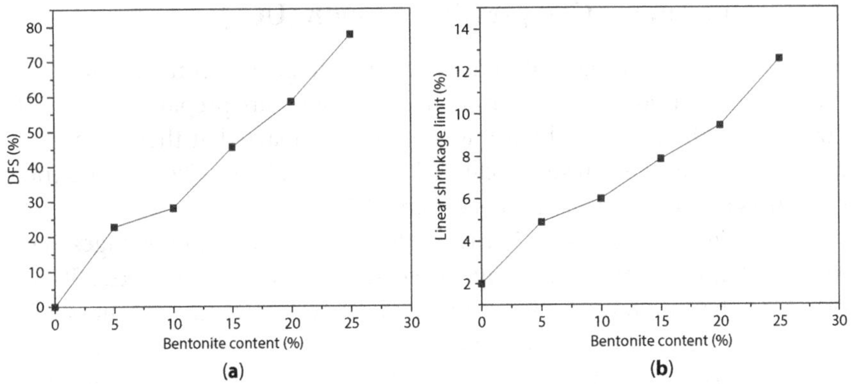

Figure 4.2 Variation of DFS and LSI of PA+B mixes with increase in bentonite content.

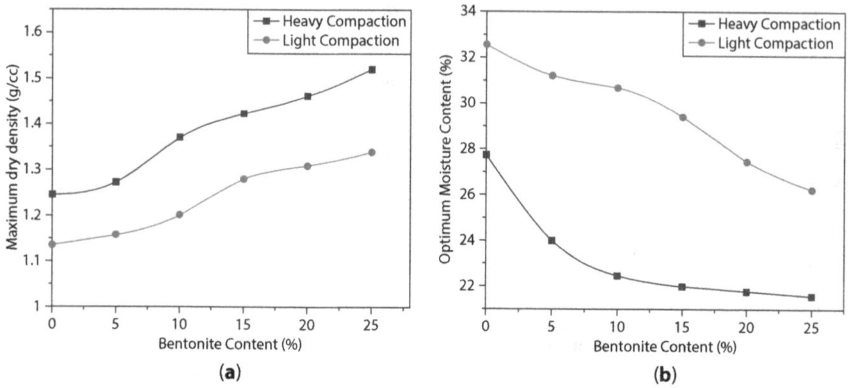

Figure 4.3 Variation of MDD and OMC of PA+B mixes with increase in bentonite content.

4.3.4 Compaction Characteristics

Compaction is generally done to modify geo-engineering behavior of soil. Compaction increases the shear strength of soil, thus increasing stability and bearing capacity. Compaction reduces compressibility and hydraulic conductivity/permeability of the soil. To determine the required compaction effort and moisture content for construction of hydraulic barrier in the field, compaction tests are conducted on PA+B mixes in the laboratory by light and heavy compaction methods. Variation of OMC and MDD of PA+B mixes with respect to bentonite content is presented in Figure 4.3.

4.3.5 Unconfined Compressive Strength (UCS)

To determine the impact of bentonite percentage and compaction effort on UCS of the pond ash-bentonite mixes, samples are prepared at 0%, 5%, 10%, 15%, 20% and 25% bentonite content compacted at their respective MDD and OMC. UCS tests are carried out for 0, 7, 15 and 28 days and their outcomes are revealed in Figures 4.4 and 4.5.

From Figures 4.4 and 4.5, the increased bentonite percentages have increased the unconfined compressive strength of PA+B mixes. It was observed that the rate of strength gained at low bentonite content is relatively less when compared to the rate of strength gained at high bentonite content. This may be due to the fact that, the addition of bentonite increased plastic nature of PA+B mixture which resulted in the gaining of strength. As the compaction effort increased, the soil particles are closely compacted and increased the strength parameters of the PA+B mix. According to the general requirement of the hydraulic barriers, the compression strength should have a value greater than 0.2 MPa.

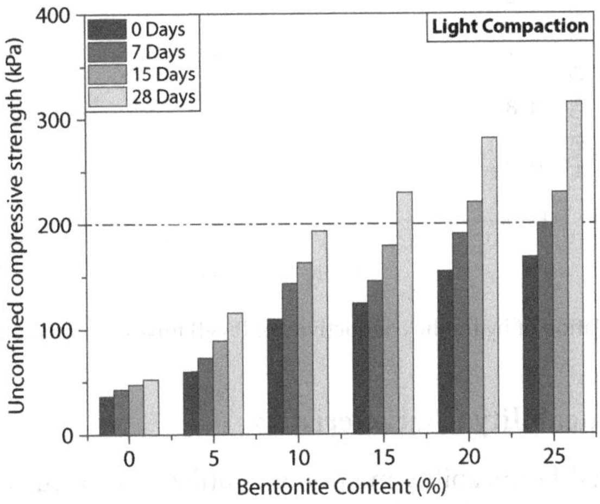

Figure 4.4 Variation of UCS of PA+B mixes with increase in bentonite content for light compaction.

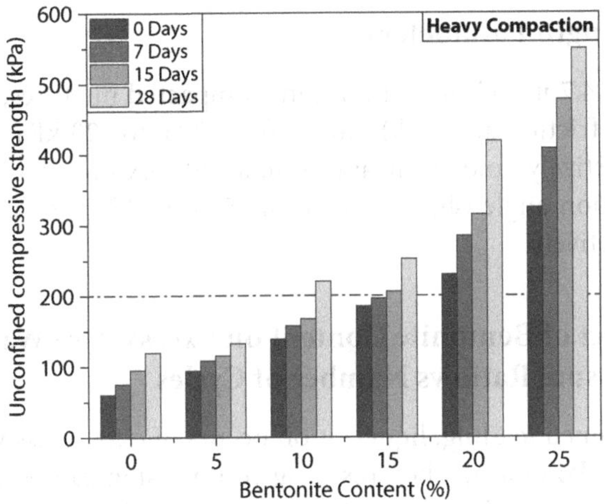

Figure 4.5 Variation of UCS of PA+B mixes with increase in bentonite content for heavy compaction.

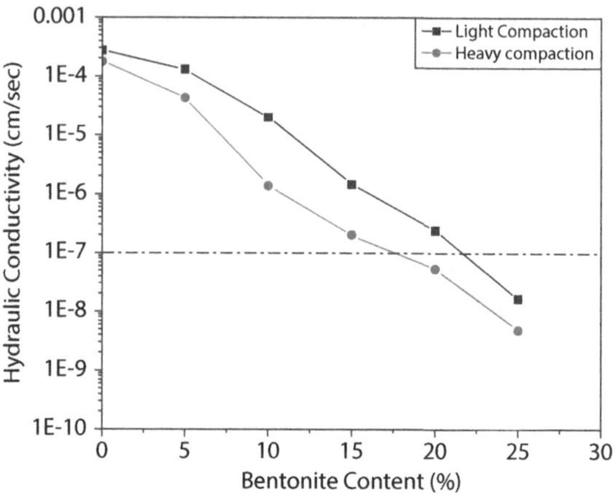

Figure 4.6 Variation of hydraulic conductivity of PA+B mixes with bentonite content.

4.3.6 Permeability Characteristics

Constant Head Permeability method was utilized to calculate the average coefficient of permeability of pond ash specimens. Pond ash has a higher permeability since it is coarser material. After adding bentonite in quantities from 0 to 25%, the permeability of compacted PA+B mixes decreased significantly as shown in Figure 4.6.

4.3.7 Strength Parameters

From Figure 4.7 it is clear that for light compacted mixtures, cohesion (c) and internal friction angle (ϕ) ranged from 28.6 to 123 kPa and 36.20° to 28.56° respectively; and for heavy compacted mixtures cohesion (c) and internal friction angle (ϕ) ranged from 45.56 to 183 kPa and 39.025° to 30.32° respectively.

4.3.8 Effect of Bentonite Content on Excess Pore Water Pressure Ratio vs Number of Cycles

Under undrained loading, liquefaction occurs due to excessive pore pressure buildup. Pond ash, when resisting compression causes compatibility issues and decreases shear strength when subjected to cyclic loading. This weakens its structural integrity, rendering it unsuitable for construction. Introducing non-plastic and plastic particles aids in generating higher pore pressure because of increased porosity. This study explores how highly

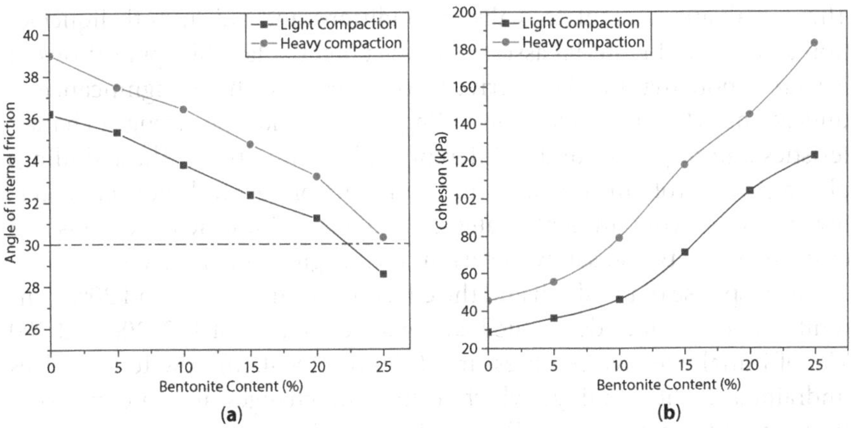

Figure 4.7 Variation of internal friction angle (ϕ) and cohesion (c) of PA+B mixes with bentonite content.

plastic bentonite influences the excess pore water pressure development in non-plastic pond ash during the cyclic loading through the undrained cyclic triaxial testing, aiming to compare their behaviors.

4.3.9 Triaxial Compression Test Unconsolidated Undrained

The triaxial test method is instrumental in determining a soil sample's compressive strength by applying loads in three perpendicular directions. Typically, soil samples, measuring 38mm in diameter and 76mm in length, are extracted from larger tubes or remolded using a split sampler. These samples undergo enclosure in rubber membrane sheaths and placed in stretchers before being positioned atop a base plate equipped with a center steel ball. The triaxial cell, filled with water and pressurized, subjects the sample to controlled strain rates. This test method allows the determination of crucial properties such as cohesion intercept, internal friction angle, and failure envelopes, all following specific procedures. Cyclic triaxial tests encompass various stages, including sample preparation adhering to ASTM D 5311, saturation, consolidation, and cyclic loading under distinct confining pressures and stress ratios. The assessment involves plotting data on excess pore pressure ratio against cycles, shedding light on the pressure generated during cyclic loading. A 13kg weight replicates pressures common in landfill scenarios. Moreover, tests performed on blends of pond ash and bentonite encompasses varying weight percentages under specific confining pressures ranging from 100 to 300 kPa.

These evaluations emulate realistic conditions and aid in both liquefaction analysis and material assessment for potential landfill applications. It is vital to note that the data derived from these tests holds significance in understanding soil behavior, providing insights into the strength characteristics and responses under different loading conditions. These findings play a pivotal role in infrastructure construction, geotechnical engineering, and waste containment strategies, facilitating informed decisions and ensuring stability and safety in various civil engineering projects.

The response of pond ash and the effects of adding 10, 15, and 20% bentonite to pond ash under cyclic loading are compared at 100, 200 and 300 kPa of initial effective cell pressures (Figures 4.8–4.10). The test involves undrained cyclic loading, which results in changes in pore pressure throughout the experiment. The pond ash exhibits an immediate development of pore pressure and liquefaction, initiated after 5 numbers of cycles. However, when a bentonite content of 10% is introduced in the pond ash, cyclic resistance significantly increased. The addition of bentonite reduces pore pressure by 20%, indicating improved resistance to liquefaction in the samples.

The response of pond ash and the effects of adding 10, 15, and 20% bentonite to pond ash under cyclic loading are compared at 100, 200 and 300 kPa of initial effective cell pressures. Further, the test involves undrained cyclic loading, which results in changes in pore pressure throughout the experiment. The pond ash exhibits an immediate formation of pore water pressure and the liquefaction initiates after 5 number of cycles. However, when the percentage of bentonite is 10% in the pond ash, cyclic resistance

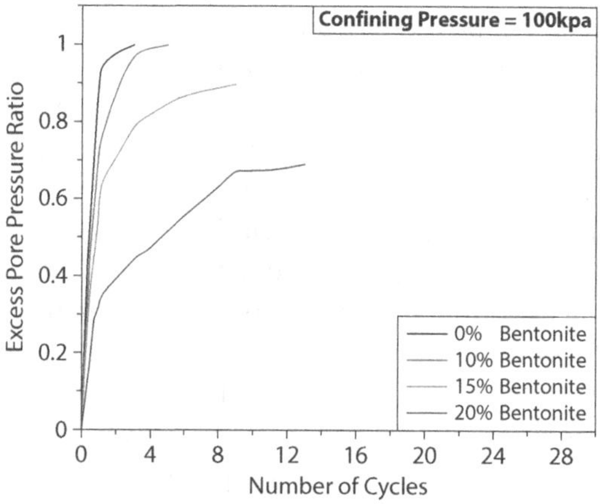

Figure 4.8 Excess pore pressure ratio variation with bentonite content.

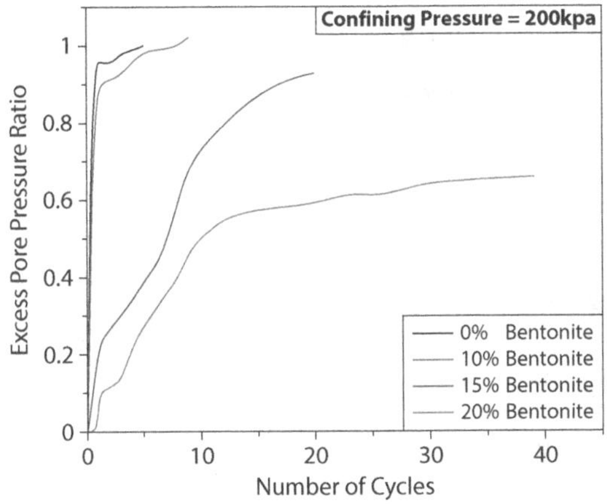

Figure 4.9 Excess pore pressure ratio variation with bentonite content.

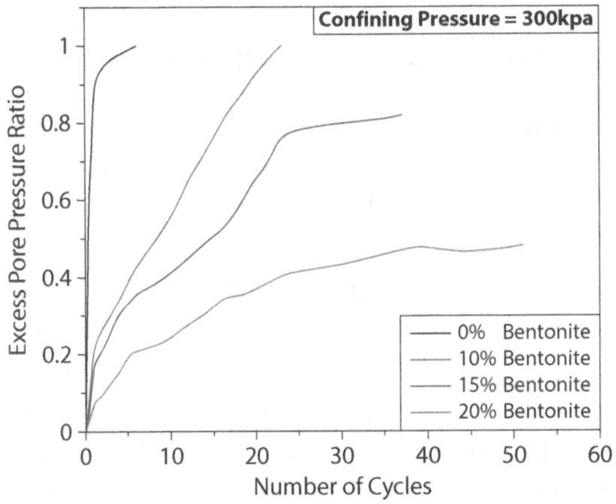

Figure 4.10 Excess pore pressure ratio variation with bentonite content.

significantly increased. The addition of bentonite reduces pore pressure by 20%, indicating improved resistance to liquefaction in the samples.

4.3.10 Effect of Plasticity Of Mix on the Cyclic Resistance

When an excess pore pressure equals the effective cell pressure or reaches a 5% double amplitude (DA) axial strain, liquefaction arises. Bentonite addition to pond ash accelerates pore pressure to rise swiftly. Beyond a specific bentonite content, cyclic resistance elevates, subsequently diminishing pore pressure.

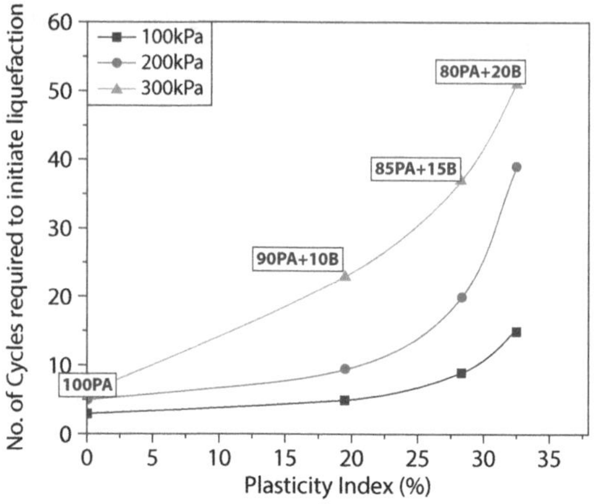

Figure 4.11 Number of cycles needed to initiate liquefaction with variation of plasticity index.

4.3.11 Typical Test Results of PA+B Mix (PA80+B20)

The pore water pressure variations were tracked in Figure 4.11 across cycles at 1kPa limiting pressure and 57 kPa cyclic shear stress. Pore pressure remained stable for 1.5 cycles, escalating slowly before sharply rising after 1.5 cycles, reaching the initial limiting pressure of 100 kPa by the 15th cycle, causing liquefaction. While the load remained constant, pore pressure didn't fluctuate. Figure 4.11 illustrates the axial strain evolution under 57 kPa cyclic shear and 100 kPa confinement. Initially stable, strain intensified after the fourth cycle and significantly after the sixth, hitting 5% at cycle 15, correlating with liquefaction onset. The pore pressure exhibited a clear upward trend after the sixth cycle, similar across multiple tests. Notably, significant strain shifts occurred after the initial liquefaction, dependent on cycle stress ratio or shear stress number.

4.3.12 Effect of Cyclic Stress Ratio (CSR) on (20% Bentonite+80% Pond Ash) Mix

The changes in cohesion with the increase in bentonite percentages is shown in Table 4.1.

Table 4.1 Cohesion alteration with the bentonite percentages.

Pond ash + bentonite (PA+B) mix	Effective confining pressure (σ3) (kPa)	CSR	2.5% Double amplitude strain	5% Double amplitude strain
80% PA+20% B	100	0.1	Not Liquefied	Not Liquefied
		0.3	16 Cycles	25 Cycles
		0.5	9 Cycles	14 Cycles
		0.7	6 Cycles	10 Cycles
		0.9	2 Cycles	7 Cycles
	200	0.1	Not Liquefied	Not Liquefied
		0.3	13 Cycles	16 Cycles
		0.5	7 Cycles	9 Cycles
		0.7	4 Cycles	6 Cycles
		0.9	1 Cycles	1.5 Cycles
	300	0.1	Not Liquefied	Not Liquefied
		0.3	10 Cycles	13 Cycles
		0.5	4 Cycles	5 Cycles
		0.7	1.5 Cycles	2 Cycles
		0.9	0.5 Cycles	1 Cycles

4.3.13 Effect of Confining Pressure on Pore Water Pressure Generation

For each of the situations that are already presented, the cycles needed to initiate the liquefaction increases with rise in 100, 200 and 300 kPa of effective cell pressures.

4.3.14 Effect of Cell Pressure on the Axial Deformation

This study explores relationships among confining pressure, cyclic shear stress, and pressure rounds impacting sample transformation. Also, the cycle numbers needed for 5% and 2.5% double amplitude (DA) strain has been analyzed. This study shows, correlations between increased restricted pressure and cyclic shear stress resulting in a 2.5% DA strain as depicted in Figure 4.12. For instance, 20, 40, and 60 kPa cyclic shear stresses generate a 2.5% DA strain for materials consolidated at 100, 200, and 300 kPa as initial effective confining pressure. Figures 4.12 to 4.15 depict axial deformation influenced by limiting pressure, cyclic shear stresses, and load cycles. There's a direct correlation between raising restricted pressure and increasing cyclic shear stress at 2.5% DA strain. Moreover, it is found that as confining pressure increases, the cyclic shear stress needed for a 5% DA strain

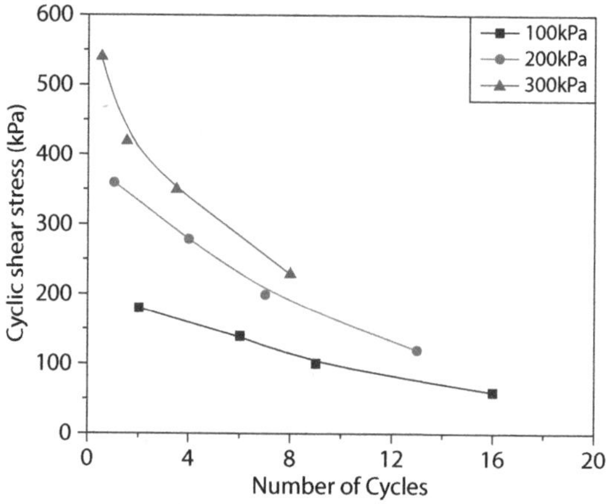

Figure 4.12 Variation of 2.5% cyclic shear stress with number of cycles.

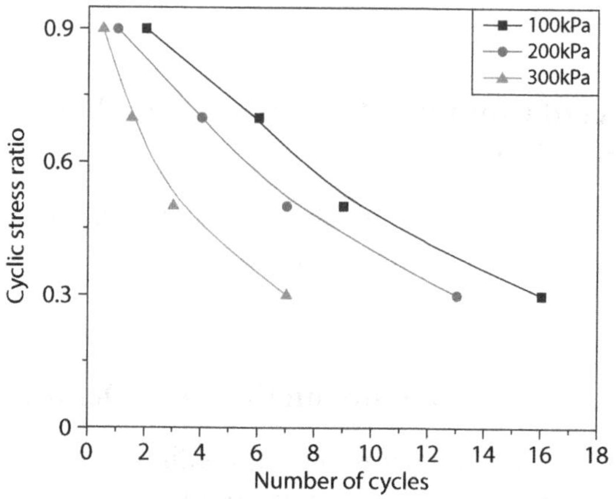

Figure 4.13 Changes in 2.5% cyclic stress ratio with cycles.

increases as depicted in Figure 4.14. The study exhibits the plots illustrating relationships between CSR, cycles required for strain generation, and initial effective confining pressure (Figures 4.13 and 4.15). Higher CSR reduces cycles needed for 2.5% DA strain. Similarly, increasing initial cell pressure reduces the CSR needed for both 2.5% and 5% DA strains. This relationship echoes the circumstances observed for the 2.5% DA strain development.

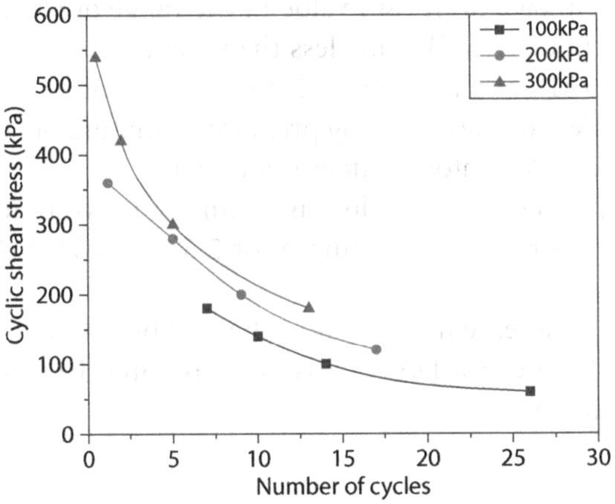

Figure 4.14 Changes in 5% cyclic shear stress with cycles.

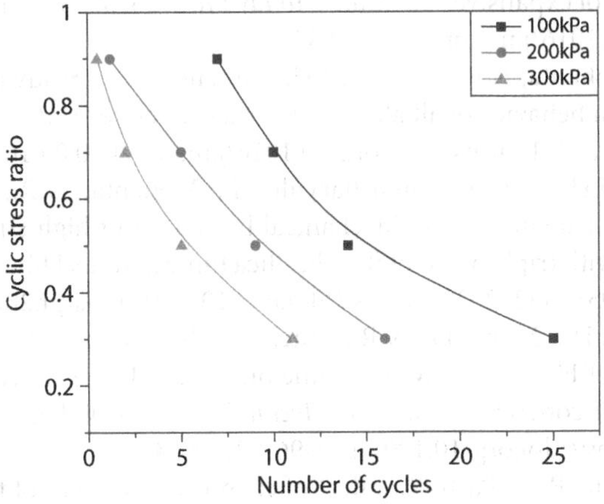

Figure 4.15 Changes in 5% shear stress ratio with cycles.

4.4 Conclusions

In this study, the feasibility of pond ash-bentonite (PA+B) mixture as an alternate liner material under dynamic loading was investigated experimentally and the following conclusions were drawn:

1. 80%PA+20%B and 75%PA+25%B heavy compacted mixture satisfying the landfill liner criteria such as permeability (k) value less than 1×10^{-7} cm/sec, unconfined compressive strength (UCS) value more than 200 kPa, PI value more than 7%, shrinkage limit (SL) value in the range of 5%-20%, linear shrinkage (LSI) value less than 10%, cohesion (c) value above 5 kPa and friction angle in the range of 20-40.
2. Regardless of the confining pressure, a mixture of 80% pond ash and 20% bentonite does not liquefy.
3. For a particular CSR value, as confining pressure increases, the number of cycles required for 2.5% and 5% DA strains decreases.
4. With an increase in CSR value, the number of cycles needed for 2.5% and 5% DA strains is decreasing, irrespective of confining pressure.

References

1. Singh, P., Dash, H.M., Samantaray, S., Effect of silica fume on engineering properties of expansive soil. *Mater. Today Proc.*, 33, 5035–5040, 2020, https://doi.org/10.1016/j.matpr.2020.02.839.
2. Ray, T., Mohanta, N.R., Kumar, M.H., Samantaray, S., Study of effect of temperature on behavior of alkali activated slag concrete. *Mater. Today Proc.*, 43, 1352–1357, 2021, https://doi.org/10.1016/j.matpr.2020.09.169.
3. Kumar, M.H., Saikrishnamacharyulu, I., Mohanta, N.R., Ashutosh, A., Mishra, P., Samantaray, S., Mechanical behaviour of high strength concrete modified with triple blend of fly ash, silica fume and steel fibers. *Mater. Today Proc.*, 65, 933–942, 2022a, https://doi.org/10.1016/j.matpr.2022.03.528.
4. Kumar, M.H., Mohanta, N.R., Patel, N., Samantaray, S., Reddy, S.V.B., Influence of Fly ash and Metakaoline on the crack resistance and shrinkage property of concrete. *Iran. J. Sci. Technol. Trans. Civ. Eng.*, 46, 2011–2026, 2022b, https://doi.org/10.1007/s40996-021-00748-y.
5. Vivek, Dutta, R.K., Parti, R., Effect of chemical treatment of the coir geotextiles on the interface properties of sand/clay-coir geotextile interface. *J. Inst. Eng. India Ser. A*, 100, 357–365, 2019.

6. Mohanta, N.R. and Samantaray, S., Study of Combined Effect of Metakaolin and Steel Fiber on Mechanical Properties of Concrete. *Pertanika J. Sci. Technol.*, 27, 3, 1381–1396, 2019, http://journals-jd.upm.edu.my/resources/files/Pertanika%20PAPERS/JST%.
7. Vivek, and Dutta, R.K., Bearing ratio behavior of sand overlying clay with treated coir geotextiles at the interface. *J. Nat. Fibers*, 19, 14, 7534–7541, 2022.
8. Jaswal, P., Vivek., Sinha, S.K., Laboratory analysis of the interface shear characteristics of chemically treated coir geotextiles and soil interface. *Int. J. Pavement Res. Technol.*, 1–14, 2023.
9. Vivek, Effects of cyclic loading on sand overlaying clay model of unpaved roads reinforced with untreated/treated coir geotextiles. *J. Text. Inst.*, 115, 9, 1694–1700, 2024.
10. Sivapullaiah, P.V. and Lakshmikantha, H., Properties of fly ash as hydraulic barrier. *Soil Sediment Contam.*, *13*, 5, 391–406, 2004.
11. Lakshmikantha, H.N. and Shivapullaiah, P.V., Liner. *Int. J. Civ. Eng. Technol.*, 7, 3, 99–106, 2006.
12. Akcanca, F. and Aytekin, M., Effect of wetting–drying cycles on swelling behavior of lime stabilized sand–bentonite mixtures. *Environ. Earth Sci.*, 66, 67–74, 2012.
13. Nayak, S., Application of coal ash-bentonite mixtures as landfill liner. *Int. J. Appl. Environ. Sci.*, 10, 3, 1033–1050, 2015.
14. Singh, S.P. and Sharan, A., Strength characteristics of compacted pond ash. *Geomech. Geoeng.*, *9*, 1, 9–17, 2014.
15. Rashid, M.A., Hossain, M.M., Sarwar, A.S.M.M.R., Rahman, M.M., Geotechnical characterization of bentonite-fly ash mixtures for their application as landfill liner. *J. Build. Eng.*, 37, 102168, 2021.
16. Ishihara, K., Yasuda, S., Ha, T., Cyclic Shear Strength of Sands with Different Clay Contents. *Soils Found.*, 29, 3, 85–97, 1989.
17. Rout, S. and Singh, S.P., Characterization of pond ash-bentonite mixes as landfill liner material. *Waste Manage. Res.*, 38, 12, 1420–1428, 2020.
18. El Mohtar, C., Anderson, D., Taha, R., Cyclic Response of Clay and Bentonite Mixtures as Landfill Liner Materials. *Geotext. Geomembr.*, 41, 50–57, 2013.
19. Puri, V. and Yazdani, S., Cyclic Response of Compacted Clay Liners in Landfill Applications: Cyclic Triaxial Tests and Analysis of Influential Factors. *Waste Manage. (Oxford)*, 88, 85–94, 2019.
20. Cai, Y., Yang, L., Liu, L., Li, S., Zhang, L., Effects of Cyclic Loading on Geotechnical Properties of Compacted Clay-GO Mixtures as Landfill Liners. *Geotext. Geomembr.*, 48, 4, 534–542, 2020.
21. IS 2720-3(2) (1980): Method of test for soil, Determination of specific gravity, Bureau of Indian Standards, 1980.
22. IS 2720-5 (1985): Methods of test for soils, Determination of liquid and plastic limit, Bureau of Indian Standards, 1985.

23. IS 2720-4 (1985): Method of test for soil, Particle size analysis, Bureau of Indian Standards, 1985.
24. IS 2720-40 (1985): Method of test for soil, Determination of free swell index of soils, Bureau of Indian Standards, 1985.
25. IS 2720-6 (1972): Methods of test for soils, Determination of shrinkage factors, Bureau of Indian Standards, 1972.
26. IS 2720-7 (1980): Determination of water content-dry density relation using light compaction, Bureau of Indian Standards, 1980.
27. IS 2720-8 (1980): Determination of water content-dry density relation using heavy compaction, Bureau of Indian Standards, 1980.
28. IS 2720-17 (1986): Method of test for soil, Laboratory determination of permeability, Bureau of Indian Standards, 1986.
29. IS 2720-11 (1993): Method of test for soil, Determination of shear strength parameters, Bureau of Indian Standards, 1993.
30. IS 2720-10 (1991): Methods of test for soils, Determination of unconfined compressive strength, Bureau of Indian Standards, 1991.
31. ASTM D-5311 Method of test for soil, Determination of pore pressure and strain.

5

Utilization of Pond Ash for Problematic Soil Remediation

Debasree and Subhashree Samantasinghar*

Odisha University of Technology and Research, Bhubaneswar, India

Abstract

The expansion of power plants in our country has made a substantial contribution to economic development, as they are the principal source of electricity generation nationwide, making it the backbone of our energy infrastructure. Indeed, the income generated by power plants accounts for a significant percentage of the Indian economy. However, with this expansion come new challenges, particularly in terms of trash management. Given the scarcity of available land and the rising costs of dumping and dusting operations, coal-based thermal plants, in particular, are confronted with a significant ash disposal challenge. Pond ash, a byproduct of power generation, has been increasingly popular in recent years due to its potential applicability in resolving the ash disposal challenge. This research addresses this issue by reporting findings of experiments conducted on specimens of clayey soil stabilized with pond ash. The experiment entailed adjusting the quantities of pond ash. The goal was to determine the effect of these admixtures on clay strength qualities and investigate their viability as a solution to the disposal dilemma. A variety of experiments were used to assess the geo-mechanical properties of the stabilized soil specimens, such as plasticity characteristics testing, compaction testing, unconfined compression testing, California bearing ratio testing and water permeability test through soil. X-ray fluorescence spectroscopy and scanning electron microscopy studies were also carried out to look at the characteristics of surface morphology. and chemical compounds present in the materials. Furthermore, the specimens underwent tests to determine their strength behavior at various curing times. to determine the efficacy of the suggested approach over time. The experiment yielded numerous noteworthy discoveries. First, the use of admixtures reduced the soil's toxicity, by raising its ideal moisture content while maintaining

**Corresponding author*: shree.singhar@gmail.com

Sandeep Samantaray, Subhashree Samantasinghar, Soumendra K. Mohanty, Vivek, Abinash Sahoo and Deba Prakash Satapathy (eds.) Waste Utilization in Geotechnical Practice, (117–148) © 2025 Scrivener Publishing LLC

its maximum dry density, which could have ramifications for compaction and engineering issues. Furthermore, the data revealed a significant improvement in the clayey soil's strength qualities, as proven by increases in tests for hydraulic conductivity, California bearing ratio and unconfined compressive strength. This shows that the suggested technology possesses the capacity to improve the technical qualities of clayey soil, thus reducing the challenges associated with its use in a variety of construction applications. In conclusion, this study provides valuable insights into the use of pond ash as a stabilizing agent for clayey soil, emphasizing its potential to address waste management issues associated with coal-based thermal power plants while also improving the soil's engineering properties. Further research and development in this field may result in novel solutions with major economic and environmental benefits for the country.

***Keywords*:** Plastic soils, pond ash, plasticity, compaction parameters, strength permeability

5.1 Introduction

The foundational material that civilizations build their buildings on soil, may be very problematic, particularly in areas where its natural qualities aren't the best. These problems can be more noticeable in regions where clay soil is the predominant soil type. Because of its weak nature, clay soil, which is characterized by small particles and strong flexibility, presents various challenges to construction endeavors. Ignoring the need for remediation of such soil can result in a host of problems when building on it, such as low shear strength, insufficient permeability, and weakened bearing capacity [1–4]. Effective ground improvement methods are, therefore, essential to reducing risks and guaranteeing the durability and stability of civil engineering constructions. There are several ways to improve soil's desired qualities so that it is more appropriate for construction. Soil stabilization is one of these methods, standing out as a practical and affordable option [5–8]. Engineers can greatly improve the soil's engineering qualities by treating it with binder materials, such as cement, lime, or other pozzolanic chemicals. Cement has long been a preferred option for stabilizing soil because of its quick strength and durability gains. Notwithstanding its efficacy, the assimilation of cement has significant limitations, especially when combined with clay soil. Stabilizing clay soil for construction purposes by the use of cement and pozzolanic materials has many advantages, but there are drawbacks as well, mainly with relation to shrinkage and cracking occurrences. In clay-rich locations, engineers can preserve the integrity and durability of civil engineering projects and successfully control the risks associated with stabilized clay soil by comprehending the underlying mechanisms driving these difficulties and putting appropriate mitigation measures in place.

Alternative strategies for stabilizing soil have been investigated recently in an effort to address the unique problems presented by substrates that are rich in clay. Using pond ash, a coal combustion by-product, to produce a pozzolanic effect in stabilizing sub-grade soil is one possible approach. The soil's durability and stability are increased by pozzolanic elements, including pond ash, reacting when there is calcium hydroxide present of moisture to generate compounds having cementitious qualities. Although adding cement quickens the stabilized soil's rate of strength development, it also introduces potential drawbacks, particularly when applied to clay soil. One notable concern is the phenomenon of shrinkage and cracking, which can compromise the integrity of the constructed base course. Clay soil, with its high plasticity and propensity to experience volumetric alterations in response to moisture percentage variations, is particularly susceptible to these issues.

The fast-paced industrialization has led to an increase in waste material output, which presents serious risks to the environment and public health [9–14]. To lessen these effects, waste management techniques that work are essential. Using waste materials for different purposes is one strategy that is gaining popularity and provides a sustainable answer to waste management issues. Nevertheless, there are significant obstacles to the efficient use of the trash generated, such as rice husk ash, fly ash, pond ash, and scrap tires. However, there are promising applications for these leftover resources in the field of civil engineering construction. Using them as admixtures for soil improvement is becoming a popular method for improving the ground, especially for troublesome soils with soft clay or swelling. Conventional construction techniques are impracticable because these soils frequently cannot support the weight of structures. Engineers can enhance the qualities of the soil and lessen the risk of unstable ground conditions by using waste materials as admixtures. This will strengthen the soil's capacity to support structures. This strategy fosters a more resilient built environment in the face of industrial waste concerns by advancing environmentally friendly construction methods and supporting sustainable waste management.

Large amounts of power plants generate waste materials called pond ash and fly ash. worldwide, which presents serious environmental problems. Millions of tons of pond ash deposits blanket almost 200 million square meters of land in India alone [15]. However, the development of leachate compounds, those which contain heavy metallic chemicals and hazardous elements raise worries about the accumulation of these ash deposits. These substances leak from ash ponds into nearby soil, contaminating surface and groundwater resources and endangering ecosystems as well as human health. The extensive pollution highlights the pressing requirement for efficient waste management techniques to lessen the negative environmental effects of disposing of fly ash and pond ash from power plants.

They [16, 17] produced additives like bitumen, cement, and lime that have historically been used to stabilize soils for building reasons. The aforementioned additions have demonstrated efficacy in markedly improving the characteristics of problematic soils, such as black cotton soils, making them appropriate for use in building projects. The cost of lime, cement, and bitumen additives has prevented their widespread use, even though they are effective. The cost of adding these compounds has been a major obstacle to their widespread use, which has limited their accessibility, especially in environments with little resources. However, there has been a discernible change in emphasis towards investigating novel methods of stabilizing soil, motivated by the need to cut expenses and lessen environmental effect. Researchers want to solve urgent environmental issues related to waste disposal as well as the cost obstacles associated with existing additives by recycling waste materials for soil stabilization. A larger dedication to sustainability and resource efficiency within the construction industry is demonstrated by the investigation of waste materials for engineering purposes. Here, efforts are increasingly directed towards optimizing the value of existing resources while reducing waste generation and environmental impact. This coordinated effort to convert industrial wastes into useful materials is a critical first step towards the creation of affordable, environmentally friendly soil stabilization solutions, opening the door for a more robust and ecologically aware building industry.

A substantial amount of research has been done over the years to investigate the possible application of waste products from industry and agriculture to improve the qualities of weak and expansive soils [18–20]. Certain research conclusions show that certain waste products have favorable qualities that, when added to soil or used alone, can improve its properties. These additives can also be utilized in conjunction with cement or lime. Researchers have shown that they may improve load-bearing capacity, reduce expansion tendencies, and improve soil stability by utilizing these waste materials, which create new opportunities for environmentally friendly soil stabilization techniques. These waste materials have become viable substitutes for traditional stabilizers, whether they are made from industrial leftovers like blast furnace slag or fly ash, or by products of agriculture, for example bagasse or rice husk ash. Furthermore, their efficacy has been further increased by their usage as admixtures with cement or lime, which has resulted in significant breakthroughs in soil improvement methods.

This coordinated research effort is a testament to the increasing awareness of the unrealized potential of waste materials in improving sustainable construction methods and solving soil-related issues. Therefore, utilizing waste materials from industry and agriculture offers a viable way to achieve environmental sustainability as well as engineering excellence in soil stabilization projects.

Several approaches and studies have been conducted to enhance the pond ash's engineering properties, with mechanical mixing of lime/cement emerging as the most effective. Additionally, research have looked into the usage of fiber reinforcement in silty sand combined with pond ash. The results of these experiments show significant improvements in a variety of metrics, including specimen ductility, peak friction angle, peak compressive strength and California bearing ratio value [21]. Furthermore, experimental results demonstrate that adding lime increases the California bearing ratio of pond ash [22]. Furthermore, research shows that increasing compactive efforts during mixing, results in a larger maximum dry density while simultaneously decreasing optimal moisture content [23]. These findings suggest intriguing pathways for optimizing pond ash's engineering qualities, perhaps providing solutions to issues connected with its use in a variety of construction applications.

Numerous research have looked into the ability of various additives and stabilization procedures to enhance the soil's engineering properties materials for geotechnical applications. One such study [24], found that a combination of clay stabilized with cement and strengthened with fibers combined with suitable amounts of pond ash and rice husk ash can efficiently function as geotechnical fill materials. This research implies that these materials have intriguing properties for use in construction projects, providing increased strength and endurance.

Another notable observation, as stated in reference [23], concerns lime being used in class f fly ash, which has shown advantages in both strength and durability. This research emphasizes the significance of proper additives in improving the performance of fly ash-based composites for construction applications. Furthermore, past research has shown that using pond ash and rice husk ash can improve the quality of sub grade for road building. The alluvial soil's compaction properties have been demonstrated to be significantly impacted by the addition of either rice husk ash or pond ash, as indicated in reference [25]. When more pond ash and rice husk ash are added, the maximum dry density of mixed soil tends to decrease, while the optimum moisture content rises. This finding emphasizes the ability of these additives to improve soil qualities for road construction projects.

Furthermore, research into the feasibility of the construction of road bases and sub-bases, stabilized pond ash has produced beneficial outcomes. A series of laboratory studies, documented in reference [26], revealed that both solo and combined use of class F pond ash with variable amounts of lime and phosphonyls can be used to produce road base and sub-base layers. This research implies that specific additive combinations can significantly improve the engineering qualities of pond ash for road building applications.

However, the field of civil engineering offers a feasible opportunity for its widespread application. A particularly effective way involves using these waste products as admixtures to improve soil qualities, which is an important feature of ground improvement methodologies. This approach is useful in remediating problematic soils, such as soft clay or swelling soil, which frequently fail to offer appropriate support for structural loads. By utilizing waste materials in this manner, not only is the burden of trash disposal minimized, but it also contributes to sustainable and environmentally friendly practices in the building and engineering industries. The dual benefit of waste reduction and soil improvement highlights the strategy's ability to promote environmental conservation and acceptable industrial practices. As the demand for infrastructure grows, incorporating waste materials into civil engineering processes becomes a tempting approach to address both environmental concerns and the need for effective waste management.

Chemical stability of residual soils is investigated by using rice husk ash and cement [27]. X-ray diffraction, strength, and Compaction are among the properties of the soil that are examined in this examination. According to test results, soil flexibility is decreased by both cement and rice husk ash. When it comes to compactability, the mixture of rice husk ash and cement reduces the maximum dry density and increases the optimal moisture content. When it comes to strength, compactness, and flexibility characteristics, and economy, the recommended ratio is 6-8% cement to 10-15% rice husk ash. They [28], investigate the effects of several compaction control parameters on the dry density of pond ash. These variables include specific gravity, tank size, mold area, moisture content, layer thickness, and compaction energy. It has been determined that pond ash has a maximum dry density of 8.40–12.25 kN/m^2, and that the ideal moisture content is between 29–46%. He tested the degree of saturation at the optimum moisture content of pond ash and discovered that it ranges between 63 and 89%. The field's moisture content and maximum dry density at every compaction energy can be found using linear empirical models. These empirical models are useful for estimating the maximum dry density and optimum moisture content of pond ash.

Using a lime proportion as low as 1.12%, they [21] examine the effects of lime stabilization on the durability and strength of class f pond ash. The specimens were preserved for 28, 45, and 90 days at 30°C with lime contents of 10 and 14%. They were tested for unconstrained compression, durability of the slake, rebound hammer, and point load strength, as well as typical rock tests. Good correlations have been established for strength characteristics obtained from unconfined compressive strength testing, tests using Schmidt rebound hammers and point load strength as well as between slake durability index and unconfined compressive strength. The bearing ratio of stabilized pond ash was examined [26] in relation to the parameters of lime content, amount of phosphogypsum and time of curing. The empirical model was developed using multiple regression analysis to ascertain the bearing ratio for stabilized blends. He offers an empirical linear relationship that may be used to calculate the stabilized pond ash's wet bearing ratio based on the unsoaked bearing ratio. The outcomes of the experiment suggest that blends of pond ash, lime, and phosphogypsum could be used as sub base and basis material for roads. They [24], presented the findings of their experiments on specimens of clayey soil that were reinforced with fibers that were dispersed randomly and stabilized with cement and pond ash. The percentage of pond ash and cement in the soil's dry weight ranged from 0 to 50 and 0 to 6%, respectively. Investigations were conducted on hydraulic chemicals, surface morphology, and clay strength properties. According to the data, adding mixes lowers the maximum dry density while raising the ideal moisture content. According to the unconfined compressive strength, sprit tensile, and pond ash test findings, the suggested approach is also extremely successful at boosting the strength of clayey soil. They [29], created and assessed the stabilized clayey soils with different proportions (up to 20%) of dust from cement kilns, rice husk ash, and their blends. Modulus of elasticity, Atterberg limits, conventional testing for proctor compaction, splitting tensile strength, unconfined compressive strength, and California bearing ratio are used to evaluate the effects of the stabilizer amounts and kinds on mechanical and fresh qualities. Additionally associated is the California bearing ratio, strength, and modulus of elasticity. The created stabilized soil combinations can be used to build affordable homes and road infrastructure because they have shown good strength and durability qualities. For such purposes, the local construction sector can benefit from sustainability through the utilization of soils that are readily available locally, e.g., rice husk ash with cement kiln dust, to produce stabilized soils. This research [30] involved the addition of fly ashes, both low-calcium (ASTM Class F Badarpur fly ash) and high-calcium (ASTM Class C Neyveli fly ash), in different weight proportions (10%-80% with an increment of 20) of black cotton soil. Investigations were conducted into the swell potential, features of compaction, and physical aspects. The results of the tests indicate that there have been significant adjustments and improvements made to the limitations of soil-fly ash combinations' expansiveness, swelling potential, and compaction properties. As the optimal moisture content rises, the maximum dry weight of mixes made of soil and fly ash decrease during the addition of fly ash to BC soil. The enhanced fly ash gradation is responsible for this. Furthermore, it has been shown that 10% of Nayeli fly ash, as opposed to 40% of Badarpur fly ash, is the optimal quantity required to lower swell potential. The effects of lime and rice husk ash on the engineering properties of clayey soils are examined [31], reporting that the application of a lime-rice husk ash (LHRA) mixture reduces plasticity in clayey soils. The plasticity index dropped dramatically from 41.25%-0.96% after combining 12-12.5% LHRA. In a similar vein, the addition of the mix resulted in a drop in the surge potential from 19.23% to 0.019%. In addition, the California Bearing Ratio increased significantly with a 6–12.5% lime–rice husk ash blend, going from 3.03% to 16.3%.They would see an increase in internal friction angle from 5.36 to 23.85 for shear strength parameters. From 54.32 to 157.19 kN/m^2, soil cohesiveness increased. From 391.12 kN/m^2 to 4131 kN/m^2, the bearing capacity rose with an increase in the shear parameter.

In conclusion, these investigations demonstrate the various methods and strategies available for increasing the engineering features of soil materials for geotechnical and building applications. Significant improvements in strength, durability, and compaction properties can be accomplished by additions of pond ash, lime, and rice husk ash, paving the way for more sustainable and efficient construction processes. Proper understanding and application of these discoveries has enormous potential for addressing infrastructure development difficulties while also ensuring the longevity and performance of developed assets.

Through a variety of techniques, pond ash's engineering properties have been greatly improved; one especially successful technique is the mechanical mixing of cement and lime. Studies investigating the addition of fibers to silty sand combined with pond ash have shown promising results. When fibers are added to soils, the experimental results show significant increases in the specimens' peak compressive strength, California bearing ratio value, peak friction angle, and ductility [21]. Furthermore, it has been noted that adding lime increases the pond ash's California bearing ratio [22]. These studies demonstrate how well these methods work to enhance the engineering qualities of pond ash and provide insightful information for real-world implementations in a range of engineering projects.

They [28], carried out a thorough investigation on the compaction properties of pond ash, employing three different kinds for examination. The study demonstrated the important effects of different compaction-controlling factors on the behavior of pond ash. The degree of saturation at the optimal moisture content varied significantly during the experiment, ranging from 63% to 89%. The researchers used multiple regression analysis to create an empirical model that enables the calculation of the dry density of pond ash, hence facilitating practical applications. Compaction energy, moisture content, and specific gravity are included as important parameters in this model. Furthermore, the maximum dry density and ideal moisture content under field conditions were predicted using linear empirical models across a range of compaction energies. With the ability to estimate maximum dry density and ideal moisture content for pond ash materials in advance and help design compaction control techniques, these predictive tools are a great resource for engineers working in the field. These kinds of insights help to improve the efficacy and efficiency of engineering methods pertaining to the use of pond ash.

The research was conducted in the laboratory on specimens of silty sand and pond ash reinforced with randomly distributed polyester fibers [32]. It is explained by the study that adding fibers to these soils improves their mechanical qualities significantly. The test findings show that the specimens' maximum friction angle and compressive strength, California bearing ratio value, and ductility have all increased. The researchers conclude from their data that pond ash samples and silty sand benefit from an ideal fiber content between 0.3% and 0.4% of the dry unit weight. These results imply that adding polyester fibers might significantly boost the overall functionality and structural properties of mixtures made of silty sand and pond ash, which may be advantageous for engineering applications that call for increased durability and strength.

A study found that the increasing compactive efforts cause the maximum dry density to increase and the optimum moisture content to decrease simultaneously [28]. According to the findings of another study, a feasible alternative for geotechnical fill materials is a blend of fiber reinforcement and cement stabilization in clay combined with the ideal amounts of pond ash and rice husk ash [24, 29]. Moreover, studies show that adding lime to class F fly ash improves its strength and durability characteristics [21]. These studies provide important insights for engineering applications by demonstrating the effectiveness of several techniques in enhancing the geotechnical properties of soil and ash mixtures.

The suitability of stabilized pond ash for use in the building of sub-bases and road bases was evaluated through laboratory experiments. The results showed that different dosages of phosphor gypsum (0.5% and 1.0%) and lime (4%, 6%, and 10%) combined with Class F Pond ash worked well for certain construction applications. According to the study, the stabilized pond ash presents possibilities for enhancing the sustainability and performance of infrastructure by acting as a dependable material for road basis and sub-base applications [26].

Using pond ash for soil stabilization can save building costs while also providing environmentally appropriate disposal options. Using these materials would reduce the need for cement, conserving energy and lowering greenhouse gas emissions. Research indicates that pond ash can significantly improve soil strength. There is a maximum amount of pond ash that can be used without noticeable improvement. Combining different admixtures is likely to boost strength significantly. The current study examines the plastic soil's geo-mechanical behavior for various combinations of clay-pond ash mixtures. The geotechnical properties of clay combined with pond ash were examined. Test specimens underwent plasticity study, compaction and strength tests. This study tries to understand soil compaction patterns which are crucial for constructions like pavements and embankments.

5.2 Materials and Experimental Methods

5.2.1 Materials Used

Two different clay soils i.e., black cotton soil and residual clay soil were taken for the present study. The residual clay was the locally available soil obtained from Rourkela, Odisha, India. The black cotton soil was procured from Chilika, Odisha, India. The current inquiry makes use of pond ash to stabilize clay soils with different plasticity i.e., clay soil with low plasticity, medium and high plasticity. For this, the two soils were mixed in such a proportion to obtain a new synthesized soil with intermediate plasticity. Industrial waste pond ash (PA) was utilized to modify the geo-mechanical properties of problematic plastic soils. The pond ash used in the study was obtained from the ash pond at Stage-I in Takua Village, which is linked with the National Thermal Power Plant (NTPC) in Angul, India. To ensure homogeneity, all obtained samples are properly combined. These samples are then sun dried before being dried further in a hot air oven, set to temperatures ranging from 105 to 110 °C for 24 hours. The dried samples are then properly preserved in airtight containers for further use in the research. This complete technique is used to treat and preserve pond ash samples for future analysis and inquiries.

Recent developments in geotechnical engineering and soil science have allowed for a greater knowledge of the characteristics and behavior of low plasticity clay. The significance of precisely characterizing these soils has been acknowledged by researchers and engineers to guarantee the efficacy and security of infrastructure projects. Low plasticity clay soils have unique engineering characteristics that need to be carefully taken into account during the design and construction phases. These soils are frequently found in areas with particular geological formations and weathering histories. Furthermore, it has been shown by recent research how important it is to take soil disturbance into account when gathering and testing data. Unrest during the sample process may have an impact on the characteristics of the soil, which could cause differences between field behavior and laboratory results. Therefore, to reduce these disparities and guarantee the accuracy of engineering studies, efforts have been made to create standardized procedures for soil sampling and testing.

In a recent study, clay, a vital component in many engineering and building applications was used. The investigation of clay soil was obtained locally and was taken from a depth of roughly 1.5 to 2.5 meters. It's significant to remember that the collection method was categorized as disturbed, which may have an effect on the qualities and actions of the soil in comparison to samples that were not disturbed. The soil is classified as low plasticity clay due to its unique engineering characteristics. In this case, the clay's behavior was best indicated by the plasticity index, which was 8.8%. The plasticity index illustrates the soil's capacity to withstand deformation without breaking or cracking by comparing its liquid limit to plastic limit. Furthermore, it was found that the swell-shrink characteristic, which gauges the soil's capacity to change volume in response to variations in moisture content, was 1.67% per percentage of moisture content. In geotechnical and construction engineering projects, knowing how the soil reacts to changes in moisture is especially important, and this parameter plays a major role in that understanding.

In the investigation, black cotton soil that had passed through a 4.75 mm IS sieve was blended in its dry form with varied percentages of pond ash by weight from 0% to 40%. Then, to conduct a traditional proctor test, the necessary amount of water was added to the mixture. Compaction was immediately started after adding water to determine the compaction characteristics of the soil-pond ash mix at the prescribed admixture percentages.

The current study aims to examine how well pond ash stabilizes two different types of clay soils with varying degrees of plasticity: residual clay soil and black cotton soil. A range of clay soils with low, medium, and high plasticity indices will be studied in order to evaluate the stabilizing ability of pond ash. To accomplish this goal, a synthetic soil with intermediate plasticity is made by combining the two clay soils in different amounts. The purpose of the study is to determine whether adding pond ash as a stabilizing agent in a ratio of 10% to 50% will enhance the stability and technical qualities of the synthetic clay soils. The goal of the research is to shed light on the viability and efficiency of using pond ash to stabilize clay soils with a range of plasticity properties through laboratory testing and analysis. The results of this study are anticipated to advance knowledge of soil stabilization methods and provide useful recommendations for engineering applications, especially in the infrastructure and construction industries, where clay soil stabilization is frequently necessary to guarantee the longevity and functionality of structures.

5.2.2 Experimental Programs

The physical, geo-engineering and micro structural analyses were conducted to characterize the raw materials i.e., clay soils and pond ash. Using three different kinds of clayey soil specimens, a series of laboratory tests were conducted, each stabilized with a different dosage and combination of pond ash as mentioned in Table 5.1 which provides an overview of the combined stabilizer combinations with soil and the related percentages of pond ash Then, the next phase of work includes the study on plastic limit (PL), liquid limit (LL)and plasticity index (PI) which are examples of plasticity characteristics, swell-shrink characteristics, i.e., linear shrinkage (LS) and free swell index (FSI)compaction characteristics, i.e., unconfined compressive strength (UCS), California bearing ratio (CBR), maximum dry density (MDD), optimal moisture content (OMC), and hydraulic conductivity which were designed to examine the soil's engineering qualities under various stabilizer conditions.

Clayey soil samples stabilized with different amounts and mixes of pond ash were used in laboratory experiments. The percentages of the mixture mass that were used ranged from 10% to 50%. These percentages were selected in accordance with the best content recommendations discovered in the research, which indicates that 30%–40% pond ash content produces positive outcomes [28]. Tests for the proctor compaction and CBR were

Table 5.1 Sample designation for clay-pond ash mixes.

Pond ash (%)	Residual soil	Synthesized soil	Black cotton soil
0	P0L	P0I	P0H
10	P10L	P10I	P10H
20	P20L	P20I	P20H
30	P30L	P30I	P30H
40	P40L	P40I	P40H
50	P50L	P50I	P50H

used to assess the stabilization's efficacy. After three days of soaking, the specimens were combined to get the ideal moisture content and maximum dry density before being tested. By evaluating the stabilized soil specimens' performance under controlled circumstances, these lab tests sought to provide light on whether pond ash is a suitable and efficient stabilizing agent for clayey soils.

The most recent standards, IS: 2720 (Part 4) 2020 [33]s for gradation and IS: 2720-Part 3, section 1 (2019) [34] for specific gravity determination, were used to evaluate the materials' gradation and specific gravity. Only 2% of the pond ash's particles were finer than 75 mm, according to analysis. The majority of the particles were medium to fine sand-sized. A consistently graded material was indicated by the values of the coefficient of curvature (Cc) and coefficient of uniformity (Cu) were determined to be 1.15 and 2.91 respectively. Additionally, the clay and combinations' LL and PL were assessed using the most recent guidelines found in IS: 2720-Part 5 (2021) [35]. These revised evaluations guarantee adherence to current standards and procedures, offering precise and trustworthy information for the study's material characterization. The swelling and shrinking characteristics of soils were determined in the laboratory in terms of free swell index and linear shrinkage respectively. The IS: 2720 Part 40 (1977) [36] describes the standard method of determining the free swell index of soil. The linear shrinkage of soil was evaluated by standard method of testing, following IS: 2720-Part 20 (1922) [37]. The representative soils were classified and identified in accordance with IS: 1498 (1970) [38].

In compliance with IS: 2720-Part 7 (1980) [39] and IS: 2720-Part 8 (1983) [40] the standard proctor compaction effort and the modified proctor compaction effort of pond ash mixtures were carefully evaluated. These experiments ensured a thorough assessment of the materials' compaction behavior under various situations, with the corresponding compaction energy levels of 595 kJ/m^3 and 2674 kJ/m^3, respectively. The ideal moisture content and the highest possible dry unit weight for clay-pond ash mixes were found through analysis of the compaction plots. These are two important factors affecting the stabilized soils' engineering characteristics and performance. The compacted clay-pond ash mixtures were then subjected to strain-controlled unconfined compression tests in accordance with IS: 2720 (Part 10) 1991 [41] criteria. A complete grasp of the stabilized soil specimens' appropriateness for engineering applications was made possible by these tests, which offered insightful information about their mechanical behavior and load-bearing capability. Through adherence to established standards and techniques, the researchers enhanced the reliability and validity of the experimental results by ensuring consistency and precision in the measurement of compaction and strength properties.

In conformity with IS code, the clays and clay-pond ash mixes compacted to the ideal moisture level and highest dry density possible of pond ash underwent unconfined compressive strength tests. Compaction tests at different compactive efforts ranging from 357 to 3488 kJ/m^3 were used to estimate the optimum moisture content and maximum dry density values. Shearing at an axial strain rate of 1.25 mm/min was applied to test objects that were cylindrical in shape and measured 50 mm in diameter and 100 mm in height. Stress-strain curves were used to calculate the specimens' unconfined compressive strengths. Additionally, samples were generated at standard or modified dry densities with various moisture levels to examine the effect of degree of saturation on the unconfined compressive strength. This extensive testing program sought to evaluate the mechanical behavior of pond ash at various moisture and compaction levels, offering important insights into the material's performance and engineering qualities.

A test for California bearing ratios were performed in compliance with IS: 2720 (Part 16) (1987) [42]. Specimens were created at each maximum dry density at the ideal moisture content using the static compaction method, utilizing a rigid metallic cylindrical mold with dimensions of 150 mm in diameter and 175 mm in height. A plunger with a 50 mm diameter and a 2.5 kg surcharge weight were used in the preparation. The recording of load data was made easier by a mechanical loading device with a moveable base that moved at a steady rate of 1.25 mm/min and an accuracy-checked with the help of a proving ring. The California bearing ratio values of the specimens were determined using the load-penetration curves, which provided vital information on the materials' engineering qualities and load-bearing capacity. The permeability of soil is an important parameter. So, the permeability of soil and soil-pond ash mix were subjected to laboratory permeability test following IS: 2720 Part 17 (1986) [43] and the permeability of soil-pond ash mixes in terms of hydraulic conductivity were reported.

5.3 Results and Discussion

For experimental study, different quantities of pond ash were mixed with clay soils that showed varying degrees of flexibility. The research included a thorough analysis of the geomechanical properties and index properties of the component materials, including the mixtures of clay and pond ash and the raw materials themselves. Following extensive testing, the outcomes were painstakingly compiled and expanded upon in the part that follows. The purpose of this study was to shed light on these mixtures' potential performance in real-world situations by offering insights into their behavior and applicability for a range of engineering applications. When clay soils and pond ash are combined, a variety of features are introduced, and to determine whether or not these combinations will work for construction and geotechnical projects, a thorough understanding of their combined behavior is required. This work adds to the body of knowledge about soil-ash mixtures by examining the geomechanical characteristics and index properties, providing useful data for both researchers and engineers. Moreover, the discourse on the results clarifies the consequences of these findings, aiding in well-informed decision-making and possibly opening the door for the most efficient application of these materials in various technical pursuits. This research project aims to improve knowledge of soil-ash mixtures by careful study and interpretation, which will promote improvements in environmentally friendly building methods and infrastructure development.

5.3.1 Raw Material Characterization

The study concentrated on three main materials: pond ash (PA), black cotton soil, and residual soil. All three materials showed consistent traits that had been previously reported in literature. Using the Indian Standard (IS) soil classification system, the clay soils were categorized as having either low or high plasticity. Table 5.2 provides comprehensive details on these soils index properties. In addition, an intermediate plasticity clay soil (CI) was created by combining residual clay with black cotton soil. The division of plasticity into three categories low, middle, and high is consistent with the results of current studies. In particular, CL, CI, and CH stood for the residual clay, mixed soil, and black cotton soil, respectively. These divisions make it easier to comprehend the behavior of the materials in a more nuanced way and offer a framework for evaluating which technical applications they are suitable for. The accurate characterization of these materials for a range of technical applications is made easier by this recent classification, which also improves our understanding of soil behavior. The consistency of these results with earlier investigations highlights the validity

Table 5.2 Geotechnical properties of raw materials.

Geotechnical parameters	Pond ash	Residual soil	Black cotton soil
Specific gravity	2.22	2.68	2.64
Particle shape	Angular	Flaky	Flaky
Sand fraction (%)	65	7	7
Silt fraction (%)	35	31	28
Clay fraction (%)	-	42	65
Coefficient of uniformity	13.2	-	-
Coefficient of curvature	0.8	-	-
Liquid limit (%)	Non plastic	27	72
Plastic limit (%)	Non plastic	18	32
Shrinkage limit (%)	Non plastic	11	12
Free swell index (%)	Non plastic	46	59
IS classification	SM	CL	CH
Maximum dry density (g/cc)	1.34	1.72	1.58
Optimum moisture content (%)	18.07	16.2	27.88
Cohesion (kN/m^2)	0	45	98
Angle of internal friction (°)	32	18	14

and applicability of the study's findings, promoting an all-encompassing comprehension of soil mechanics and engineering principles.

Data on the elemental composition of the pond ash and the low plasticity clay are shown in Table 5.3, which was acquired using X-ray fluorescence (XRF) analysis. The ability to precisely determine a material's chemical composition has been made possible by recent developments in analytical techniques like XRF [44]. This technique allowed for an accurate assessment of the chemical makeup of the clay soil sample and the pond ash soil sample. Insights into the fundamental components of these materials are offered by this current analysis, which is vital knowledge for a variety of applications in disciplines spanning from material engineering to environmental science and geology. Researchers and practitioners can ensure compatibility with intended objectives and minimize risks by making informed judgements on the utilization of low plasticity clay and PA by knowing their elemental composition.

Figures 5.1 show scanning electron microscopic (SEM) observations of all the raw resources utilized in this investigation, i.e., pond ash, residual clay soil, and black cotton soil. The SEM pictures show various properties in particle form and surface details. Notably, pond ash particles have a spherical shape, as seen in figure. These particles have glossy surfaces with no obvious dust collection. In contrast, the clay particles have an angular to sub-round shape. SEM micrograph of naturally occurring black cotton soil and residual clay soils shows a surface morphology that is scaly and has

Table 5.3 Chemical composition of raw materials.

Chemical compounds	Pond ash	Residual soil	Black cotton soil
SiO_2	58.2	54.8	72.81
Al_2O_3	16.78	30.91	11.25
Fe_2O_3	7.27	4.56	5.2
CaO	1.78	0.05	1.34
MgO	0.45	0.42	2.9
Na_2O	0.54	0.4	0.1
K_2O	1.2	5.12	0.61
LOI	13.78	3.74	5.79

Figure 5.1 Scanning electron microscopic image of raw materials.

a well-developed network of cracks, indicating changes in microstructure and drying-induced inter grain porosity.

5.3.2 Plasticity Characteristics

5.3.2.1 Plasticity Index

The variance in the plastic limit (PL), plasticity index (PI), and liquid limit (LL) of both pure pond ash and PA-clay soil mixtures is shown in Figure 5.2. Notably, as the figure illustrates, a decrease in LL, PL, and PI values is correlated with an increase in pond ash concentration. As the pond ash concentration increases from 10% to 30%, the LL, PL, and PI show virtually linear patterns, with values falling from 17% to 32%, 32% to 58%, and 52% to 90%, respectively. The pond ash has gaps within its particles, which are partially filled by the additional pond ash, that is thought to be the cause of this decrease in LL for PA mixes. This decreases the amount of material available to coat particle surfaces. Lower LL, PL, and PI values for PA mixtures are the outcome of the absence of such gaps in sand, which causes the full mass of bentonite to be employed either for coating particle surfaces or filling inter-particle void spaces. Pond ash has a reduced specific gravity and increased specific surface area than sand, which is further explained by this occurrence.

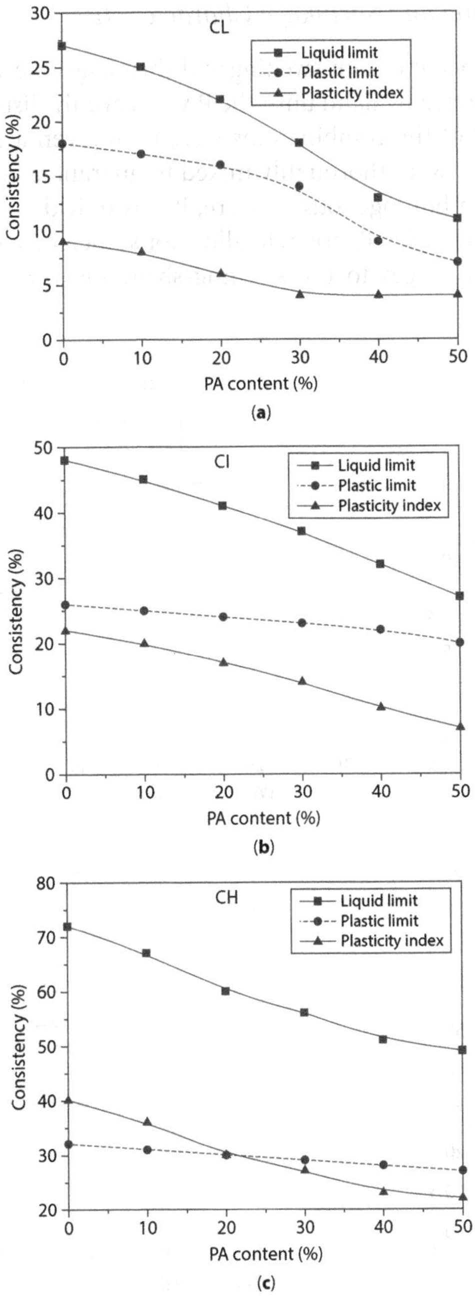

Figure 5.2 Variation in plastic properties of (a) low, (b) intermediate and (c) high plastic clays with pond ash content.

5.3.2.2 Swelling and Shrinkage Characteristics

To prepare the specimens for swelling and shrinkage, the soil and pond ash were carefully mixed by hand until the PA was evenly distributed throughout the soil. All of the combinations were then given a 16 percent water content, and they were thoroughly mixed to guarantee homogeneity [45]. After achieving a homogenous mixture, it was quickly placed into plastic bags, sealed, and left for 48 hours to allow for sufficient stabilization. There are two separate stages to the swelling-shrinking process, i.e., swelling

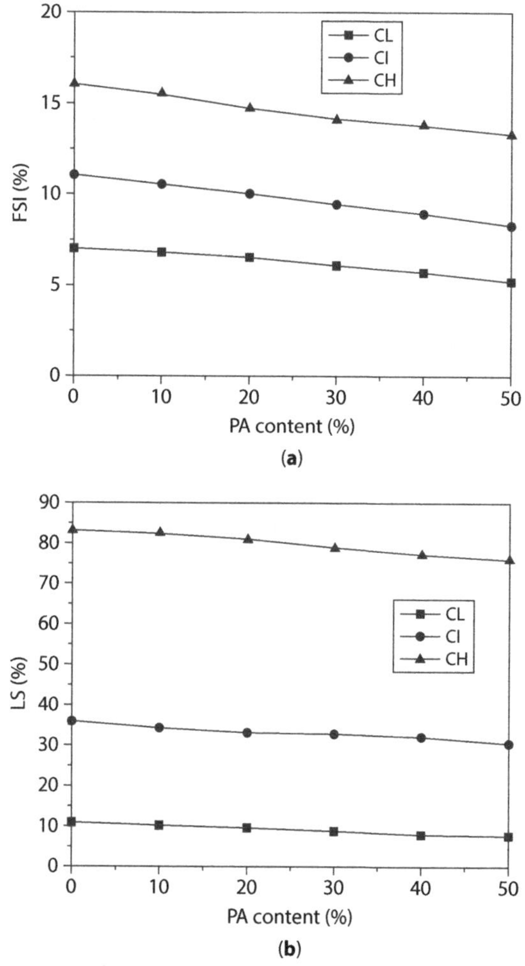

Figure 5.3 Variation in (a) free swell value and (b) linear shrinkage value with pond ash content.

and shrinkage [46]. Each specimen was a compaction in an unconfined compressive strength mold to ensure a single-layer compaction that complied with typical compaction standards for the swell test. A thin layer of Vaseline was put inside the consolidation cells cutting ring to help with easy specimen preparation. After gently inserting the cutting ring into the specimen, any excess material was scraped off the top and bottom surfaces with a spatula to leave a flat, smooth surface.

The results of the swelling-shrinkage experiments are shown in Figure 5.3, which also illustrate the differences in shrinkage strain, swell-shrink index, and swelling strain after varying the amount of pond ash added. Significantly, shrinkage strain, swell-shrink index, and swelling strain were all reduced upon the inclusion of PA. The reduction in swell-shrink characteristics can be ascribed to the replacement of clay, which has a strong affinity for water, with PA, which is mainly made up of non-expansive silica, and is hence less water-responsive. Moreover, the angular shape of PA particles promotes a strong interlocking mechanism in the soil that prevents soil expansion. Thus, with low plasticity clay, the use of PA offers resistance against heaving and shrinking. These results [47, 48] demonstrated the validity and consistency of the test results.

5.3.3 Compaction Parameters

An extensive investigation is conducted into the connection between the dry unit weight of pond ash-clay mixtures and the molding moisture content compacted conventionally at both the original and modified proctor compaction efforts. The variations between the maximum dry density and ideal moisture content of clay soils containing pond ash are shown in Figure 5.4. The outcome demonstrates that raising the pond ash content in a clay-pond ash mix raises the optimum moisture content, lowering the maximum dry density up to the optimum pond ash content for a certain compactive effort. This drop in dry density may be explained by the fact that solid pond ash is having a low specific gravity than clay particles. Above this point, the highest possible moisture and dry density show very little variation, when more pond ash is added. Plastic soil particles, which are usually clay-sized, are combined with pond ash, which comprises fine to medium-sized sand particles. When compaction occurs, this mixing results in a well-graded mix that alters particle fractions and increases particle packing. The ideal level of water content increases as the maximum dry density decreases. However, as clay concentration rises, there is very little volume gain because the clay paste is forced into the intra-particle gaps in the pond ash by the matrix's expanding pressure.

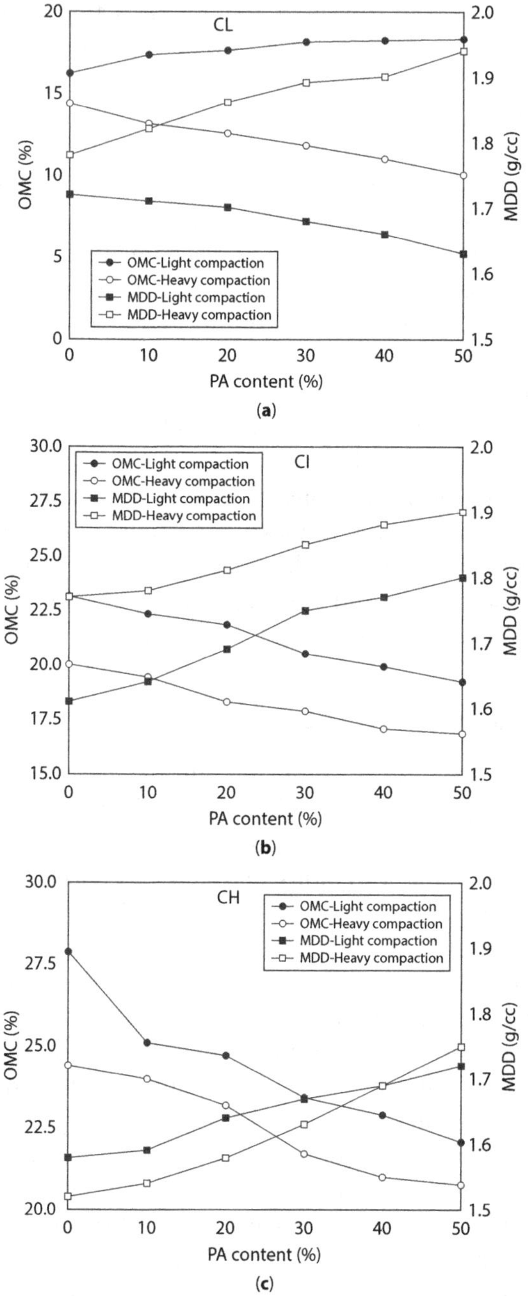

Figure 5.4 Compaction parameters for (a) low, (b) intermediate and (c) high plastic clays with pond ash content.

With an emphasis on the PA-clay soil mixture, compaction tests were conducted for modified proctor compaction testing. The results of testing several combinations of PA-soil mixtures are shown in Figures 5.4. The maximum dry density and optimum moisture content changes for various PA dose levels from 0% to 50%, are depicted in these figures. Remarkably, when the PA dosage is increased, both MDD and OMC show a steady upward trend. This finding raises the potential for a relationship between the soil's compaction properties combination and the PA dosage. The results highlight how PA affects compaction characteristics and point to a possible use for PA to improve soil stability and compactability. Engineering techniques can benefit greatly from such insights, especially in projects where soil alteration and compaction are essential for maintaining structural integrity and performance.

An examination of Figure 5.4 indicates a pattern in which the OMC increases simultaneously with the dose increase of pond ash in mix mode stabilization. The reason behind this rise in OMC is the faster rate of pozzolanic interaction between PA and soil elements. Significantly, specimens stabilized with PA show a more marked rise in OMC, which is thought to require more water because of its higher fineness, and hence improved hydration. Moreover, a discernible reduction in maximum dry density is noted concomitant with an increase in OMC. This MDD reduction could be caused by PA's relatively lower specific gravity values, as they may also cause a decrease in PA. This, in turn, causes the mixture's void ratio to rise and its overall density to decrease. These detailed observations highlight the complex interactions that exist between stabilizing agents and soil properties, clarifying the complex mechanisms that underlie variations in compaction characteristics and emphasizing the value of these understandings in enhancing soil stabilization methods for a range of engineering applications.

5.3.4 Unconfined Compressive Strength

The unconfined compressive strength test results are displayed in Figure 5.5 on clay stabilized with pond ash and different amounts of cement. According to analysis, increasing the proportion of PA by up to 50% causes a notable rise in UCS values in all types of soil. The UCS, for example, increases from 110 to 140 kPa in CL soil, from 160 to 240 kPa in CI soil, and from 254 to 270 kPa in CH soil. The cohesive characteristics of PA, when combined with clay, enable the simultaneous mobilization of both cohesion and frictional components, which are principally responsible for the observed improvement in UCS. Furthermore, the admixtures included

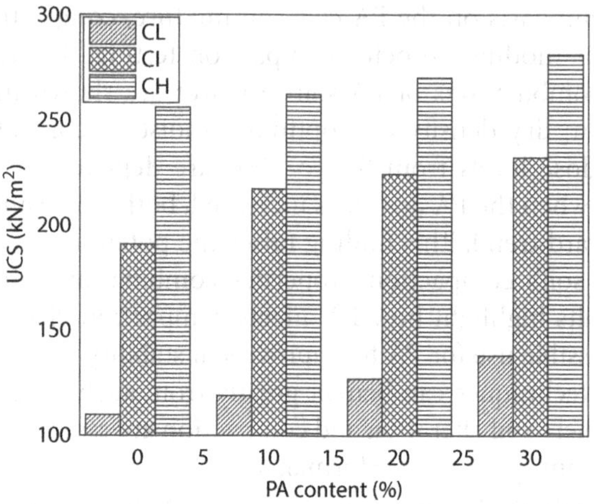

Figure 5.5 Variation in UCS value with pond ash content for different plastic soils.

in PA enhance particle packing, which increases the stabilized clay's overall strength. These outcomes fairly concur with earlier studies on clay stabilization, especially when it comes to the use of Fly Ash as a stabilizing agent, as reported in the literature [49].

The compilation of recent research highlights the continued development of soil stabilization techniques overall, with PA emerging as a useful and efficient supplement to improve the engineering qualities of clay soils. Sustained research endeavors focused on refining mixture designs, comprehending basic mechanics, and investigating innovative additives exhibit considerable potential for propelling the domain of soil stabilization and tackling modern engineering predicaments.

5.3.5 California Bearing Ratio

The clay soils of varying plasticity were modified by replacing with pond ash, and the CBR values of the resulting mixtures were evaluated. The unsoaked and soaked California bearing ratio values of clay stabilized with PA are shown in Figure 5.6, which indicates that the addition of PA raises the CBR amount. Although the main mechanism by which clay resists deformation is cohesion, the addition of PA creates a frictional component that strengthens this mechanism. Thus, when the PA content rises, so does the load-carrying capacity. Nevertheless, the data also shows that load-carrying capability is not significantly increased by adding PA in increments greater than 40%. This indicates that the best dosage for increasing

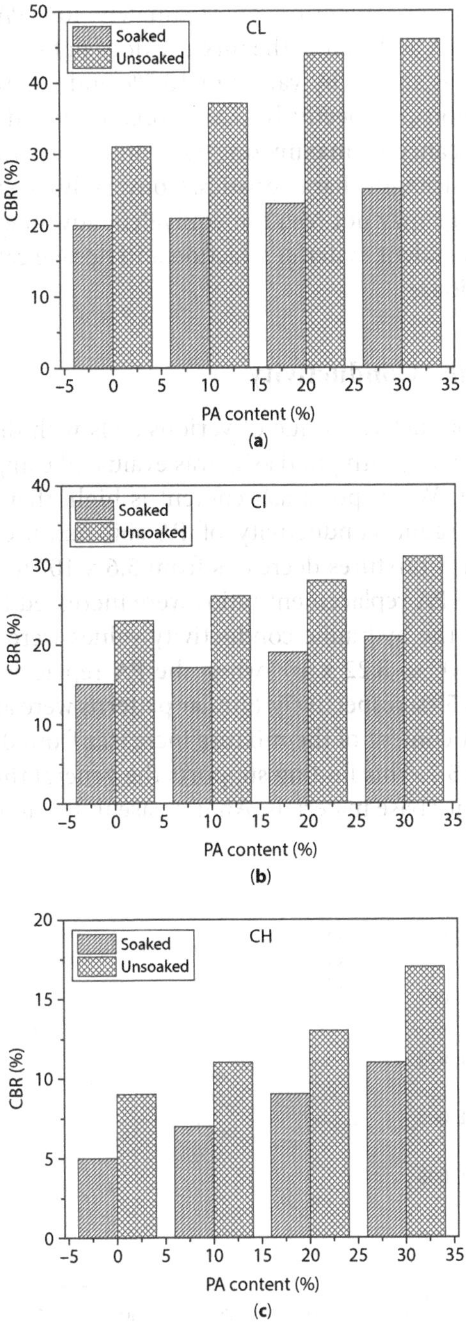

Figure 5.6 Variation in CBR of (a) low, (b) intermediate and (c) high plastic clays with content of pond ash content.

the PA-clay mixture's strength is approximately 40% PA content. Above this point, more PA addition to the mixture does not significantly increase the CBR values, as shown by values of 18.3% and 13.7% for soaked and unsoaked situations, respectively. The aforementioned discovery highlights the significance of maximizing PA content to attain the intended enhancement in strength characteristics. Conversely, an excess of PA above the optimal dosage may not result in equivalent advantages, underscoring the necessity of prudent material selection and dosage calculation in engineering applications.

5.3.6 Hydraulic Conductivity

The hydraulic conductivity value of various soils with different plasticity, modified by replacing with pond ash, was evaluated using the falling head permeability test. When pond ash content is high, there is a discernible drop in the hydraulic conductivity of PA-soil mixtures. The hydraulic conductivity of PA mixtures decreases from 5.6 x 10^{-8} to 1.1 x 10^{-7}m/s for CL soil when the PA replacement ratios were increased from 10% to 50% respectively. But the hydraulic conductivity values were increased for CI soil from 2.8 x 10^{-8} to 8.22 x 10^{-8} when the PA replacements were increased from 105 to 50% respectively. Similar patterns were also found for CH soil when the PA content of the mixture increases from 0% to 50%, represented in Figure 5.7. This finding supports the general theory that less PA is required for finer base materials. An increase in the amount of pond ash

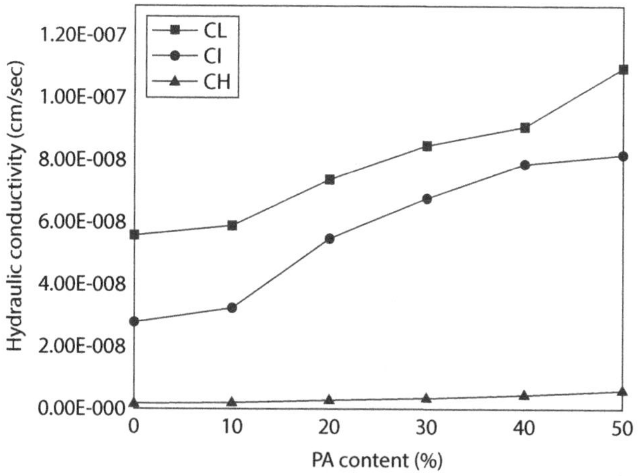

Figure 5.7 Hydraulic conductivity of clay-pond ash mixes.

causes the micro pores in the compacted mixture to shrink, which in turn controls the flow. Hydraulic conductivity decreases as a function of micro pore size because of their decreasing impact on flow. This relationship has ramifications for engineering applications where the control of fluid flow through compacted materials is essential for stability and performance, as it highlights the vital role that bentonite content plays in regulating the hydraulic conductivity of PA-soil mixes.

5.4 Conclusions

Various plastic soils were characterized for their geotechnical applications. The pond ash was utilized to modify clay soils' geotechnical characteristics with different plasticity. The geo-engineering properties were assessed by conducting the laboratory experiments following various soils' testing standards. The following conclusions are drawn

1. Black cotton soil has increased flexibility and is more prone to swelling and shrinking in response to variations in moisture levels than residual soil, which has decreased plasticity, swelling, and shrinkage properties.
2. When pond ash was added, the plasticity of plastic clay soils reduced, and these soils' liquid limits likewise fell—regardless of how plastic they had been before. Furthermore, mixing various types of clay with pond ash resulted in a drop in the plastic limit. As the amount of pond ash in the clay soils rose, the linear shrinkage values dropped. Additionally, the free swell value in the plastic soils dropped as the pond ash content increased.
3. Pond ash was found to increase the ideal moisture content in plastic soils, but it also caused the maximum dry density to drop under both heavy and light compaction energies.
4. The ideal moisture content was lower when the material was compressed with high compactive effort as contrasted to low compactive effort. Furthermore, after applying strong compactive force to all plastic soil-pond ash mixes, a rise in maximum dry densities was observed.
5. It was discovered that the addition of pond ash to the soils, increased the unconfined compressive strength of all the plastic soil-pond ash mixtures. The low plastic soils, i.e., the

residual clay soil exhibited a higher strength compare to that for high plastic clays for all the pond ash contents in soils.

6. When pond ash was added, the plastic soils' California bearing ratio showed an increase. Furthermore, in every soil-pond ash combination, the unsoaked California bearing ratio values were higher than the soaked values.
7. Plastic clays are less permeable. The incorporation of pond ash to the clays raises the permeability of pond ash-clay mixes. After adding of pond ash to the clay, the hydraulic conductivity values increased.

References

1. Jaswal, P., Vivek, Sinha, S.K., Improvement in the performance of two layered model pavement with treated coir geotextile at the interface. *J. Ind. Text.*, *52*, 15280837221114161, 2022.
2. Vivek, Shafi Mir, M., Sehgal, R., Studies of Modulus of Resilience on Unpaved Roads Reinforced with Untreated/Treated Coir Geotextiles. *J. Nat. Fibers*, *19*, 16, 13563–13573, 2022.
3. Vivek, Shafi Mir, M., Sehgal, R., Study on bearing capacity of unpaved roads reinforced with coir geotextiles using finite element method (fem). *J. Nat. Fibers*, *19*, 15, 11735–11748, 2022.
4. Jaswal, P. and Vivek, Experimental study on monotonic behaviour of two-layered unpaved road model reinforced with treated coir geotextiles. *Int. J. Pavement Res. Technol.*, 17, 1377–1388, 2023.
5. Vivek, Dutta, R.K., Parti, R., Effect of chemical treatment of the coir geotextiles on the interface properties of sand/clay-coir geotextile interface. *J. Inst. Eng. India Ser. A*, 100, 357–365, 2019.
6. Mohanta, N.R. and Samantaray, S., Study of Combined Effect of Metakaolin and Steel Fiber on Mechanical Properties of Concrete. *Pertanika J. Sci. Technol.*, 27, 3, 1381–1396, 2019, http://journals-jd.upm.edu.my/resources/files/ Pertanika%20PAPERS/JST%.
7. Vivek, and Dutta, R.K., Bearing ratio behavior of sand overlying clay with treated coir geotextiles at the interface. *J. Nat. Fibers*, 19, 14, 7534–7541, 2022.
8. Vivek, Effects of cyclic loading on sand overlaying clay model of unpaved roads reinforced with untreated/treated coir geotextiles. *J. Text. Inst.*, 115, 9, 1694–1700, 2023.
9. Singh, P., Dash, H.M., Samantaray, S., Effect of silica fume on engineering properties of expansive soil. *Mater. Today Proc.*, 33, 5035–5040, 2020, https://doi.org/10.1016/j.matpr.2020.02.839.

10. Kumar, M.H., Mohanta, N., Samantaray, S., Kumar, N.M., Combined effect of waste glass powder and recycled steel fibers on mechanical behavior of concrete. *SN Appl. Sci.*, 3, 1–18, 2021a, https://doi.org/10.1007/s42452-021-04422-w.
11. Kumar, M.H., Macharyulu, I.S., Ray, T., Mohanta, N.R., Jain, M., Samantaray, S., Sahoo, A., Effect of water absorption and curing period on strength and porosity of triple blended concrete. *Mater. Today Proc.*, *43*, 2162–2169, 2021b, https://doi.org/10.1016/j.matpr.2020.12.092.
12. Ray, T., Mohanta, N.R., Kumar, M.H., Samantaray, S., Study of effect of temperature on behavior of alkali activated slag concrete. *Mater. Today Proc.*, *43*, 1352–1357, 2021, https://doi.org/10.1016/j.matpr.2020.09.169.
13. Kumar, M.H., Saikrishnamacharyulu, I., Mohanta, N.R., Ashutosh, A., Mishra, P., Samantaray, S., Mechanical behaviour of high strength concrete modified with triple blend of fly ash, silica fume and steel fibers. *Mater. Today Proc.*, *65*, 933–942, 2022a, https://doi.org/10.1016/j.matpr.2022.03.528.
14. Kumar, M.H., Mohanta, N.R., Patel, N., Samantaray, S., Reddy, S.V.B., Influence of Fly ash and Metakaoline on the crack resistance and shrinkage property of concrete. *Iran. J. Sci. Technol. Trans. Civ. Eng.*, 46, 2011–2026, 2022b, https://doi.org/10.1007/s40996-021-00748-y.
15. Gupta, D. and Kumar, A., Performance evaluation of cement-stabilized pond ash-rice husk ash-clay mixture as a highway construction material. *J. Rock Mech. Geotech. Eng.*, *9*, 1, 159–169, 2017.
16. Ola, S.A., Stabilization of lateritic soils by extensible fibre reinforcement. *Eng. Geol.*, *26*, 2, 125–140, 1989.
17. Balogun, J.A., Akomolafe, C.T., Amusa, L.O., Grip strength: effects of testing posture and elbow position. *Arch. Phys. Med. Rehabil.*, *72*, 5, 280–283, 1991.
18. Phanikumar, B.R., Sharma, R.S., Rao, A.S., Madhav, M.R., Granular pile anchor foundation (GPAF) system for improving the engineering behavior of expansive clay beds. *Geotech. Test. J.*, *27*, 3, 279–287, 2004.
19. Oriola, F. and Moses, G., Groundnut shell ash stabilization of black cotton soil. *Electron. J. Geotech. Eng.*, *15*, 1, 415–428, 2010.
20. Bairwa, M., Rajput, M., Sachdeva, S., Modified Kuppuswamy's socioeconomic scale: social researcher should include updated income criteria, 2012. *Indian J. Community Med.: Off. Publ. Indian Assoc. Preventive Social Med.*, *38*, 3, 185, 2013.
21. Chand, S.K. and Subbarao, C., Strength and slake durability of lime stabilized pond ash. *J. Mater. Civ. Eng.*, *19*, 7, 601–608, 2007.
22. Sarkar, R., Abbas, S.M., Shahu, J.T., Comparison Study of Geotechnical Behaviour of Lime Reinforced Pond Ashes Available in Delhi Region. *GEOMATE J.*, *3*, 5, 273–279, 2012.
23. Bera, A.K., Ghosh, A., Ghosh, A., Behaviour of model footing on pond ash. *Geotech. Geol. Eng.*, *25*, 315–325, 2007a.

24. Gupta, D. and Kumar, A., Strength characterization of cement stabilized and fiber reinforced clay–pond ash mixes. *Int. J. Geosynth. Ground Eng.*, *2*, 1–11, 2016.
25. Roy, T.K. and Chattopadhyay, B.C., A study on the effect of cement on alluvial soil strengthened with pond and rice husk ash for construction of road subgrade. *The 12th international conference of international association for computer methods and advances in geomechanics*, October 2008, pp. 3508–3513.
26. Ghosh, A., Compaction characteristics and bearing ratio of pond ash stabilized with lime and phosphogypsum. *J. Mater. Civ. Eng.*, *22*, 4, 343–351, 2010.
27. Basha, E.A., Hashim, R., Mahmud, H.B., Muntohar, A.S., Stabilization of residual soil with rice husk ash and cement. *Constr. Build. Mater.*, *19*, 6, 448–453, 2005.
28. Bera, A.K., Ghosh, A., Ghosh, A., Compaction characteristics of pond ash. *J. Mater. Civ. Eng.*, *19*, 4, 349–357, 2007b.
29. Anwar Hossain, K.M., Stabilized soils incorporating combinations of rice husk ash and cement kiln dust. *J. Mater. Civ. Eng.*, *23*, 9, 1320–1327, 2011.
30. Mir, B.A. and Sridharan, A., Physical and compaction behaviour of clay soil–fly ash mixtures. *Geotech. Geol. Eng.*, *31*, 1059–1072, 2013.
31. Muntohar, A.S. and Hantoro, G., Influence of rice husk ash and lime on engineering properties of a clayey subgrade. *Electron. J. Geotech. Eng.*, *5*, 2000, 1–13, 2000.
32. Kumar, R., Kanaujia, V.K., Chandra, D., Engineering behaviour of fibre-reinforced pond ash and silty sand. *Geosynth. Int.*, *6*, 6, 509–518, 1999.
33. IS: 2720 Part 4 (1985) Methods of Test for Soils: Grain Size Analysis, Bureau of Indian Standards, New Delhi.
34. IS: 2720 Part 3, section-1 (1980) Determination of Specific Gravity of Soil, Bureau of Indian Standards, New Delhi.
35. IS: 2720 Part 5 (1985) Determination of Liquid Limit and Plastic Limit of Soil, Bureau of Indian Standards, New Delhi.
36. IS: 2720 Part 40 (1977) Methods of Test for Soils: Determination of free swell index of Soil, Bureau of Indian Standards, New Delhi.
37. IS: 2720 Part 20 (1922) Methods of Test for Soils: Determination of linear shrinkage, Bureau of Indian Standards, New Delhi.
38. IS: 1498 (1970) Classification and identification of soils for general engineering purposes, Bureau of Indian Standards, New Delhi.
39. IS: 2720 Part 7 (1980) Methods of Test for Soils: Determination of Water Content-Dry Density Relation Using Light Compaction (second revision), Bureau of Indian Standards, New Delhi.
40. IS: 2720 Part 8 (1983) Methods of Test for Soils: Determination of Water Content-Dry Density Relation Using Heavy Compaction (second revision), Bureau of Indian Standards, New Delhi.
41. IS: 2720 Part 10 (1991) Methods of Test for Soils: Determination of Unconfined Compressive Strength for Soil, Bureau of Indian Standards, New Delhi.

42. IS: 2720 Part 16 (1987) Methods of Test for Soils: Laboratory determination of CBR, Bureau of Indian Standards, New Delhi.
43. IS: 2720 Part 17 (1986) Methods of Test for Soils: Laboratory Determination of Permeability (second revision), Bureau of Indian Standards, New Delhi.
44. Ogundalu, A.O. and Oyekan, G.L., Mineralogical and geotechnical characterization of maidaguri black cotton soil by X-ray diffraction (XRD), x-ray photoelectron (XPS) and scanning electron spectroscopy (SEM). *Int. J. Eng. Technol.*, *4*, 6, 345–353, 2014.
45. Li, J., Zou, J., Bayetto, P., Barker, N., Shrink-swell index database for Melbourne. *Aust. Geomech. J.*, *51*, 3, 61–76, 2016.
46. Fityus, S.G., Cameron, D.A., Walsh, P.F., The shrink swell test. *Geotech. Test. J.*, *28*, 1, 92–101, 2005.
47. Blayi, R.A., Sherwani, A.F.H., Ibrahim, H.H., Faraj, R.H., Daraei, A., Strength improvement of expansive soil by utilizing waste glass powder. *Case Stud. Constr. Mater.*, *13*, e00427, 2020.
48. Yaghoubi, E., Yaghoubi, M., Guerrieri, M., Sudarsanan, N., Improving expansive clay subgrades using recycled glass: Resilient modulus characteristics and pavement performance. *Constr. Build. Mater.*, *302*, 124384, 2021.
49. Pandian, N.S., Krishna, K.C., Sridharan, A., California bearing ratio behavior of soil/fly ash mixtures. *J. Test. Eval.*, *29*, 2, 220–226, 2001.

6

Utilization of Steel Slag in the Construction of Granular Sub-Base Layer

Chitaranjan Dalai*, Subhashree Samantasinghar, Debiprasad Panda, Sai Sumiran Panda and Deba Prakash Satapathy

Department of Civil Engineering, Odisha University of Technology and Research, Bhubaneswar, India

Abstract

Water can damage soil, so good drainage is essential for soil durability. To make a pavement to last longer, it is important to consider drainage as water can damage it. Recent studies have suggested the use of non-conventional materials such as industrial waste and dolochar and geotextiles as support layers to hold the aggregate. Steel bracing is also good for smaller classes because it can support heavy loads without moving or settling, making it cost-effective for large parking spaces. Incorporating scrap steel into construction projects is sustainable because it reduces waste and conserves natural resources. It is also more wear-resistant than traditional materials and requires less maintenance. It is generally popular and sustainable for construction projects. This research study investigates the use of non-traditional aggregates as drainage layer materials in the subgrade for flexible pavement design. These aggregates, such as steel slag and dolochar, are industrial wastes from iron ore processing and sponge iron processing industries, respectively. The grading of these aggregates meets ASSTHO (grade 2)/MoRTH (grade V & VI) specifications. Geosynthetics, specifically jute and woven polypropylene, are also utilized in accordance with MoRTH specifications (type II) to provide additional support for permeability value. The best combination is found to be with woven polypropylene geotextile on both the middle and bottom layer of the subgrade comprising iron slag, resulting in a satisfactory coefficient of permeability value of 12.37 meters/day. The use of this combination also leads to a 0.96% reduction in coefficient of permeability value when compared to the subgrade layer consisting solely of steel slag. It's resulting in a satisfactory permeability

**Corresponding author*: dr.chitaranjandalai@gmail.com

Sandeep Samantaray, Subhashree Samantasinghar, Soumendra K. Mohanty, Vivek, Abinash Sahoo and Deba Prakash Satapathy (eds.) Waste Utilization in Geotechnical Practice, (149–178) © 2025 Scrivener Publishing LLC

value. The study takes into account moderate installation conditions and storm water levels in the area. Using steel slag in road construction has multiple benefits, including reducing the need for landfills for disposal and saving natural resources. Additionally, the costs associated with disposing of the slag can be saved, and the pollution caused by dumping it can be reduced. Another advantage is that using geosynthetics and inexpensive steel slag can create a more economical and durable drainage layer in sub-grade sections. This can prevent subgrade materials from washing away and causing pavement cracks, ultimately leading to lower maintenance costs.

Keywords: Soil, dolochar, steel slag, geotextile, CBR, permeability

6.1 Introduction

The construction industry faces a steady challenge in finding feasible materials for different framework ventures. Geo materials like soil are generally adopted for road construction [1–5]. The over-the-top transfer of mechanical by-products, such as steel slag, presents a natural concern and a missed opportunity for asset utilization. Steel slag, a by-product of steel fabricating, has picked up consideration as a potential fabric for utilization within the development granular sub-base layer. This chapter explores the utilization of steel slag in the construction granular sub-base layer, focusing on its properties, benefits, challenges, and best practices. For pavement construction or pavement design; drainage of water is one of the major considerations, because of its adverse effect on pavement materials. Initially, its effect on the pavement is negligible but subsequently, it deteriorates the material strength, causing pavement failure. This will escalate to pavement resurfacing or complete removal of damaged pavement for construction. Free water can enter the pavement by several means such as infiltration, defective workmanship, heavy traffic load, or traffic volume in a water-prone area. In comparison to other forms of water, free water can be removed vertically by a subgrade layer or laterally by a lateral drainage system.

Several studies have been conducted on adoption of waste materials in construction [6–12]. Researches show the effectiveness of utilization of industrial wastes as road pavement construction materials in an environmentally and economically sustainable way [13–17]. They [18, 19] studied on recycled materials in pavement construction and concluded its efficacious application as pavement material. They [20], experimented on California bearing ratio, permeability and durability study and find the optimum slag content. About 50% slag with natural aggregate proved to be effective for granular subbase construction. They [21], studied strength and durability performances of electric arc furnace slag and incorporated cementitious material for road construction. Partial replacement of natural aggregate by slag produces material with greater strength and durability. Further, to enhance the load carrying capacity and stability of road sub-base, it is necessary to provide proper drainages which facilitates passage of water out of the layer if not so, the water may be present within the layer which in turn may reduce the stability and bearing capacity. Provision of geotextiles proved to be an alternate filtration method [22–26]. Most of the past study presented the use of slag for production of cementitious materials to be used for road construction and use of geosynthetics for drainage. This chapter aims to utilize dolochar in granular sub bases with enhanced strength, stability and drainage properties.

- **Benefits of Using Steel Slag in Granular Sub-Base Layer**

The utilization of steel slag in granular sub-base layers presents a multitude of compelling benefits, offering a sustainable, cost-effective, and high-performance alternative in construction practices [20]. Firstly, steel slag, as a by-product of steel production, significantly reduces the environmental burden by repurposing industrial waste into a valuable construction material, thereby diminishing landfill waste and conserving natural resources. Its innate characteristics, including high density, angularity, and excellent load-bearing capacity, enhance the structural integrity and stability of the sub-base layer, thereby reducing settling and deformation under heavy loads. Moreover, steel slag's chemical composition promotes good drainage, mitigating the risk of water accumulation and associated structural damage. Durability and weather resistance extend the life of and reduce maintenance requirements, contributing to long-term cost savings. In addition, the use of reinforced concrete in the bottom layer promotes sustainability by turning waste into functional parts of the structure, contributing to the circular economy. Overall, the incorporation of scrap steel has many benefits, including environmental sustainability, improved structural performance, and the economic viability of gravel pavement construction.

- **Best Practices for Utilizing Steel Slag in Sub-Base Layers**

The optimal use of steel concrete in basements must comply with a set of good practices to ensure the correct installation and performance of steel in construction projects. First, comprehensive characterization and testing of steel scrap, including physical, chemical and mechanical properties, is essential to determine the suitability and appropriate input design. Adequate quality control procedures during procurement and steel scrap processing ensure application consistency. The inclusion of steel scrap in the base layer must be carefully graded and combined with other aggregates or binding materials to achieve the required mechanical properties such as stiffness, stability and transport resistance. If proper moisture addition and management techniques are implemented during the construction, that will ensure a very good and solid foundation. Regular monitoring and inspection of the manufacturing process will help ensure performance expectations are met and identify deviations that may require correction. Furthermore, dissemination of knowledge and awareness regarding the benefits and proper utilization techniques of steel slag among construction stakeholders remains integral to fostering its widespread adoption and realizing the full potential of this sustainable construction material.

6.2 Materials and Methodology

6.2.1 Materials Used

The materials employed in the utilization of steel slag within the construction of the granular sub-base layer encompass a range of components tailored to enhance structural integrity, stability, and sustainability. Firstly, steel slag, a by-product of the steelmaking process, serves as the primary material due to its potential as an alternative aggregate. Additionally, the soil, chosen for its cohesive properties and ability to improve workability, is combined with steel slag to optimize the mix design. Iron slag, another by-product rich in iron oxide, is utilized to enhance the binding properties and overall strength of the sub-base layer. Dolochar, a waste material from coal gasification, is introduced to improve the layer's resistance to moisture and provide better load-bearing capabilities. Jute, in the form of natural fibers, is integrated to reinforce the sub-base layer, contributing to its tensile strength and preventing cracking. Geotextiles, used as a separation and filtration layer, assist in restraining the intermixing of soil and aggregates, promoting drainage, and enhancing overall stability. The combined utilization of these diverse materials aims to create a durable, high-performing granular sub-base layer while simultaneously addressing sustainability and environmental concerns within construction practices.

6.2.1.1 Soil

Soil is the basic requirement of the experiment as any kind of pavement needs to transmit its strength solely to the soil. Soil sample was collected

Table 6.1 Soil properties.

Description	Test value
Specific gravity	2.65
Liquid limit	24.2 %
Plastic limit	2.43 %
Maximum dry density	2.057 gm/cc
Optimum moisture content	12.01 %
California Bearing ratio	0.91 %

from OUTR campus. The geotechnical properties of soil are tabulated in Table 6.1.

6.2.1.2 Dolochar

Dolochar is the residual waste product of sponge iron industry. Dolochar can be used as a replacement of natural aggregate in pavement construction and can also be used as drainage material in pavement. Dolochar was collected from Ganesh sponge iron factory of Talcher, Anugul, Odisha. The chemical and physical properties of dolochar are tabulated in Tables 6.2 and 6.3 respectively.

Table 6.2 Chemical properties of dolochar.

Major constituents	Values (%)
Fe_2O_3	51.10
SiO_2	20.35
Al_2O_3	12.37
MgO	11.80

Table 6.3 Physical properties of dolochar.

Physical property	Values (%)
Fineness modulus	6.87
Specific gravity	2.64
Aggregate crushing value	30.23
Aggregate impact value	29.91
Water absorption (%)	6.14

6.2.1.3 Steel Slag

Steel slag is the waste product obtained from iron and steel industry. After processing of iron ore, residual slag obtained from the furnace can be used for various purposes. Generally, it is used as a replacement for normal granite aggregate for construction of pavement. The chemical composition of steel slag can vary depending on the steelmaking process, potentially leading to variations in performance. Careful quality control and testing are essential. The porous nature of steel slag promotes efficient water drainage, reducing the risk of water-related damage. In our experiment, iron slag is

Table 6.4 Chemical properties of steel slag.

Major constituents	Values (%)
CaO	41.7
SiO_2	33.8
MgO	7.4
Al_2O_3	6.4
Fe_2O_3	7.4
SO_3	0.8
MnO	0.3
P_2O_5	<0.1

Table 6.5 Physical properties of steel slag.

Physical property	Values
Specific gravity	3.18
Water absorption	2 %
Aggregate impact value	4.98 %
Aggregate crushing value	16.4 %
Unit weight (kg/m^3)	1600.19

used as a drainage layer in pavement. Iron is obtained from Rourkela steel plant, Rourkela, Syndergaard, Odisha. The chemical and physical properties of steel slag are tabulated in Tables 6.4 and 6.5 respectively.

6.2.1.4 Jute

Jute is a natural occurring fiber collected from ribbon, i.e., the outer skin of jute plant. It is also considered to be the strongest among all other natural fibers. Jute occupies second place, next to cotton in world's production of natural fiber. In our experiment we used Jute as a medium of passing of water through it without clogging. Jute was collected from local market of Cuttack, Odisha. The physical and chemical properties of jute fiber are summarized in Tables 6.6 and 6.7 respectively.

Table 6.6 Physical properties of jute fiber.

Physical property	Values
Diameter	0.035-0.14 mm
Specific gravity	1.48 gm/cc
Tenacity	2.7-5.3 gm/tex
Breaking elongation	0.8-1.8%
Moisture regained	13%

Table 6.7 Chemical properties of Jute fiber.

Chemical property	Values (%)
Cellulose	58 – 63
Hemi cellulose	20 – 22
Lignin	12 – 14.5
Wax & fats	0.4 – 0.8
Pectin	0.2 – 0.5
Protein	0.8 – 2.5

6.2.1.5 Geotextile

This is normally used alongside soil, rock or in any geotechnical materials for civil engineering application. It can be polymeric, planar, synthetic, natural, permeable, woven, nonwoven or knitted. geotextiles form one of the two largest groups of geo synthetics. The raw material, i.e., polypropylene, Polyethylene and Polyamide comes from polyester for making geotextile. Geotextile is applicable to various engineering sector, including civil engineering and especially in construction site. Generally, to strengthen the

Table 6.8 Properties of woven polypropylene geotextile.

Properties	Values
Specific gravity	0.91
Unit weight	500 g/m^2
Thickness	2mm
Color	White
Breaking strength	16 kN/m
Elongation Break	25 – 10 (%)
Tear strength	0.42 kN
Opening size	0.07-2 mm

soil, geotextiles are placed in the tension area. Geotextile was purchased from Sun textiles private limited from Kolkata. In our experiment we used Woven polypropylene type of geotextile. The properties of woven polypropylene geotextile are summarized in Table 6.8.

6.2.2 Methodology

The methodology for the utilization of steel slag in the construction of the granular sub-base layer involves several key steps aimed at assessing the feasibility, suitability, and effectiveness of incorporating steel slag into this layer of construction. Initially, comprehensive literature review and research will be conducted to gather existing knowledge, data, and findings related to the characteristics, properties, and performance of steel slag as a construction material. This includes the study of physical, chemical, and mechanical properties, as well as possible environmental effects. Later, laboratory tests and other tests are conducted to analyze the special characteristics of the steel plate in terms of deformation, strength, permeability and durability of the steel plate when used in the subgrade layer. It evaluates various designs and proportions to determine the best combination. of scrap steel and other aggregates or binders to achieve the desired mechanical properties. Field tests and on-site tests will also be conducted to validate the laboratory results and evaluate the actual performance of the microsatellite platform in terms of payload capacity, stability and long-term stability. A lot of time, a comparative analysis with conventional materials is performed to determine the economic and environmental benefits of using steel scrap in construction. Overall, this method aims to systematically investigate and demonstrate the potential benefits and advantages of incorporating steel slag into gravel subgrades to provide a sustainable and efficient construction solution. In this experiment iron slag is used as a drainage layer of subgrade layer of flexible pavement and with the combination of geotextile or jute or combination of both jute and geotextile in different layers of subgrade, permeability is determined. The combinations of placement of geotextile are as follows:

1. On top
2. On bottom
3. On middle
4. On both top and bottom
5. On both top and middle
6. On both middle and bottom
7. On top, middle and bottom

6.3 Results and Discussion

In the previous section experiments were mentioned about in brief were performed. In this section, the outcome of the performed experiments is duly explained with all numerical data and graphs. After determining the results of each experiment, results were discussed and the best suitable combination is determined.

6.3.1 Tests on Soil

6.3.1.1 *Specific Gravity of Soil*

The results were obtained when the soil was tested to find its specific gravity in accordance with IS: 2720 (Part-III):1980 [27]. The average specific gravity of soil at room temperature is 2.65.

6.3.1.2 *Sieve Analysis of Soil*

The representative soil sample underwent mechanical analysis by conducting sieve analysis, performed in accordance with IS: 2720 (Part-IV):1985 [28]. From the graph in Figure 6.1, the value of coefficient of curvature (Cc) and coefficient of uniformity (Cu) was determined.

$D_{10} = 0.014$
$D_{30} = 0.28$
$D_{60} = 0.76$
$Cu = D_{60} / D_{10} = 54.28$
$Cc = D_{30}^{2} / D_{10} \times D_{60} = 7.386$
75μ passing percentage finer is 17.43%, which is > 12%.
$I_P = 0.73(W_L\text{-}20) = 3.139$.

Considering the above calculation determined from experiments, and analyzing these with IS code IS 1498:1970 [29] (Classification of soil for general engineering purpose), the type of soil we get is SM (silty sandy soil).

6.3.1.3 *Liquid Limit of Soil*

The liquid limit test was performed on the soil sample in accordance with IS: 2720 (Part-V):1985 [30]. From the flow curve in Figure 6.2, the liquid limit value of the soil corresponding to 25 blows was found to be 24.2%.

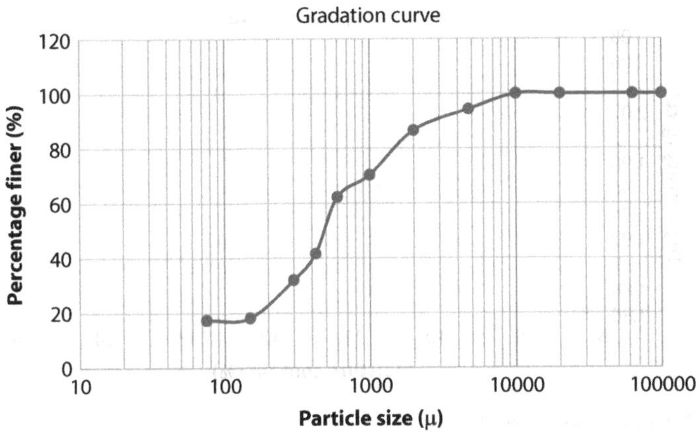

Figure 6.1 Grain size distribution curve for soil.

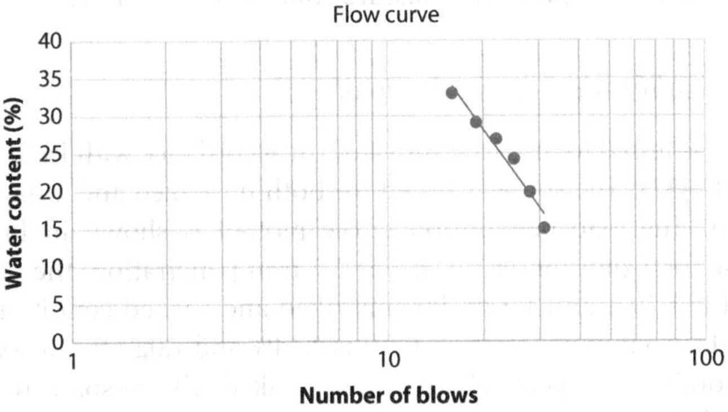

Figure 6.2 Liquid limit flow curve.

6.3.1.4 *Plastic Limit of Soil*

The plastic limit test was performed on the soil sample in accordance with IS: 2720 (Part-V):1963 [30]. The average plastic limit of the 3 observations was found to be 2.43%. So, the plastic limit of the soil obtained was 2.43%.

6.3.1.5 *Maximum Dry Density and Optimum Moisture Content of Soil*

The standard proctor test was performed on the soil sample in accordance with IS: 2720 (Part-VIII):1983 [31]. When the compaction curve was plotted for various water content and corresponding dry density values as

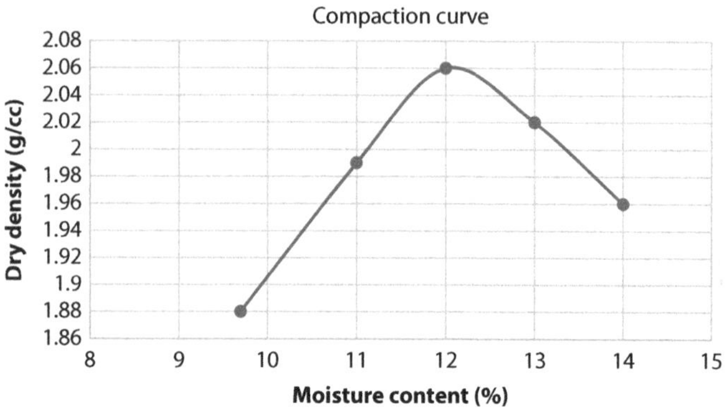

Figure 6.3 Compaction curve for soil.

shown in Figure 6.3, it was found that the max dry density of the soil was 2.06 g/cm^3 and the optimum moisture content was 12.01%.

6.3.1.6 California Bearing Ratio of Soil

The soil was tested to find its CBR value in accordance with IS: 2720 (Part-XVI):1987 [32]. The soil was tested for both unsoaked and 4 days-soaked condition. The load-penetration curve plotted is shown in Figure 6.4. The value for 1 division is 1.172. For 2.5 mm penetration, the CBR value obtained is 2.39% and 1.32% for unsoaked and soaked condition respectively and for 5 mm penetration CBR is 2.14% and 1.02%for unsoaked and soaked condition respectively. So, the unsoaked CBR of soil is reported as 2.39% and the soaked CBR as 1.32%.

6.3.2 Tests on Aggregates

6.3.2.1 Impact Test of Dolochar

Using dolochar as aggregate, the impact test was done according to IS: 2386 (Part-IV):1963 [33]. The average impact value of the aggregates used was found to be 35.89%.

6.3.2.2 Impact Test of Steel Slag

Using steel slag as aggregate in the sub base, the impact test was done according to IS: 2386 (Part-IV):1963 [33]. The average impact value is 4.987%.

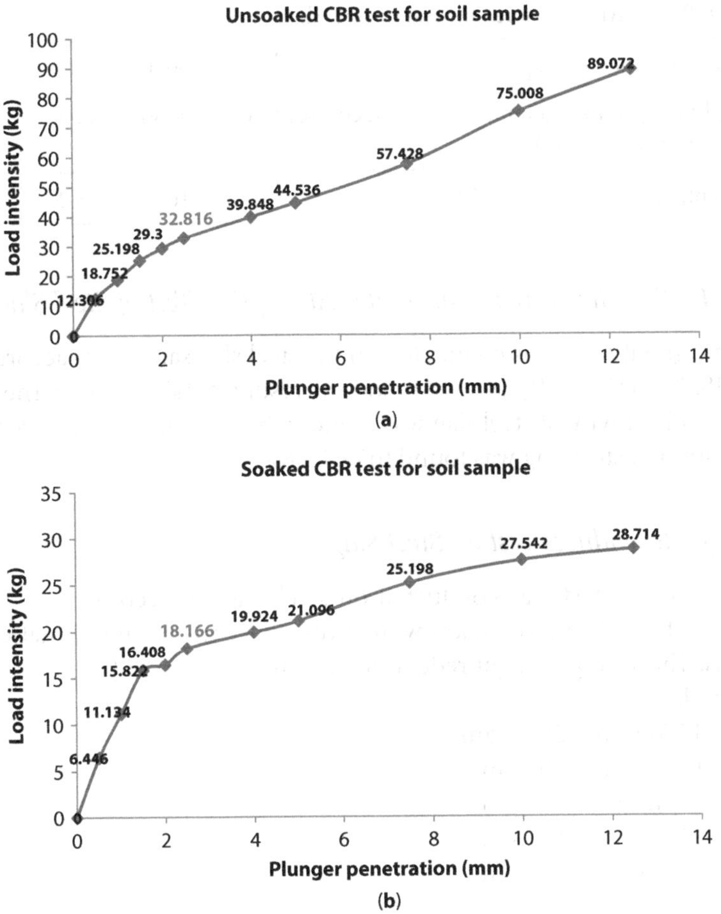

Figure 6.4 Load-penetration curve of soil for (a) Unsoaked and (b) Soaked condition.

6.3.2.3 Los Angeles Abrasion Test of Steel Slag

Using steel slag as aggregate in the sub base, the impact test was done according to IS: 2386 (Part-IV):1963 [33]. The results obtained are shown in Table 6.9. Here, we used Grade A samples. The percentage wear was found to be 16.1%.

Sample size- Grade (A)
(20-12.5) mm = 1250 gm
(25-20) mm = 1250 gm
(40-25) mm = 1250 gm
(63-40) mm = 1250 gm

Table 6.9 Los Angeles abrasion test of steel slag.

Wt. of specimen (W_1) g	5000	5000	5000
Weight of specimen after abrasion test coarser than 1.70mm IS sieve (W_2) g	4180	4240	4165
Percentage wear (W_1-W_2/W_1)×100	16.4	15.2	16.7

6.3.2.4 *Specific Gravity and Water Absorption Test of Steel Slag*

Specific gravity test was conducted on the steel slag sample in accordance with IS: 2386 (Part-III):1963 [34] and the results obtained were: The average specific gravity of steel slag was found to be 3.33 and the average water absorption of steel slag was found to be 0.78%.

6.3.2.5 *Soundness Test on Steel Slag*

The soundness test was conducted on steel slag in accordance with IS: 2386 (Part-V):1963 [36].To know their resistance to extreme climatic conditions, The average weight reduction percentage was 0.83 %.

Sample size-
(10-4.75) mm = 200.5 gm
(20-12.5) mm = 201 gm
(25-20) mm = 206.5 gm
(40-25) mm = 201.5 gm
(63-40) mm = 210.5 gm

6.3.2.6 *CBR Test on Steel Slag*

The steel slag was tested for both unsoaked CBR and soaked CBR in accordance with IS: 2720 (Part-XVI):1987 [32]. The load-penetration curves for soaked and unsoaked CBR test are shown in Figure 6.5. The value for 1 division was 1.172.

6.3.2.7 *Constant Head Permeability Test on Steel Slag*

The steel slag was tested to know its permeability in accordance with IS: 2720 (Part-XXXIV):1986 [36]. The coefficient of permeability of slag was found to be 13.69 m/day.

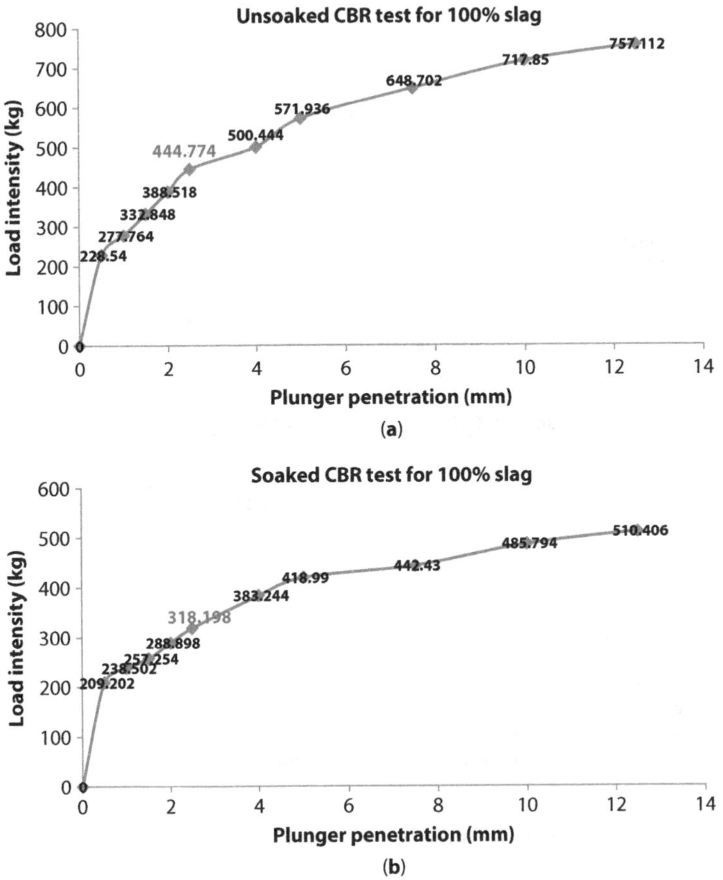

Figure 6.5 Load-penetration curve of slag for (a) Unsoaked and (b) Soaked condition.

6.3.3 Tests on Soil and Aggregate Combined

6.3.3.1 Unsoaked CBR Tests Using Soil and Varying Proportion of Aggregate

Soil and varying proportions of steel slag were tested for unsoaked CBR in accordance with IS: 2720 (Part-XVI):1987 [35]. Value for 1 division is 1.172. The results obtained are plotted and is shown in Figure 6.6.

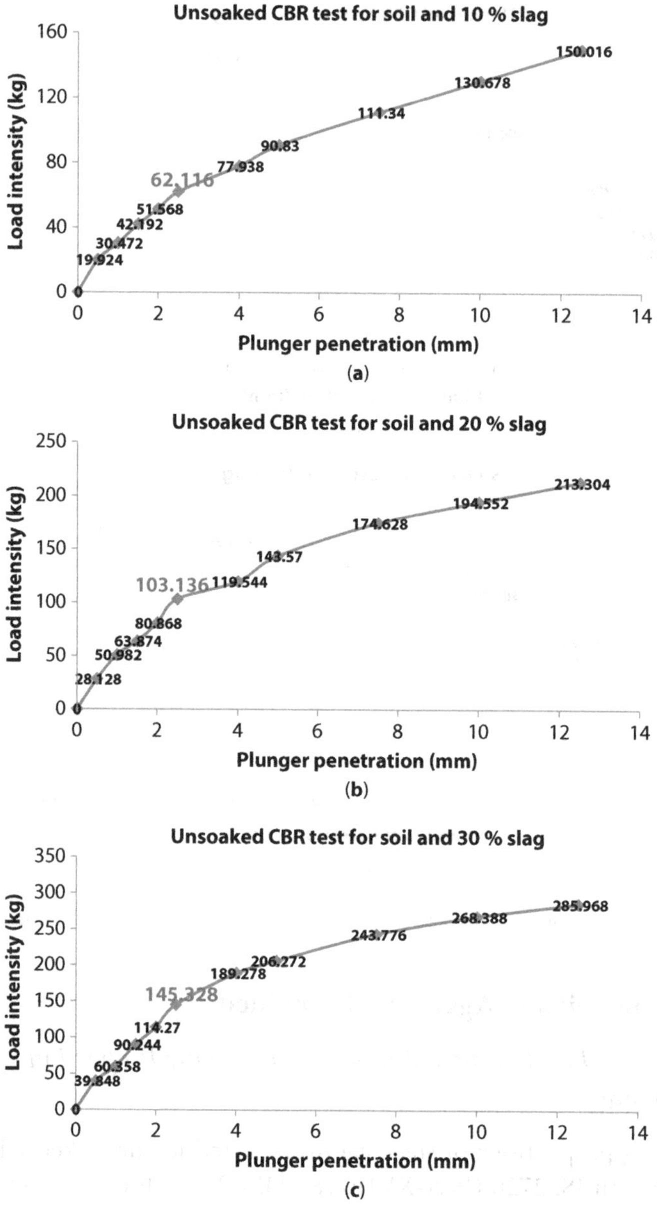

Figure 6.6 Unsoaked CBR test graph for penetration vs load intensity (a) 10% slag and soil (b) 20% slag and soil, (c) 30% slag and soil. *(Continued)*

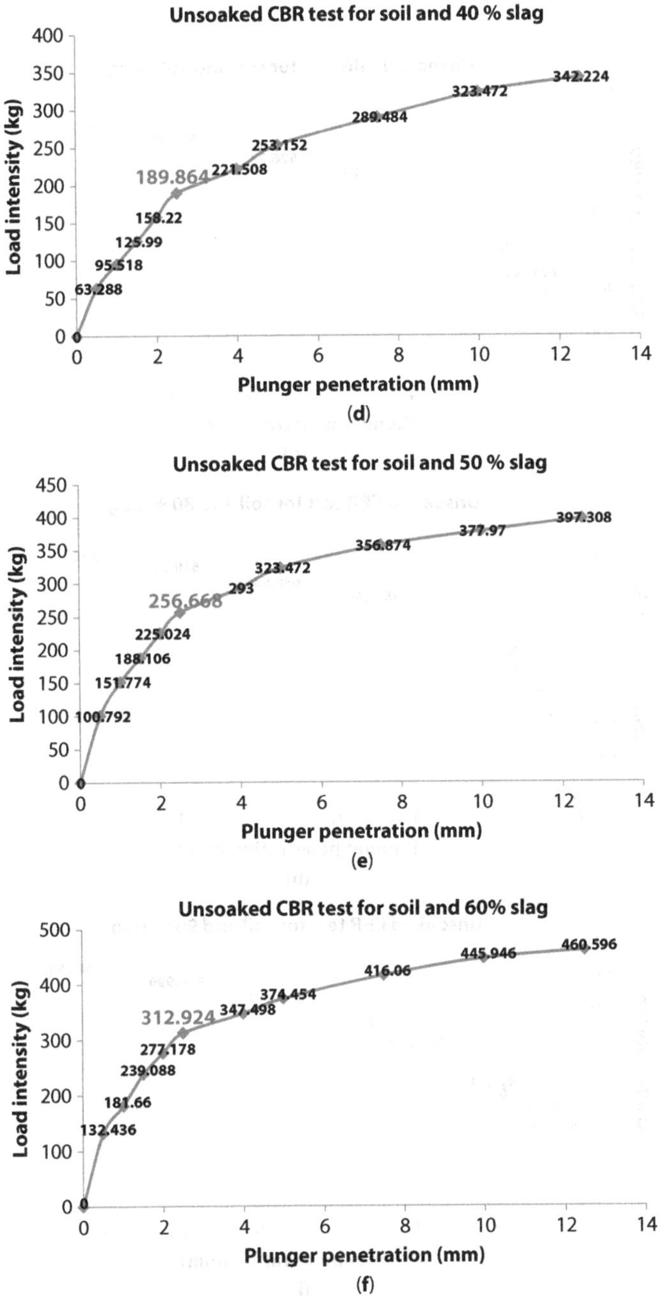

Figure 6.6 (Continued) Unsoaked CBR test graph for penetration vs load intensity (d) 40% slag and soil (e) 50% slag and soil (f) 60% slag and soil. *(Continued)*

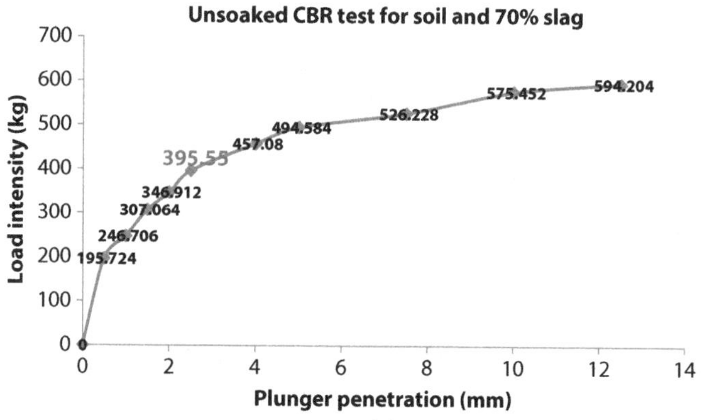

(g)

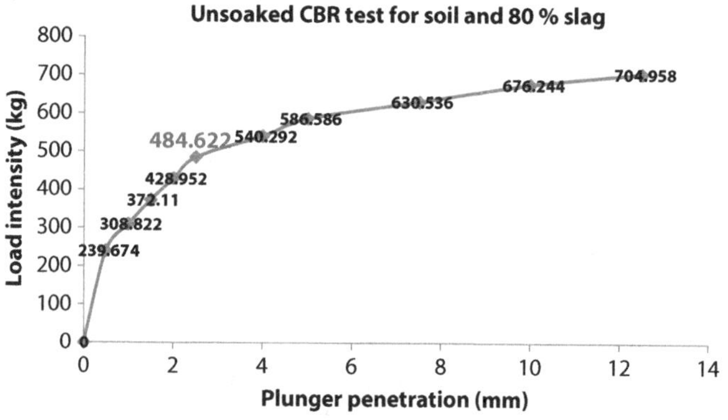

(h)

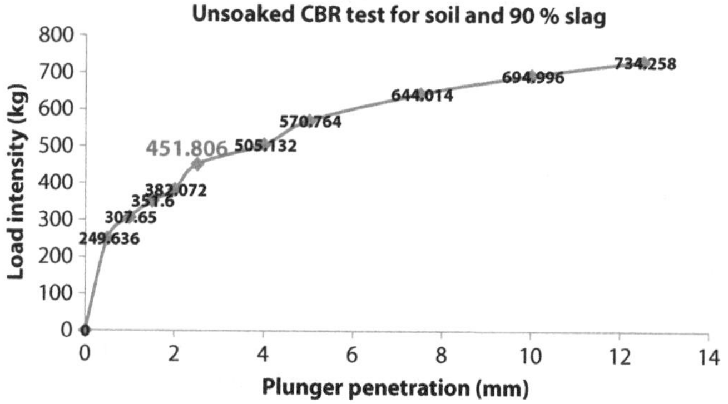

(i)

Figure 6.6 (Continued) Unsoaked CBR test graph for penetration vs load intensity | (g) 70% slag and soil, (h) 80% slag and soil (i) 80% slag and soil.

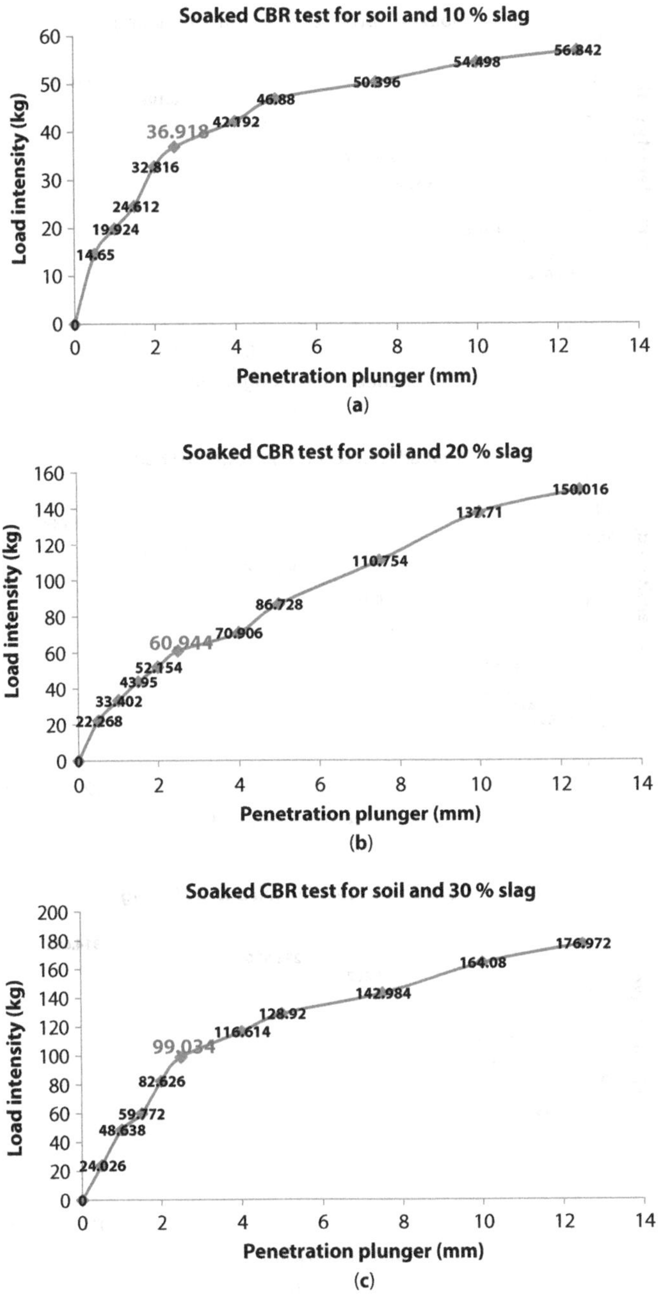

Figure 6.7 Soaked CBR test graph for penetration vs load intensity (a) 10% slag and soil (b) 20% slag and soil, (c) 30% slag and soil. *(Continued)*

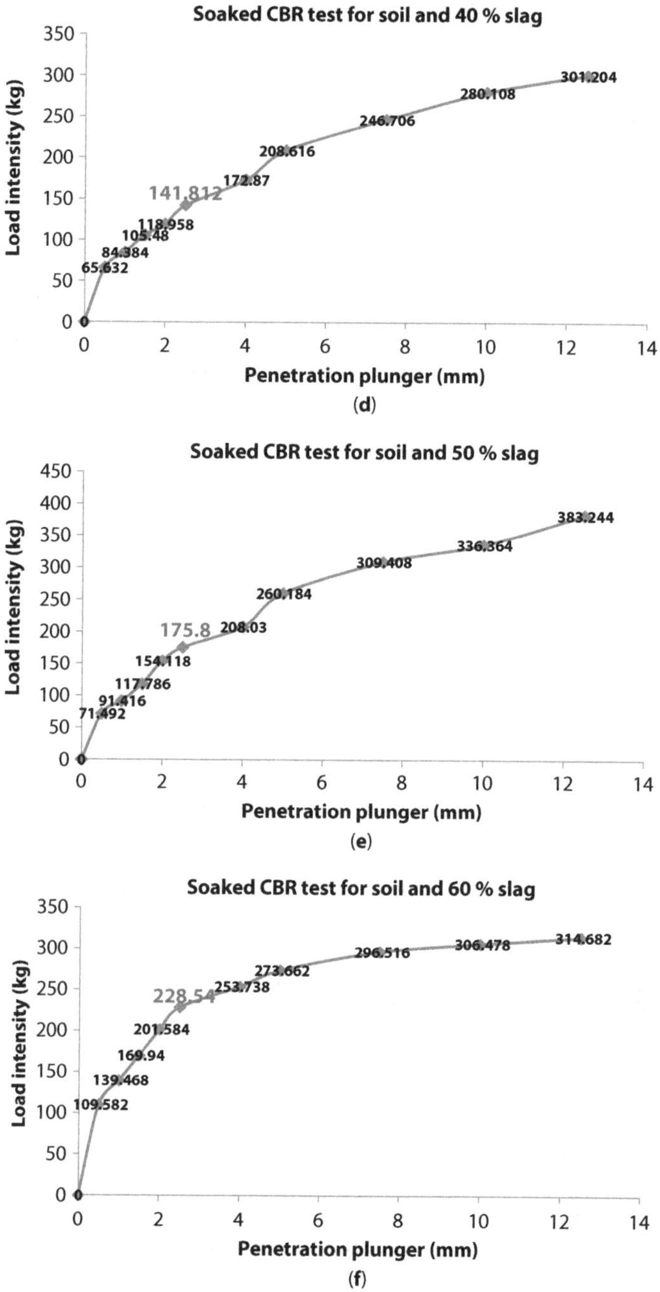

Figure 6.7 (Continued) Soaked CBR test graph for penetration vs load intensity (d) 40% slag and soil (e) 50% slag and soil (f) 60% slag and soil. *(Continued)*

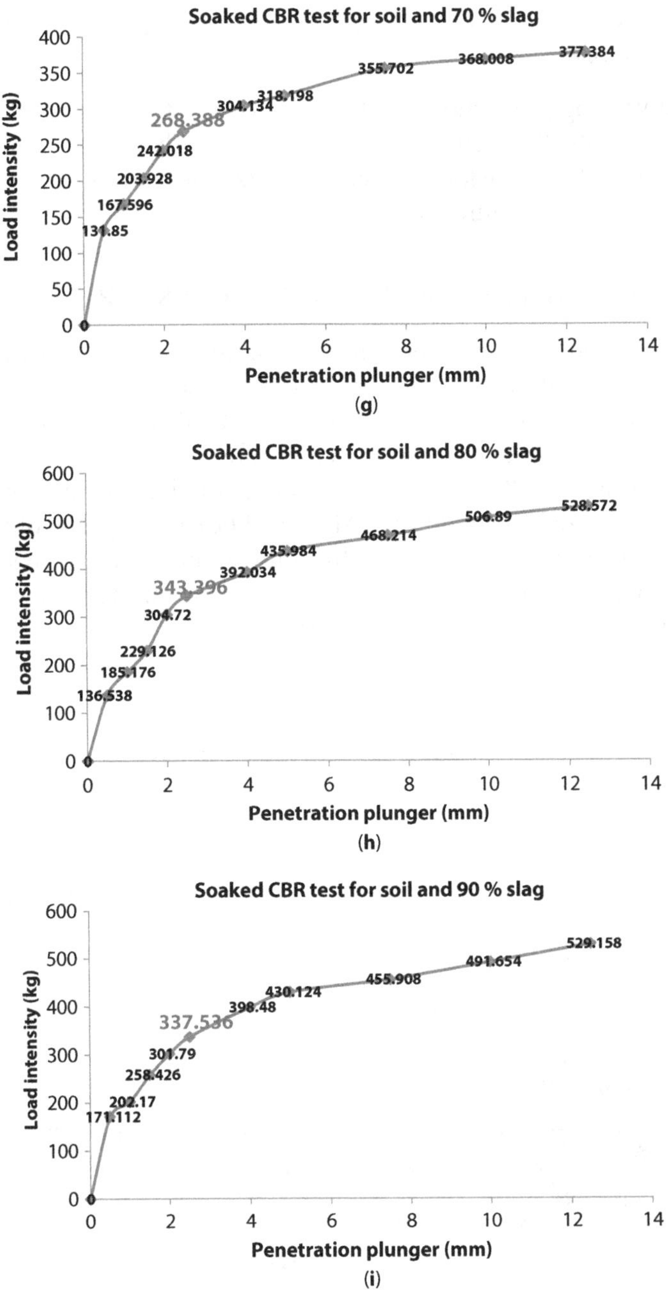

Figure 6.7 (Continued) Soaked CBR test graph for penetration vs load intensity (g) 70% slag and soil, (h) 80% slag and soil and (i) 90% slag and soil.

6.3.3.2 *Soaked CBR Tests Using Soil and Varying Proportions of Slag*

Soil and varying proportions of steel slag were tested for soaked CBR in accordance with IS: 2720 (Part-XVI):1987 [32]. The results obtained are shown below. The value for 1 division is 1.172. The load-penetration graph plotted is shown in Figure 6.7.

6.3.4 Proportions of Steel Slag with Soil in Sub Base

After doing the CBR tests for soil-slag mixtures under unsoaked and 4 days soaking conditions, in both unsoaked and soaked conditions, it was observed from the graph in Figure 6.8 that the CBR values go on increasing as we increase the slag content up to 80% after which there is a decrease. The CBR values of 80% slag and 20% soil is the highest in case of both soaked and unsoaked conditions. Also, load bearing capacity of this proportion is also the highest as seen from the graphs in Figure 6.9. Hence, for further tests to analyze the drainage conditions using jute, we use the proportion containing 80% slag and 20% soil.

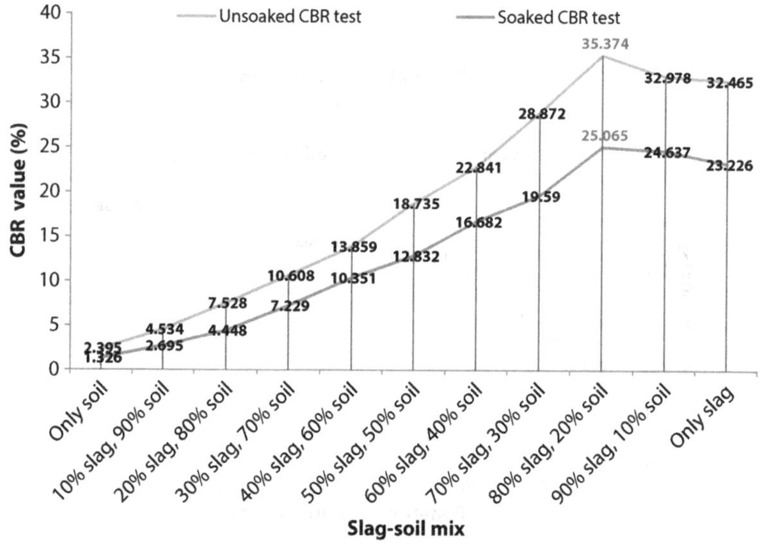

Figure 6.8 Soaked vs unsoaked CBR comparison.

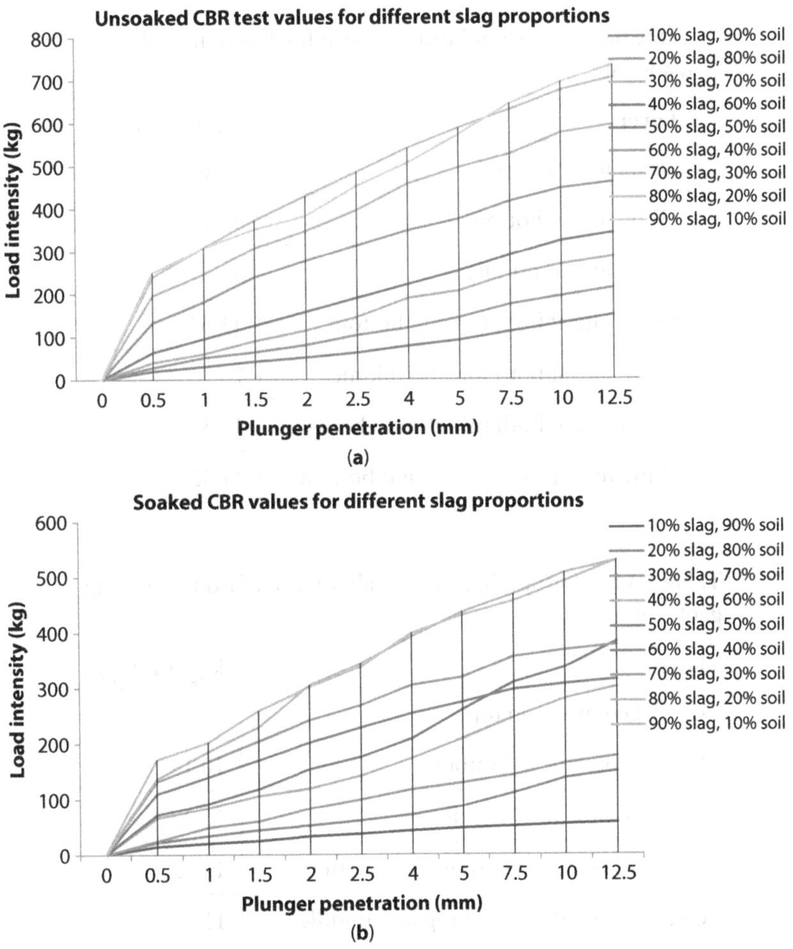

Figure 6.9 Comparison of (a) Unsoaked and (b) Soaked CBR values (Plunger penetration in mm vs Load intensity in kg).

6.3.5 Comparative Study

Dolochar is used as a drainage layer of subgrade medium in flexible pavement. Different combinations of jute, geotextile and combinations of jute and geotextile are placed in top, bottom and bottom layer with all combinations that can be established were all performed. Analysis of the results was specified in Table 6.10, Table 6.11 and Table 6.12 for average permeability of jute, geotextile and combined jute-geotextile reinforcement respectively in soil-slag mixture and comparative studies in terms of graph were drawn in Figure 6.10.

Table 6.10 Constant head permeability test of iron slag using jute.

Layer	k_{avg} (m/day)
Using jute on top	9.10
Using jute at bottom	11.77
Using jute in middle	11.27
Using jute at both top and bottom	11.53
Using jute at both top and middle	7
Using jute at both middle and bottom	11.36
Using jute at top, middle and bottom	11.32

Table 6.11 Constant head permeability test of iron slag using geotextile.

Layer	k_{avg} (m/day)
Using geotextile on top	12.15
Using geotextile at bottom	10.18
Using geotextile in middle	12.22
Using geotextile at both top and bottom	12.12
Using geotextile at both top and middle	11.92
Using geotextile at both middle and bottom	12.37
Using geotextile at top, middle and bottom	11.59

The better combination is found by providing jute at bottom of the sub base layer of pavement containing 80% steel slag and 20% silty sandy soil, which has the coefficient of permeability value to be 11.77 meters/day. Further tests can be carried out with geotextiles to see if they can provide more strength and better drainage facilities.

Table 6.12 Constant head permeability test of iron slag using both jute and geotextile.

Layer	k_{avg} (m/day)
Using jute and geotextile on top	8.72
Using jute and geotextile at bottom	10.12
Using jute and geotextile in middle	11.10
Using jute and geotextile at both top and bottom	11.24
Using jute and geotextile at both top and middle	10.54
Using jute and geotextile at both middle and bottom	10.72
Using jute and geotextile at top, middle and bottom	11.12

6.4 Conclusion

The utilization of steel slag in construction granular sub-base layers offers a sustainable and cost-effective solution that addresses environmental concerns while enhancing project performance. By understanding the properties, benefits, challenges, and best practices associated with steel slag, construction professionals can harness its potential to create more resilient and eco-friendly infrastructure. Further research and collaboration between the construction and steel industries will continue to advance the use of steel slag in construction applications. Based on the tests conducted on soil and steel slag, it is found that both have a decent physical and mechanical property to be used in subgrade layer. The soil sample was found to be silty sandy soil with a maximum dry density of 2.05 g/cc with optimum moisture content of 12.015%. The steel slag shows better result when compared with conventional aggregate, having specific gravity of 3.33 and water absorption of 0.78% to be specific. While conducting constant head permeability test of iron slag, geotextile (jute and woven polypropylene) was placed in different positions. It is observed that the woven polypropylene geotextile on both middle and bottom layer together with combination of iron slag in subgrade layer shows a better coefficient of permeability value of 12.37 m/day when compared with subgrade comprising steel slag alone (coefficient of permeability of 13.69 m/day). Also, a reduction in 0.96% in coefficient of permeability value is observed in combination of woven polypropylene and steel slag than steel slag alone. The use of natural fabric jute is not suitable because of its water holding

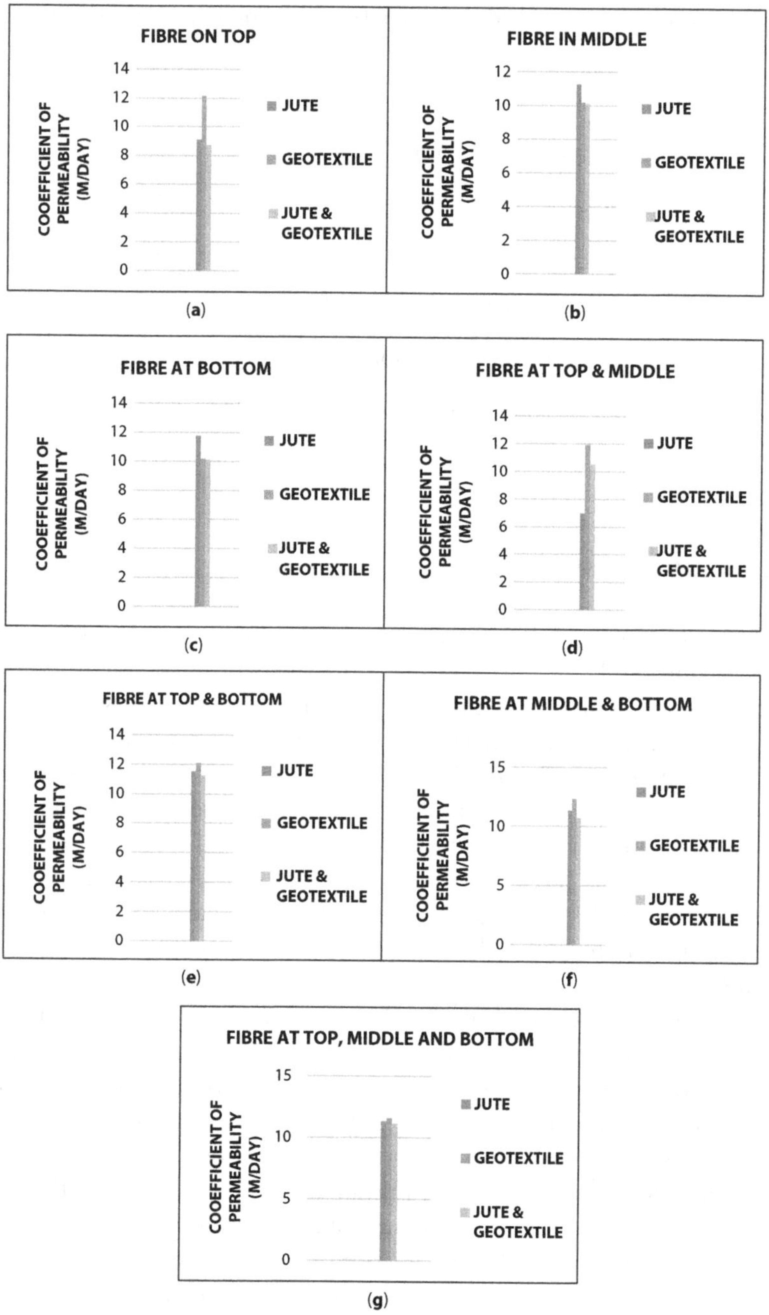

Figure 6.10 Constant head permeability variation with fiber at different layer configuration (a) top (b) middle (c) bottom (d) top and middle (e) top and bottom (f) middle and bottom (g) top, middle and bottom.

ability (coefficient of permeability value of 11.77 m/day when placed at bottom) which will deteriorate the subgrade layer and durability subsequently. Dolochar, upon sudden and gradual load shows poor result when compared with steel slag and it is also observed that the pulverized particle can be a viable cause for clogging off the voids of drainage layer.

References

1. Vivek, Dutta, R.K., Parti, R., Effect of chemical treatment of the coir geotextiles on the interface properties of sand/clay-coir geotextile interface. *J. Inst. Eng. India Ser. A*, 100, 357–365, 2019.
2. Mohanta, N.R. and Samantaray, S., Study of Combined Effect of Metakaolin and Steel Fiber on Mechanical Properties of Concrete. *Pertanika J. Sci. Technol.*, 27, 3, 1381–1396, 2019, http://journals-jd.upm.edu.my/resources/files/Pertanika%20PAPERS/JST%.
3. Vivek, and Dutta, R.K., Bearing ratio behavior of sand overlying clay with treated coir geotextiles at the interface. *J. Nat. Fibers*, 19, 14, 7534–7541, 2022.
4. Jaswal, P. and Vivek, Laboratory analysis of the interface shear characteristics of chemically treated coir geotextiles and soil interface. *Int. J. Pavement Res. Technol.*, 2023, 1–14, 2023, https://doi.org/10.1007/s42947-023-00369-w.
5. Vivek, Effects of cyclic loading on sand overlaying clay model of unpaved roads reinforced with untreated/treated coir geotextiles. *J. Text. Inst.*, 115, 19, 1694–1700, 2023.
6. Singh, P., Dash, H.M., Samantaray, S., Effect of silica fume on engineering properties of expansive soil. *Mater. Today Proc.*, 33, 5035–5040, 2020, https://doi.org/10.1016/j.matpr.2020.02.839.
7. Kumar, M.H., Mohanta, N., Samantaray, S., Kumar, N.M., Combined effect of waste glass powder and recycled steel fibers on mechanical behavior of concrete. *SN Appl. Sci.*, 3, 1–18, 2021a, https://doi.org/10.1007/s42452-021-04422-w.
8. Kumar, M.H., Macharyulu, I.S., Ray, T., Mohanta, N.R., Jain, M., Samantaray, S., Sahoo, A., Effect of water absorption and curing period on strength and porosity of triple blended concrete. *Mater. Today Proc.*, *43*, 2162–2169, 2021b, https://doi.org/10.1016/j.matpr.2020.12.092.
9. Kumar, M.H., Saikrishnamacharyulu, I., Kumar, U., Jena, S., Mohanta, N.R., Samantaray, S., Coupling effect of fly ash, metakaoline and different types of steel fibers on mechanical performance of concrete. *AIP Conf. Proc.*, 2417, 1, 2021c, https://doi.org/10.1063/5.0072573.
10. Ray, T., Mohanta, N.R., Kumar, M.H., Samantaray, S., Study of effect of temperature on behavior of alkali activated slag concrete. *Mater. Today Proc.*, *43*, 1352–1357, 2021, https://doi.org/10.1016/j.matpr.2020.09.169.

11. Kumar, M.H., Saikrishnamacharyulu, I., Mohanta, N.R., Ashutosh, A., Mishra, P., Samantaray, S., Mechanical behaviour of high strength concrete modified with triple blend of fly ash, silica fume and steel fibers. *Mater. Today Proc., 65*, 933–942, 2022a, https://doi.org/10.1016/j.matpr.2022.03.528.
12. Kumar, M.H., Mohanta, N.R., Patel, N., Samantaray, S., Reddy, S.V.B., Influence of Fly ash and Metakaoline on the crack resistance and shrinkage property of concrete. *Iran. J. Sci. Technol. Trans. Civ. Eng.*, 46, 2011–2026, 2022b, https://doi.org/10.1007/s40996-021-00748-y.
13. Lee, J., Edil, T.B., Benson, C.H., Tinjum, J.M., Building environmentally and economically sustainable transportation infrastructure: green highway rating system. *J. Constr. Eng. Manage., 139*, 12, A4013006, 2013.
14. Jaswal, P., Vivek, Sinha, S.K., Improvement in the performance of two layered model pavement with treated coir geotextile at the interface. *J. Ind. Text., 52*, 15280837221114161, 2022.
15. Vivek, Shafi Mir, M., Sehgal, R., Studies of Modulus of Resilience on Unpaved Roads Reinforced with Untreated/Treated Coir Geotextiles. *J. Nat. Fibers, 19*, 16, 13563–13573, 2022.
16. Vivek, Shafi Mir, M., Sehgal, R., Study on bearing capacity of unpaved roads reinforced with coir geotextiles using finite element method (fem). *J. Nat. Fibers, 19*, 15, 11735–11748, 2022.
17. Jaswal, P. and Vivek, Experimental study on monotonic behaviour of two-layered unpaved road model reinforced with treated coir geotextiles. *Int. J. Pavement Res. Technol.*, 1–12, 2023.
18. Arulrajah, A., Piratheepan, J., Disfani, M.M., Bo, M.W., Geotechnical and geoenvironmental properties of recycled construction and demolition materials in pavement subbase applications. *J. Mater. Civ. Eng.*, 25, 8, 1077–1088, 2013.
19. El-Badawy, S.M., Gabr, A.R., Abd El-Hakim, R.T., Recycled materials and by-products for pavement construction, in: *Handbook of ecomaterials*, pp. 1–22, Springer, New York, 2018.
20. Gonawala, R.J., Kumar, R., Chauhan, K.A., Usage of crushed EAF slag in granular sub-base layer construction, in: *Innovations in Infrastructure: Proceedings of ICIIF 2018*, pp. 257–266, Springer Singapore, 2019.
21. Autelitano, F. and Giuliani, F., Electric arc furnace slags in cement-treated materials for road construction: Mechanical and durability properties. *Constr. Build. Mater., 113*, 280–289, 2016.
22. Palmeira, E.M. and Gardoni, M.G., Drainage and filtration properties of non-woven geotextiles under confinement using different experimental techniques. *Geotext. Geomembr., 20*, 2, 97–115, 2002.
23. Muthukumaran, A.E. and Ilamparuthi, K., Laboratory studies on geotextile filters as used in geotextile tube dewatering. *Geotext. Geomembr., 24*, 4, 210–219, 2006.
24. Vivek, Dutta, R.K., Parti, R., Effect of chemical treatment on the tensile strength behavior of coir geotextiles. *J. Nat. Fibers, 17*, 4, 542–556, 2020.

25. Jaswal, P., XXX, V., Sinha, S.K., Investigation on tensile strength characterisation of untreated and surface treated coir geotextiles. *J. Ind. Text.*, *52*, 15280837221118847, 2022, https://doi.org/10.1177/15280837221118847.
26. Vivek, Dutta, R.K., Kumari, A., Effect of chemical treatment on the durability behavior of coir geotextiles. *J. Nat. Fibers*, *19*, 8, 3127–3146, 2022.
27. IS: 2386 (Part-III):1963, Determination of specific gravity and water absorption of coarse aggregates or stone, Bureau of Indian standards, New Delhi, India, 1963.
28. IS: 2386 (Part-IV):1963, Aggregate impact test, Bureau of Indian standards, New Delhi, India, 1963.
29. IS: 1498:1970, Classification and identification of soil for general purpose, Bureau of Indian standards, New Delhi, India, 1970.
30. IS: 2720 (Part-V):1985, Methods of test for soils: Determination of liquid and plastic limit of soil, Bureau of Indian standards, New Delhi, India, 1985.
31. IS: 2720 (Part-VIII):1983, Methods of test for soils: Determination of water content-dry density relation using heavy compaction, Bureau of Indian standards, New Delhi, India, 1983.
32. IS: 2720 (Part-XVI):1987, Methods of test for soils: Laboratory determination of CBR, Bureau of Indian standards, New Delhi, India, 1987.
33. IS: 2386 (Part-IV):1963, Methods of tests for aggregates for concrete: Mechanical properties, Bureau of Indian standards, New Delhi, India, 1963.
34. IS: 2386 (Part-III):1963, Methods of tests for aggregates for concrete: Specific gravity, density, voids, absorption and bulking, Bureau of Indian standards, New Delhi, India, 1963.
35. IS: 2386 (Part-V):1963, Methods of tests for aggregates for concrete: Soundness, Bureau of Indian standards, New Delhi, India, 1963.
36. IS: 2720 (Part-XXXIV):1986, Methods of test for soils: Laboratory determination of permeability of granular soil (Constant head), Bureau of Indian standards, New Delhi, India, 1986.

7

Multi Stabilizers Optimization for Expansive Soil Treatment Using Taguchi Approach

Doleswar Nayak[1], Biswajit Majhi[1]* and Saleema Panda[2]

[1]*Department of Civil Engineering, Odisha University of Technology and Research, Bhubaneswar, India*

[2]*Department of Process, Energy and Transport Engineering, Munster Technological University, Cork, Ireland*

Abstract

The goal of the current study is to determine how well the Taguchi method can be used to optimize the use of many stabilizers for expansive soil treatment. Sawdust (SD), Silica fumes (SF) and Ground granulated blast furnace slag (GGBFS) were among the stabilizers which were added with soil. The optimized parametric levels were established in the Taguchi planned experiment study, which yielded the results for Differential free swell index (DFS) and California bearing ratio (CBR). The response data were also investigated while using an interaction of super-efficiency models and grey relational analysis and the results revealed that 10% SD, 40% SF, and 20% GGBFS were the best additive combinations for simultaneously optimizing the expansive soil.

Keywords: Taguchi approach, sawdust, silica fumes, ground granulated blast furnace slag, expansive soil, California bearing ratio

**Corresponding author*: bmajhi@outr.ac.in

Sandeep Samantaray, Subhashree Samantasinghar, Soumendra K. Mohanty, Vivek, Abinash Sahoo and Deba Prakash Satapathy (eds.) Waste Utilization in Geotechnical Practice, (179–208) © 2025 Scrivener Publishing LLC

7.1 Introduction

Soil stabilization is generally understood as the way of improving the technical qualities of the soil to increase its strength, durability and bearing capacity [1–4]. There are various methods and techniques used for soil stabilization like mechanical, chemical, geosynthetics, electrokinetic, etc. It's important to note that the different soil types and project requirements may influence the choice of soil stabilization. Expansive soil, also known as shrinking soil or swelling clay, is a form of soil that expands and contracts in the action of changes on moisture level [5–8]. This soil has the ability to expand as it receives water and contract when it loses water. This behavior can pose a variety of problems for structures and infrastructures built on or near the expansive soils. Most sections of the world have several types of expansive soils. Because of the remarkable growth in road infrastructure in developing countries such as India, soil stabilization has emerged as a significant obstacle in the construction industry. Building runways and highways requires stabilization. Increased strength and durability are implied by stabilization, which may be a tactic for managing materials that are readily available to provide low-cost road design and construction, with an emphasis on the effective utilization of waste byproducts such as GGBS, fly ash and silica fumes to decrease development costs [9–15]. It has poor shear strength, undesired settlement, and high compressibility. Because of these issues, researchers have been compelled to devise mitigation measures to address the persistent issues linked with it. The Taguchi technique, which is an offline quality optimization strategy, was proposed by Genichi Taguchi, an engineer and quality assurance expert from Japan. The majority of process parameters and process conditions can be improved. It aids in the design of trials as well as searching for the best parameter set by utilizing orthogonal arrays (OA). Fewer partial factorials tests are done when utilizing the OA in Taguchi method, resulting in reduction in the time and energy required to evaluate the optimal set of parameters for a process utilizing the full multifactorial, an additive used to boost the soil's engineering properties. For process optimization, the Taguchi technique introduces a new variable called as SNR (signal to noise ratio). The SNR is used to understand experimental data. Throughout the Taguchi method's application, three primary analyses are performed. The first step is to compute the average of the levels to identify the ideal value of process parameters. Finally, using the ideal value of the process parameters, the optimum value for the response is computed. GRA is a mathematical tool for analyzing the interactions and correlations between several elements or variables.

It is widely utilized in decision-making processes, especially when dealing with complex systems or when data is limited or unclear. GRA was established by Professor Deng Julong as part of grey system theory in the early 1980s.

Grey connection analysis is especially useful when there is limited or confusing data, making it difficult to create precise mathematical models. It allows for the comparison and evaluation of multiple factors based on their relative significance or influence on a certain outcome. The fundamental goal of GRA is to assess the similarity between a reference series and a comparison series. The comparison series depicts the current state or performance, while the reference series represents the ideal or intended situation. It is widely utilized in decision-making processes, especially when dealing with complex systems or when data are limited or unclear. A grey relational grade can be evaluated by calculating the grey relational coefficient between each pair of linked values in the reference and comparison sets. The degree of connection or resemblance between the two series is shown by the grey relational grade. It expresses how closely the comparative series resembles the pattern of the reference series. The stronger the correlation between the two series, the higher the grey relational grade.

The DEA is non-parametric approach to calculate the comparative efficiency of a collection of DMUs that convert numerous inputs into several outputs. In operations research and management science, DEA is extensively used to analyze the effectiveness of organizations such as school, hospitals, banks, and industrial enterprises. The main principle behind DEA is to evaluate the performance of DMUs by building an efficient frontier or set of production possibilities. DMUs on the efficient frontier are regarded technically efficient, which means they produce the most feasible output for a given set of inputs. DMUs located below the border are considered inefficient and have the potential to improve. DEA assigns efficiency scores to each DMU, reflecting its relative efficiency in relation to the others in the dataset. Traditional DEA models, on the other hand, only consider relative efficiency, therefore they cannot distinguish between DMUs that are all technically efficient. This constraint led to the development of super-efficiency models. Super-efficiency DEA models were created to identify DMUs that are not only somewhat efficient but also have the potential to exceed efficient units. These models are used to evaluate DMUs and identify those that are very efficient or performing extraordinarily well. The basic DEA model is modified to generate super-efficiency DEA models by removing the DMU being evaluated from the collection of efficient units. This change enables the detection of super-efficient units, which are not only efficient but also outperform other efficient units. Super-efficiency models provide more precise information on DMU performance and are especially valuable for benchmarking and establishing best practices.

They [16], stated that the rough texture of sawdust, a byproduct of the production of wood, may provide a significant level of friction between the fibers and the soil matrix, greatly enhancing bearing capacity and shear strength. The concept of increasing soil quality by randomly dispersing natural fibers was driven by the organic phenomenon of plants or grass that, when grown on a slope, may efficiently stabilize the slope. They [17], stated that when the amount of saw dust increases, the plastic limit reduces, which will aid in the improvement of the subgrade. The findings reveal that as the percent of saw dust increases, the plasticity index decreases, assisting in the improvement of the subgrade. The results demonstrate that the fraction of saw dust increases the CBR value; the highest value is found to be 21.93% and 11.45%, respectively. According to the data, increasing saw dust percentage increases UCS up to 8.5% and then it decreases. UCS rises from 63.54 KN/m^2 to 133.12 KN/m^2. He [18], studied the multiple responses that were utilized to optimize the Taguchi method. In the Taguchi technique, the AA models and DEA were utilized to address the multiple responses. Based on the findings and outcomes of this investigation, when admixed with TBC, a mixture of 20% quarry dust, 20% saw dust ash, and 5% Portland cement improved the UCS, CBR, and DFS significantly.

They [19], studied the design experiments taken by the Taguchi orthogonal array of mixed levels, L18 (6 x 3). For the UCS, 8% PC, 20% SDA, and 20% QD; the CBR, 8% PC, 20% SDA, and 20% QD; and the DFS, 3% PC, 20% SDA, and 20% QD were shown to achieve the best results when utilizing simply the Taguchi optimization approach. The ideal outcome produced by combining GRA and Taguchi was found to be 8% PC, 20% QD, and 20% SDA. They [9], stated that soil stabilization was accomplished using varying amounts of silica fumes from 10% to 100%. The Atterberg limit, DFS, CBR, UCS, UU testing are all performed on natural soil and measured in soil stabilized with silica fume. The results show that adding SF reduces the DFS. According to the compaction curves, adding silica fume improves the OMC value but it decreases the MDD value. Stabilizers (such as silica fumes) improve C values in UCS, resulting in increased cohesion. The ideal value of silica fumes was shown to be 50%. They [19], studied on expansive soil and stated that the expansive soil was established using varied amounts of polypropylene fiber, 0.25%, 0.5%, 10% and SF (2%, 4%, and 8%) and by the dry soil's weight. The results demonstrate that the presence of SF and PP fiber raises the CBR value and reduces the shrinkage value. The experimental results show that adding SF and PP fibers reduces the Atterberg limit; the CBR value of segment BC+0.25%PP+8.00%SF increased by 107.89%. Increasing the CBR value enhances the road surface's load-bearing capability. The strongest reduction in free swelling index was seen with 8% SF content.

They [20], evaluated the strength and durability of fly ash-slag based geopolymer modified soil and concluded that the slag based geopolymer as an effective and durable stabilizer. They [21], evaluated the effect of GGBS and quicklime on the strength properties of locally available soils. To determine their impact on the soil's saturated CBR strength and UCC strength, GGBS was combined with locally available clayey sand (SC soil) and fast lime. To find the appropriate proportion of lime and GGBS to be added to clayey soil to achieve the best UCC strength with immersed CBR, the percentage of quicklime varied from 2%, 4%, 6%, 8% and the percentage of GGBS from 20%, 25%, 30%, 35%, etc. has changed to 40%. They [22], explored the properties of BCS and adjusted by stabilizing the soil using stabilizers such as GGBS 30%, lime 6% and nano silica 0.3%, 0.6%, and 0.9%. Soil increases geotechnical properties like WL, PL, compaction characteristics, UCS, CBR and swelling characteristics. An experiment was carried out to look at the impact of the combination on the strength properties. They [23], found that the amount of GGBS affects the strength improvement. To determine the microstructural investigation, a soil XRD test was also carried out for both treated and untreated soil. Work continues to determine the variance of soil strength with and without GGBS as a function of intensity properties. The ideal concentrations of GGBS for sodium bentonite clay, organic clay, calcium bentonite clay and BC soil are 12%, 25%, 24% and 24% respectively on the basis of stress performance testing. They [24], indicated that adding up to 40% GGBS enhances UCS strength of specimen. Above 30%, however, the improvement in UCS strength is minor. The UCS intensity rose from 230.1 kPa for (0 %GGBS+ 100% BC) to 443.3 kPa for (30% GGBS+70% BC). As a consequence, they determined that 30% was the ideal dose of his GGBS in soil. They [25], included experiments to investigate the impact of GGBS on BCS treated with tiny doses of lime. It has been demonstrated to be extremely beneficial to the intended purpose, and also good to the environment by repurposing industrial waste and lowering carbon footprint by eliminating the need for cement. They [26], used pyroclastic dust, a by-product of the rock quarrying process, to affect the geotechnical parameters of an expanding soil. When pyroclastic dust is applied to the soil, the plasticity index dropped dramatically. In addition, an optimal 8% pyroclastic dust was applied to the expanding soil; a notable increase was found in both the CBR and the UCS. They [27], analysed the Taguchi method and optimised the process parameters to get a greater impact over the process's performance, efficiency, and cost. Offline quality assurance is regarded to be an effective technique to improve product quality at a low cost. The ANOVA (analysis of variance) approach is used to determine the effect of process factors on the entire process of machining.

The Taguchi method, the ANOVA (analysis of variance) and the signal-to-noise (S/N) ratio are used to examine the performance characteristics. They [28], stabilized the BC soil with different ratio of GGBS and found that the amount of GGBS used determines how much strength is improved. Additionally, it was demonstrated that as GGBS content increases, the initial tangent modulus values typically increase as well. The inclusion of GGBS enhanced the strength by up to 20%. GGBS content for the 7 and 14-day and up to 40% for the 28 days curing periods. The strength of the soil-GGBS mixture is reduced by the addition of GGBS. He [29], examined the soil stabilization using varying volumes of GGBS and fly ash applied individually. Physical and strength performance tests, such as specific gravity, Atterberg limits, standard proctor test, and California Bearing Ratio test are used to assess the stabilized soil performance at an optimal moisture content. The optimal values for fly ash and GGBS are 15% and 20%, respectively. They [30], conducted CBR and UCS experiments to evaluate SDA potential for improving the soil. According to the experimental investigation, adding SDA significantly raises both CBR and UCS. While its performance may be enhanced by mixing it with other bonding materials, such as lime, it can also be used as an alternative to industrial waste to lower the cost of building roads, especially in the nation's rural areas. The optimum value of SDA is found at 4% where CBR and UCS value increases. They [31], found that the following benefits of employing GGBS as a stabiliser are available: waste management, cost savings, and availability. Additionally, it was shown that adding additional slag results in more stable soil than by just adding lime. They [32], reported that fly ash-based geopolymer might be used to stabilize BCS for highway subgrade and sub-base preparation, as well as the viability of using fly ash geopolymer to stabilize BCS. The strength provided by FA geopolymer stabilized BCS at the end of the 7-day curing period well exceeds the minimum strength requirement for sub-base specified in IRC SP37 (2012) [33].

The Taguchi approach with super efficiency model identifies the most suitable soil stabilization processes, additives, and their optimal proportions to acquire the intended technical qualities of the stabilized soil. The goal is to maximize soil strength and durability while minimizing the use of additives, reducing costs and minimizing negative environmental impact.

7.2 Materials

7.2.1 Expansive Soil

The soil utilized in this investigation was extracted at a depth of 1.5 m in Kantabanji, Balangir District, Orissa, India (20.4639°N, 82.9206°E). The soil sample and different materials are shown in Figures 7.1-7.4. The geotechnical characteristics of soil are shown in Table 7.1. From this, it was found that it belongs to the CI classification (clay with moderate plasticity).

Table 7.1 Geotechnical characteristics of soil.

Sl. no.	Properties	Value	Confirming IS code
1	Specific gravity (G)	2.6	IS 2720: Part 3: Sec 1: 1980
2	Maximum dry density (MDD)	1.9 g/cc	IS 2720: Part VII: 1980
3	Optimum moisture content (OMC)	20%	IS 2720: Part VII: 1980
4	Liquid limit	48.10%	IS 2720: Part 5: 1985
5	Plastic limit	26.09%	IS 2720: Part 5: 1985
6	Shrinkage limit	25%	IS 2720: Part 5: 1985
7	Differential free swell	48%	IS 2720: Part XL: 1977
8	Classification	CI	IS 1498 (1970)

Figure 7.1 Soil sample.

Figure 7.2 Silica fumes material.

7.2.2 Silica Fumes

The silica fumes were collected from local sources. The manufacturing of silicon metal and ferrosilicon alloys produces silica particles. When SiO_2 is vaporized in an electric arc furnace to create these minerals, silica particles,

Table 7.2 Chemical components of silica fumes.

Chemical component	Results
Silica	99.886%
Alumina	0.043%
Ferric oxide	0.040%
Calcium oxide	0.001%
Calcium oxide	0.001%
Magnesium oxide	0.000%
Potassium oxide	0.001%
Sodium oxide	0.003%
Loss of ignition	0.015%
Heavy metal	
Lead	0.000%
Arsenic	0.000%

are created. The majority of the particles in silica fumes have a diameter of less than one micron. Silica fumes are frequently employed as a supplemental cementitious material in concrete and cementitious products because of their distinctive physical and chemical qualities. The chemical combination of silica fumes is presented in Table 7.2.

7.2.3 Sawdust

The saw dust was gathered from the nearby sawmill in Ghatikia, BBSR. Sal wood was cut into pieces, and sawdust was collected from that process. Saw dust is really a by-product of sawmills that is created when cutting wood. It is the splintered materials or wood chips that are produced when cutting wood into useful pieces. Sawdust consists mainly of carbohydrates and lignin, with small amounts of carbon, hydrogen, oxygen and nitrogen. Clean sawdust, containing little bark or other organic matter, was collected and air-dried at room temperature. The lumps, gravels, and other soil-damaging components were subsequently removed from the sawdust by sieving it through 600micron sieves.

Figure 7.3 Saw dust material.

Figure 7.4 GGBS material.

Table 7.3 Chemical parameters of GGBS (provided by Toshali Cement Pvt. ltd.).

1	SiO_2	33.43 %
2	Al_2O_3	20.52 %
3	Fe_2O_3	0.85 %
4	CaO	35.32 %
5	MgO	7.45 %
6	SO_3	0.82 %
7	Chlorides	0.015 %
8	Loss of Ignition	1.63 %

7.2.4 GGBS

GGBS is an outcome from iron and steel industry, especially from the blast furnace process. It is a supplemental cementitious substance used in the construction sector to partially substitute Portland cement in concrete production. GGBS is made by quenching molten blast furnace slag with water or air, which results in the formation of a glassy granular material. GGBS has gained significant attention as a sustainable and eco-friendly alternative to traditional cement due to its various beneficial properties. The GGBFS used in the study was obtained from Toshali Cements Pvt Ltd. and the obtained product satisfies all requirements according to IS 12089:1987. The chemical properties of GGBS are presented in Table 7.3.

7.3 Methodology

The soil geotechnical properties such as specific gravity, granulometry, Atterberg limit, compaction properties, CBR, UCS, and DFS are measured in the laboratory, and were done in compliance with the directives established according to the Indian Standards. The planned experiments were derived from mixed planar orthogonal Taguchi arrays were used to assess responses (CBR, UCS, DFS) and were performed using standard Indian procedures. The soil optimization data obtained from the experiments were analyzed by utilizing an approach of super efficiency model and grey relational analysis. After the optimization, optimum parameter can be found out and for further study a confirmatory test can be evaluated.

7.3.1 Specific Gravity

In accordance with IS: 2720 (Part 3)-1980, the specific gravity test was carried out by the density bottle mechanism. It was calculated by dividing the weight of a substance with an equivalent amount of distilled water at that equal temperature. A 20gm sample of raw soil was used for the test, which was carried out using distilled water at room temperature.

7.3.2 Grain Size Analysis

A typical laboratory test for investigating soil mechanics is by particle size analysis. In particle size analysis, sieves can be used to measure particles with sizes varying from 0.075 mm to 100 mm to ascertain the sieve analysis. The distribution of particle sizes was found by sieve size analysis according to IS.2386 (Part 1) – 1963. In this 1 kilogram of the sample, the raw soil is put in different sizes of sieves that were arranged in the order of 2mm to 0.0073mm. Then the sieves were put in sieve shaker; and then, the particle size that was retained and passed through the sieves were calculated.

7.3.3 Atterberg Limits

7.3.3.1 Liquid Limit

The liquid limit of expansive soil was evaluated according to IS.2720 (Part 5) - 1985 as shown below. The moisture level at which grooves in a raw soil sample close after 25 hits is known as the liquid limit. Liquid limits were determined with the Casagrande apparatus.

7.3.3.2 Plastic Limit

The plastic limit of expansive soil was evaluated according to IS.2720 (Part 5) - 1985 as shown below. The water content at which the soil starts to break when stretched into a thread with a diameter of 3 mm is known as the plasticity limit. It was measured by rolling a soil sample mixed with water on a glass plate. After oven drying, the water content of each sample was determined individually.

7.3.4 Compaction Characteristics

Compaction properties, i.e., OMC and MDD of the materials were evaluated by standard and modified proctor tests in accordance with IS: 2720 (Part 7) - 1980 for normal soil and soil with additives. About 3 kg of sample was prepared for standard proctor test and about 5 kg for modified proctor test. In standard compaction test, a 2.6 kg rammer was used to compact the sample with 25 blows. (Light compaction). For modified compaction (Heavy compaction), a 4.6 kg rammer was used with 56 blows In the dry density versus moisture content graphs, the compaction characteristics are shown.

7.3.5 California Bearing Ratio

The CBR of raw soil and soil with additives according to Taguchi design experiments, was determined in accordance with IS: 2720(Part 16): 1987. The test material used for the retread samples passed through the 19 mm IS sieve and was retained on the 4.75 mm IS sieve. The sample was mixed with optimum value of their water content and was placed on CBR mold to perform compaction process. Then the sample is placed on CBR test apparatus to determine the CBR value. Soil CBR values were measured in both wet and dry circumstances. The percentage of force required per unit area to penetrate a mass of soil at a speed of 1.25 mm/min using a standard 50 mm diameter rotating piston is known as the California strength ratio (CBR) of soil. This ratio is compared to the force needed to penetrate a ground mass of standard material. Commonly, penetration depths between 2.5 and 5 mm are used to compute this ratio. The 5mm ratio is utilized when it consistently outpaces the 2.5mm ratio; if not, the 2.5mm ratio is employed.

7.3.6 Unconfined Compression Strength Test

The UCS test of soil was carried out in accordance with IS: 2720 (Part 10) - 1991. The samples were evenly arranged in three layers in a cylinder with a diameter of 50 mm and 180 mm height, with 25 impacts per layer. A steel pipe with a diameter of 38 mm and 100 mm height was then pushed and injected into the soil mixture using a hydraulic jack and the extruded sample used for testing.

7.3.7 Differential Free Swell

DFS is a specification used to identify expansive soils. For this test, two 10g soil samples, having passed through BS No. 425μ were taken and carefully poured into two different 100cc graduated cylinders. One contains water and the other kerosene. Both assemblies were allowed to sit for 24 hours after the initial volume was recorded. The final volume of soil samples in both liquids was recorded and the extent of swelling was assessed.

The differential free swelling is calculated as,

$$DFS= ((V_d-V_k)/V_k)\ x100$$

where V_d - Volume in distilled water,

V_k - Volume in kerosene

The differential free swelling value of raw soil and soil with different additives as described by Taguchi design experiment was calculated.

7.3.8 Plan of Optimization

The experimental data were examined utilizing a series of GRA and super-efficiency models. In this study planned the experiment using the Taguchi $L_8\left(4^\wedge 1^* 2^\wedge 2\right)$ mixed-level orthogonal array, with silica fumes allocated four levels and Saw dust and GGBS allocated three levels each, as shown in Table 7.4. Table 7.5 shows the Taguchi design experiments and Table 7.6 shows the arrangement of the experiment was designed with both actual and coded values. He [34], used a technique which was somewhat changed in this study. It was divided into six parts. The Taguchi technique was utilized to get quality loss in step one. GRA theory was employed in steps 2–4 to generate the grey relational coefficient, while the AP model was used in step 5 to evaluate the efficiency of the grey relational coefficients, which are suitably defined as DMU_s in this study. Finding the average efficiency for all DMU_s operating at the same parameter level was done in the sixth and final stage. The parameter ideal level was determined to be the one that increased the level of efficiency to its maximum. The techniques for concurrently optimizing additives are described in the steps that follow.

Step 1: Evaluation of Quality loss

The first step is to evaluate the quality loss $L_i(k)$, for each response. It uses suitable function provided below. Equation (7.3) is used for the larger, the better value; Equation (7.2) is used for the nominal the best value and Equation (7.1) is used for the STB (Smaller the better) value.

$$L_i\left(k\right)=h\left[\frac{1}{w}\sum_{t=1}^{w}\left(y_{it}^{\ 2}\right)\right] \tag{7.1}$$

$$L_i(k) = h\left(\frac{s_i^2}{\overline{y}_i^2}\right) \tag{7.2}$$

$$L_i(k) = h\left[\frac{1}{w}\sum_{t=1}^{w}\left(\frac{1}{y_{it}^2}\right)\right] \tag{7.3}$$

where i = 1, …, n (n indicate the number of experiments), h indicate the quality loss coefficient for each response, t = 1, …., w (w indicates the number of repeats response), k = 1, …, m (m is the number of responses), $L_i(k)$ signifies the loss function for the kth response in the ith experiment, s_i indicates the response propagates standard deviation, $\overline{y}_i$ indicates the average of the response repeats, y_{it} shows the response replicated.

Step 2: Grey relation generation
This is the stage at which the $L_i(k)$ is normalized. The normalized cc $L_i(k)$ values vary from 0 to 1. and are also affected by the response quality features. The objective of the optimization is to reach a higher target value in the answer, where equation (7.4) is employed, whereas when obtaining a lower target value in the response in the optimization process, equation (7.5) is employed.

$$R_{ik} = \frac{L_i(k) - \min L_i(k)}{\max L_i(k) - \min L_i(k)} \tag{7.4}$$

$$R_{ik} = \frac{\max L_i(k) - L_i(k)}{\max L_i(k) - \min L_i(k)} \tag{7.5}$$

R_{ik} represents the normalized $L_i(k)$ value for the kth response in the ith experiment.

Step 3: Evaluation of deviation sequence
Equation (7.6) is used to determine the deviation sequence, which is the third stage.

$$\gamma_{0i}(k) = R_0(k) - R_i(k) \tag{7.6}$$

where $\gamma_{0i}(k)$ indicates the total variation between the benchmark sequences of $R_0(k)$ and the sequence of comparison $R_i(k)$.

Step 4: Determine Gray's Relation Coefficient

The fourth stage is the link between the best normalized value $L_i(k)$ and the real normalized $L_i(k)$. It is expressed numerically, as seen in the equation.

$$\Delta_{ik} = \frac{\gamma_{min} - \varnothing\gamma_{max}}{\gamma_{0i}(k) - \varnothing\gamma_{max}} \tag{7.7}$$

where Δ_{ik} denotes the coefficient of grey relational, γ_{min} indicates the minimum value of $\gamma_{0i}(k)$, γ_{max} symbolizes the maximum value of $\gamma_{0i}(k)$, $\varnothing$ symbolizes the discriminant coefficient $0 \leq \varnothing \leq 1$ for the interval modification of Δ_{ik}.

Step 5: Evaluations of efficiency

In this phase, the efficiency of each DMU_i is determined. The AP model is used to estimate the efficiency of each DMU. The Δ_{ik} value is set as output and input value set as 1.

Step 6: Calculating Efficiency Level

The final step calculates the average efficiency of each DMUs that operate at the equal parameter level. The ideal level of a parameter is determined by the level at which level efficiency is maximized.

Table 7.4 Parameter level.

Sl. no.	Process parameters	Level 1	Level 2	Level 3	Level 4
1	Silica fumes	10	20	30	40
2	Saw dust	5	10	-	-
3	GGBS	10	20	-	-

Table 7.5 Taguchi design experiments.

Taguchi design experiment	Additive combination mixed with soil		
	SF	SD	GGBS
T1	10	5	10
T2	10	10	20
T3	20	5	10
T4	20	10	20
T5	30	5	20
T6	30	10	10
T7	40	5	20
T8	40	10	10

Table 7.6 Orthogonal array layout coded and actual value of parameters.

Experimental runs	Coded values of parameter			Actual values of parameters		
	A	B	C	Silica fumes	Saw dust	GGBS
1	1	1	1	10	5	10
2	1	2	2	10	10	20
3	2	1	1	20	5	10
4	2	2	2	20	10	20
5	3	1	2	30	5	20
6	3	2	1	30	10	10
7	4	1	2	40	5	20
8	4	2	1	40	10	10

7.4 Results and Discussion

7.4.1 Experimental Results

7.4.1.1 Specific Gravity

A density bottle was used to quantify the specific gravity of the soil. The normal range for expansive soil specific gravity is 2.65 to 2.75.

7.4.1.2 Particle Size Analysis

The distribution of soil's particle sizes as shown, can be observed that from the grain size distribution curve of soil the soil consists primarily of clay and silt, with just a small percentage of gravel and sand. The finer percentage of particles based on the sieve sizes is depicted in Figure 7.5. Black cotton soil has a coefficient of curvature (Cc) and co-efficient of uniformity (Cu) values of 2.81 and 44 respectively. Figures 7.6 depict water content against the number of blows and dry density against water content which is shown in Figure 7.7. These results indicate that the soil has received a good grade.

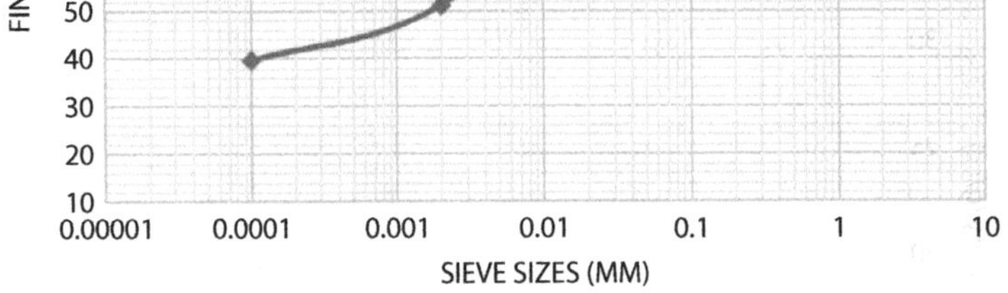

Figure 7.5 Particle size vs. percentage finer graph.

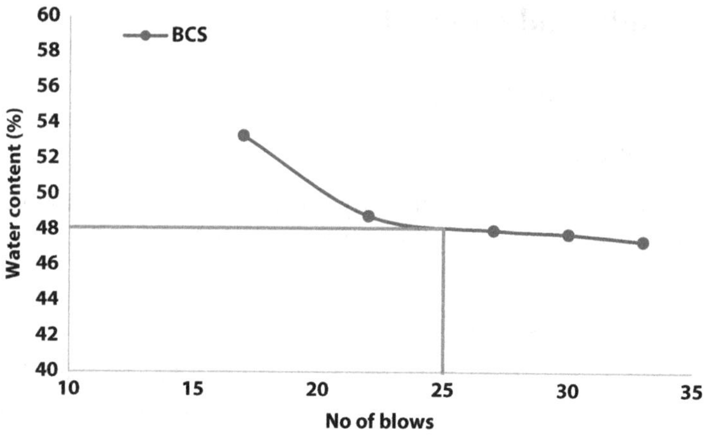

Figure 7.6 Graph of water content vs. no. of blows.

7.4.1.3 *Liquid Limit*

The results of LL for soil are presented in Figure 7.6. The liquid limit was found to be 48%.

7.4.1.4 *Plastic Limit*

The PL has been found to be 26.09%. Plastic index may be defined as the difference between the LL and PL. The plasticity index has been found to be 21.94.

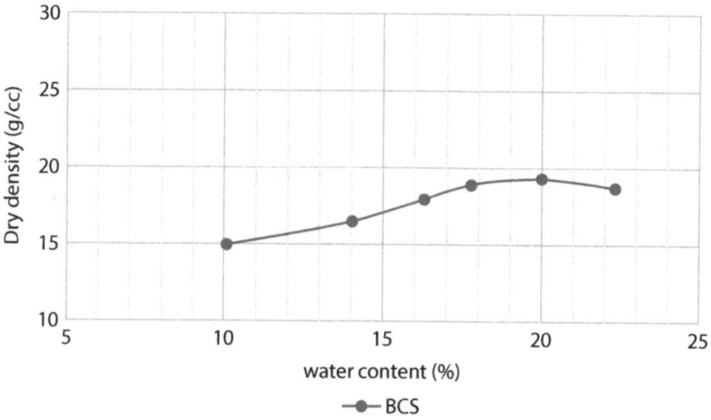

Figure 7.7 Dry density versus water content curve.

7.4.1.5 Compaction Characteristics

The MDD was found to be 19.33(kg/m3) and OMC was 20% from the graph, as shown in Figure 7.7.

7.4.1.6 California Bearing Ratio

The CBR of unmodified soil and soil with various additives according to the Taguchi design experiment are listed below in Table 7.7. It stated that CBR value of soil in unsoaked condition was found to be 3.12% and in soaked condition it was found to be 2.71%. Soil mixed with additives shows higher CBR vale than plain soil.

The graph presented in Figure 7.8 shows that the CBR of soil is increasing while mixed with different additive combination.

7.4.1.7 Unconfined Compression Strength Test

The UCS of soil and soil with additives according to the Taguchi design experiments are tabulated below in Table 7.7, and represented in Figure 7.9. The UCS of raw soil was found to be 115.27 kPa.

7.4.1.8 Differential Free Swell

The DFS index was carried out for the unmodified soil and soil with additives according to the Taguchi design experiments. It shows that the free

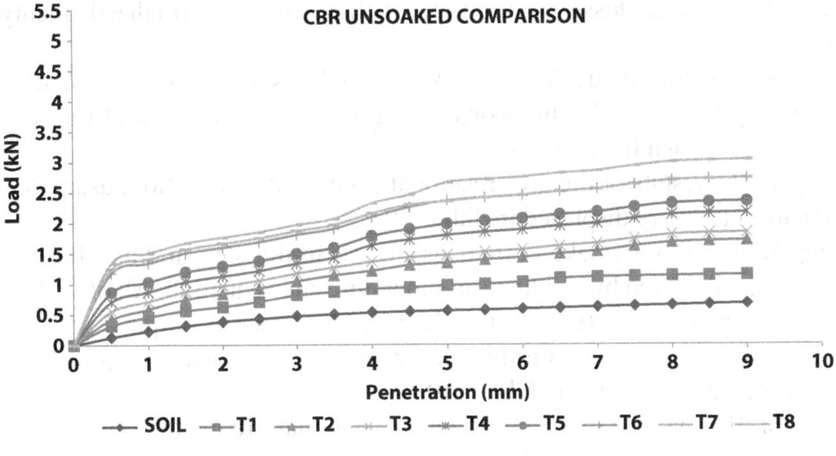

Figure 7.8 Load vs penetration graph.

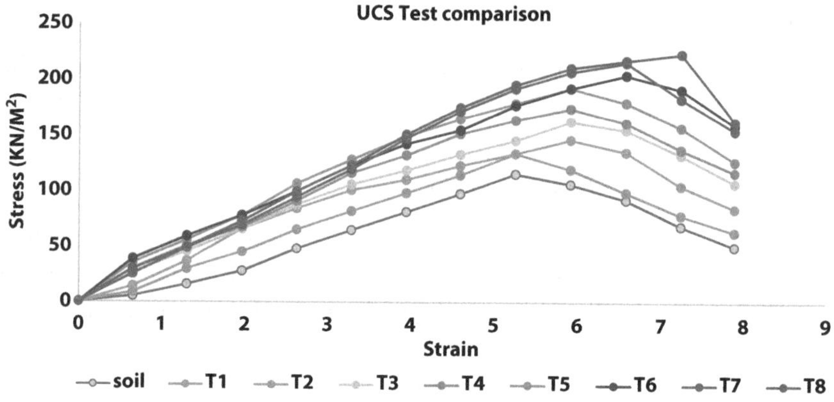

Figure 7.9 UCS test graph.

swelling index decreases while the additive mixed with soil shows different % according to Taguchi design experiment.

7.4.2 Procedure to Determine Grey-Super-Efficiency

Step 1: By using Table 7.7 data, the quality losses, L_{ik} (k = 1, 2, and 3) can be calculated. Table 7.8 displays the results of the calculations made for each response using the suitable function in Equation (7.1-7.3). Accepting that the quality loss coefficient for every response increased to one allowed us to calculate the quality losses.
Step 2: Using the proper approach from Eq. (7.4, 7.5), normalizing the quality losses observed during step 1. Table 7.8 displays the results for each response's quality losses and compares them to the normalized quality losses.
Step 3: The absolute deviation γ_{0i}(k), for each response was calculated by using Eq. (7.6) shortly after normalizing the quality losses, and the outcomes are shown in Table 7.9.
Step 4: The results of the grey relational coefficients calculation, using Δ_{ik}. Equation (7.7) are displayed in Table 7.9.
Step 5: Each test was taken care of as a DMU in this step, and its effectiveness was calculated by tackling the super efficiency equation. All of DMU's inputs were set to unity while the Δ_{ik} values were set as outputs. The efficiency of each DMU was estimated using the DEA solver program. The outcomes are displayed in Table 7.10.
Step 6: The average of all DMU efficiencies that are on the same factor level was calculated in the last stage to establish the best combination of

additives, and the calculated level with the excellent averaged productivity was chosen as the ideal level. Table 7.11 shows the parameter impacts on the efficiencies, and Figures 7.10, 7.11, and 7.12 show the parameter impact on the efficiencies of SF, SD and GGBS respectively.

7.4.3 Experimental Data

Table 7.7 Experimental data of UCS, CBR, and DFS.

Experiment no.	UCS	CBR	DFS
1	134.16	71.96	41
2	145.6	93.12	30
3	163.4	104.77	33
4	174.8	122.29	28
5	193	147.87	22
6	204.86	169.91	17
7	223.96	186.1	20
8	214.9	175.1	15

Table 7.8 Grey relation generation and quality loss for the responses.

	Quality loss			Normalized quality loss		
Exp. no.	L_i (1) UCS	L_i (2) CBR	L_i (3)DFS	R_i (1)	R_i (2)	R_i (3)
1	5.55799E-05	0.00019314	1888.666667	1	1	1
2	4.71782E-05	0.000115327	900	0.7642879	0.526296	0.40573
3	3.74542E-05	9.11053E-05	1089	0.4914786	0.378843	0.519335
4	3.27331E-05	6.68689E-05	784	0.3590254	0.231299	0.336005
5	2.68469E-05	4.57342E-05	484	0.1938857	0.102637	0.15568
6	2.38338E-05	3.46376E-05	289	0.1093534	0.035084	0.038469
7	1.9936E-05	2.88745E-05	400	0	0	0.105189
8	2.16538E-05	3.26162E-05	225	0.0481929	0.022779	0

Table 7.9 Deviation sequence, grey relational coefficients for the response.

	Deviation sequence			Grey relational coefficients		
Exp. no.	γ_{0i} **(1)**	γ_{0i} **(2)**	γ_{0i} **(3)**	Δ_i **(1)**	Δ_i **(2)**	Δ_i **(3)**
1	0	0	0	1	1	1
2	0.235712	0.473704	0.59427	0.80925	0.678563	0.627246
3	0.508521	0.621157	0.480665	0.662901	0.616843	0.675372
4	0.640975	0.768701	0.663995	0.609394	0.565387	0.600963
5	0.806114	0.897363	0.84432	0.553675	0.527047	0.542205
6	0.890647	0.964916	0.961531	0.52892	0.508928	0.509806
7	1	1	0.894811	0.5	0.5	0.527757
8	0.951807	0.977221	1	0.512346	0.50576	0.5

Table 7.10 Efficiency for the response.

DMU 1	**DMU 2**	**DMU 3**	**DMU 4**	**DMU 5**	**DMU 6**	**DMU 7**	**DMU 8**
0.508	0.797	0.812	0.885	0.953	0.992	1.024	1.023

Table 7.11 Efficiency for the level parameter.

	Level value			
Parameter	**1**	**2**	**3**	**4**
SILICA FUMES	0.6526	0.8485	0.9725	1.0235
SAW DUST	0.824	0.841	-	-
GGBFS	0.833	0.914	-	-

From Table 7.11 we found the maximum value of efficiency which is considered as optimum level or combination. The efficiency for various parameter is -

a) 1.0235 for SF at level 4 which is 40% SF
b) 0.841 forb SD at level 2 which is 10% SD
c) 0.914 for GGBS at level 2 which is 20% GGBS

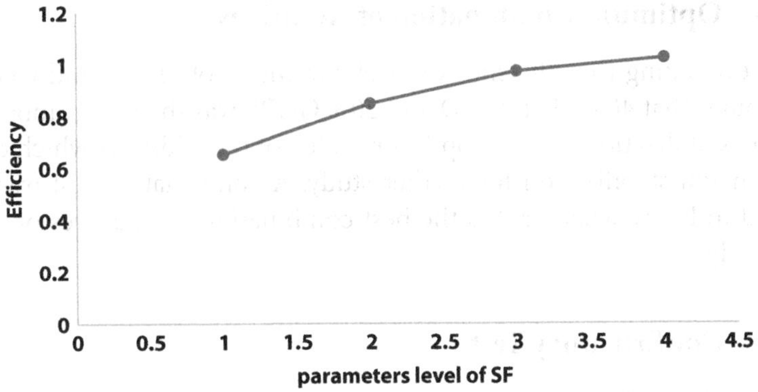

Figure 7.10 Parameters effect on efficiency of SF.

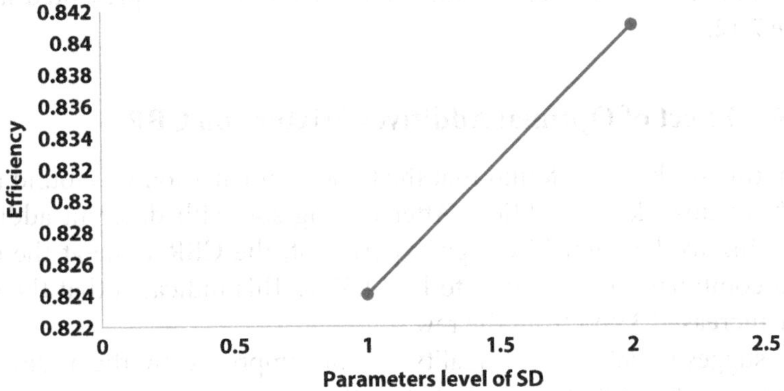

Figure 7.11 Parameters effect on efficiency of SD.

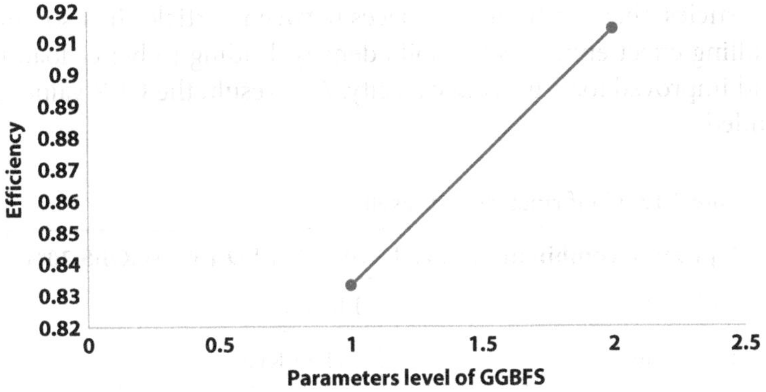

Figure 7.12 Parameters effect on efficiency of GGBS.

7.4.4 Optimum Combination of Additives

After evaluating the efficiency of level parameter of SF, SD and GGBS, it was found that 40% SF, 10% SD and 20% GGBS was the best combination for soil stabilization. It is the optimum value of the additives which can be used in soil stabilization for further study. A confirmatory test was conducted to know whether it is the best combination of additives or not in this study.

7.4.5 Confirmatory Test

After testing of the soil with optimum combination, it was found that the CBR and UCS value increased from the other combination and DFS value decreased, which indicates that the combination used was the optimum combination for the soil. The outcomes of the tests are presented in the Table 7.12.

7.4.6 Effect of Optimal Additives Mixture on CBR

From this study it was found that the CBR of normal soil was found to be 3.12% in unsoaked condition. After mixing soil with different additives according to the Taguchi design experiment, the CBR value at the optimum combination was found to be 14.35%. This indicated that the CBR value increased 359% from the raw soil.

It suggests that the soil quality can be improved by the addition of additives. When GGBS is included to the soil, it responds with the free lime present within the soil to form calcium silicate hydrate (CSH) gel, which provides binding properties and stabilizes the soil. This stabilization process can lead to an increase the CBR value, because silica fumes include tiny particles, that can fill in the spaces between particles in a soil matrix. This filling effect enhances the soil's density, leading to better load transfer and improved load-bearing capacity. As a result, the CBR value can be expanded.

Table 7.12 Confirmatory test result.

Optimum combination – SOIL +SF 40%+ SD 10% +GGBS 20%	
CBR Value	14.35%
UCS Value	234.25 KPa
DFS Value	13%

7.4.7 Effect of Optimal Additives Mixture on UCS

From this study it was found that the UCS of normal soil was found to be 115.23 KPa. After mixing soil with different additives according to Taguchi design experiment, the UCS value at optimum combination was found to be 234.25 KPa. This is because silica fumes have a pozzolanic effect, implying that when water is present, they react with calcium hydroxide to create new cementitious compounds. This response results in the development of stronger bonds and increased cohesion within the soil, resulting in higher UCS values. GGBS also impact on increase on UCS value.

7.4.8 Effect of Optimal Additives Mixture on DFS

From this study it was found that the DFS value of normal soil was found to be 48 %. After mixing soil with different additives according to Taguchi design experiment, the UCS value at optimum combination was found to be 13%, which shows that the value decreased dramatically. This is because silica fume can chemically react with the soil constituents, forming stable compounds. These compounds can reduce the swelling value of the soil by altering its mineralogy or changing the properties of the clay minerals present. GGBS has a high content of amorphous silica and alumina, which can act as pozzolanic materials. When GGBS is added to soil, it can react with lime and other alkalis present in the soil, forming stable compounds. This reaction reduces the availability of free lime and alkalis, which are responsible for the swelling of certain clay minerals. Consequently, the addition of GGBS can reduce the swelling behavior of the soil. Adding sawdust can also help mitigate the extreme swelling behavior of such soils.

7.5 Conclusion

The goal of the current work was to simultaneously optimize three additives by utilizing grey relational analysis and the Taguchi method's super-efficiency model Different levels of the additives SF, SD and GGBS were used. The following conclusions are possible to determine from experimental data in this study and the confirmatory test conducted in our current investigation.

1. The best additive combination resulted in the substantial enhancements among the various tests like UCS, CBR, and DFS at 40% SF, 20% GGBS, and 10% SD.

2. The optimal additive mixture resulted in a substantial boost in the soil UCS value. GGBS reacts with the free lime present in the soil to produce calcium silicate hydrate (CSH) gel, which provides binding properties. Silica fumes have a pozzolanic effect; it forms cementitious compound when reacting with water. It develops stronger bond with soil, resulting in the increase in strength of soil.
3. When the optimal Additives mixture was introduced with the soil, its CBR value increased dramatically. The rise in CBR of the soil additive admixtures was attributable to the micro-filler impact observed when the SF conferred on mechanical attributes of the soil. Furthermore, the impact of exchange cations that happened in the GGBS when it was exposed in water and form additional cementitious compounds, resulted in improvement in soil's CBR value after it was mixed with additives.

The application of the optimal supplement mixture dramatically lowered the DFS of the raw soil. This decrease is likely due to the formation of cement bonds (CSH) during GGBS hydration and the effect of SF micro-filler actions. Sawdust also acts as a filler, reducing the clay's ability to absorb water. The Taguchi technique is able to link and integrate efficiency models and grey relational analysis to optimize multiple additives for extended soil remediation.

References

1. Jaswal, P., Vivek, Sinha, S.K., Improvement in the performance of two layered model pavement with treated coir geotextile at the interface. *J. Ind. Text.*, 52, 15280837221114161, 2022.
2. Vivek, Shafi Mir, M., Sehgal, R., Studies of Modulus of Resilience on Unpaved Roads Reinforced with Untreated/Treated Coir Geotextiles. *J. Nat. Fibers*, 19, 16, 13563–13573, 2022.
3. Vivek, Shafi Mir, M., Sehgal, R., Study on bearing capacity of unpaved roads reinforced with coir geotextiles using finite element method (fem). *J. Nat. Fibers*, 19, 15, 11735–11748, 2022.
4. Jaswal, P. and Vivek, Experimental study on monotonic behaviour of two-layered unpaved road model reinforced with treated coir geotextiles. *Int. J. Pavement Res. Technol.*, 17, 1377–1388, 2023.

5. Vivek, Dutta, R.K., Parti, R., Effect of chemical treatment of the coir geotextiles on the interface properties of sand/clay-coir geotextile interface. *J. Inst. Eng. India Ser. A*, 100, 357–365, 2019.
6. Mohanta, N.R. and Samantaray, S., Study of Combined Effect of Metakaolin and Steel Fiber on Mechanical Properties of Concrete. *Pertanika J. Sci. Technol.*, 27, 3, 1381–1396, 2019, http://journals-jd.upm.edu.my/resources/files/Pertanika %20PAPERS/JST%.
7. Vivek, and Dutta, R.K., Bearing ratio behavior of sand overlying clay with treated coir geotextiles at the interface. *J. Nat. Fibers*, 19, 14, 7534–7541, 2022.
8. Vivek, Effects of cyclic loading on sand overlaying clay model of unpaved roads reinforced with untreated/treated coir geotextiles. *J. Text. Inst.*, 115, 1694–1700, 2023.
9. Singh, P., Dash, H.M., Samantaray, S., Effect of silica fume on engineering properties of expansive soil. *Mater. Today Proc.*, 33, 5035–5040, 2020, https://doi.org/10.1016/j.matpr.2020.02.839.
10. Kumar, M.H., Mohanta, N., Samantaray, S., Kumar, N.M., Combined effect of waste glass powder and recycled steel fibers on mechanical behavior of concrete. *SN Appl. Sci.*, 3, 1–18, 2021a, https://doi.org/10.1007/s42452-021-04422-w.
11. Kumar, M.H., Macharyulu, I.S., Ray, T., Mohanta, N.R., Jain, M., Samantaray, S., Sahoo, A., Effect of water absorption and curing period on strength and porosity of triple blended concrete. *Mater. Today Proc.*, 43, 2162–2169, 2021b, https://doi.org/10.1016/j.matpr. 2020.12.092.
12. Kumar, M.H., Saikrishnamacharyulu, I., Kumar, U., Jena, S., Mohanta, N.R., Samantaray, S., Coupling effect of fly ash, metakaoline and different types of steel fibers on mechanical performance of concrete. *AIP Conf. Proc.*, 2417, 24–31, 2021, https://doi.org/10.1063/5.0072573.
13. Ray, T., Mohanta, N.R., Kumar, M.H., Samantaray, S., Study of effect of temperature on behavior of alkali activated slag concrete. *Mater. Today Proc.*, 43, 1352–1357, 2021, https://doi.org/10.1016/j.matpr.2020.09.169.
14. Kumar, M.H., Saikrishnamacharyulu, I., Mohanta, N.R., Ashutosh, A., Mishra, P., Samantaray, S., Mechanical behaviour of high strength concrete modified with triple blend of fly ash, silica fume and steel fibers. *Mater. Today Proc.*, 65, 933–942, 2022a, https://doi.org/10.1016/j.matpr.2022.03.528.
15. Kumar, M.H., Mohanta, N.R., Patel, N., Samantaray, S., Reddy, S.V.B., Influence of Fly ash and Metakaoline on the crack resistance and shrinkage property of concrete. *Iran. J. Sci. Technol. Trans. Civ. Eng.*, 46, 2011–2026, 2022b, https://doi.org/10.1007/s40996-021-00748-y.
16. Medina-Martinez, C.J., Sandoval Herazo, L.C., Zamora-Castro, S.A., Vivar-Ocampo, R., Reyes-Gonzalez, D., Use of Sawdust Fibers for Soil Reinforcement: A Review. *Fibers*, 11, 7, 58, 2023.

17. Rashid Rather, M. and Gupta, S., Enhancing the Index and Engineering Properties of Virgin Soil by the Addition of Saw Dust. *Int. J. Res. Publ. Rev.*, 3, 2381–2389, 2022.
18. Ikeagwuani, C.C. and Nwonu, D.C., Multi-additive optimization for expansive soil treatment using grey-super-efficiency model integrated in Taguchi method. *Indian Geotech. J.*, 51, 6, 1166–1181, 2021.
19. Tiwari, N., Satyam, N., Patva, J., Engineering characteristics and performance of polypropylene fibre and silica fume treated expansive soil subgrade. *Int. J. Geosynth. Ground Eng.*, 6, 2, 18, 2020.
20. Samantasinghar, S. and Singh, S.P., Strength and durability of granular soil stabilized with FA-GGBS geopolymer. *J. Mater. Civ. Eng.*, 33, 6, 06021003, 2021.
21. Lakshmi, S.M., Geetha, S., Selvakumar, M., Sivanesan, N., Sreedharan, K.R., Application of lime and GGBS to improve the strength of clayey sand. *IOP Conf. Ser.: Mater. Sci. Eng.*, 989, 1, 012028, November 2020.
22. Vastrad, M., Karthik, M., Dhanavandi, V., Shilpa, M.S., Stabilization of black cotton soil by using GGBS, lime and nano-silica. *Int. J. Res. Eng. Sci. Manage.*, 3, 9, 1–7, 2020.
23. Narendra, K. and Prasad, A.D., Effect of ground granulated blast furnace slag on expansive soils under static loading. *VFSTR J. Sci. Technol. Eng. Manage.*, 3, 61–65, 2017.
24. Sameeuddin, M., Azeemuddin, M., Sohail, M.S., Rather, M.U.D., Girija, K.M., Stabilization of Black Cotton Soil by Ground Granulated Blast Furnace Slag (GGBS), *Int. Res. J. Eng. Technol.*, 5, 804–809, 2018.
25. Manjunath, K.V., Shekhar, H., Kumar, M., Kumar, P., Kumar, R., Stabilization of black cotton soil using ground granulated blast furnace slag, in: *Proc., Int. Conf. on Advances in Architecture and Civil Engineering (AARCV 2012)*, pp. 387–390, June 2012.
26. Ene, E. and Okagbue, C., Some basic geotechnical properties of expansive soil modified using pyroclastic dust. *Eng. Geol.*, 107, 1–2, 61–65, 2009.
27. Karna, S.K. and Sahai, R., An overview on Taguchi method. *Int. J. Eng. Math. Sci.*, 1, 1, 1–7, 2012.
28. Kumar Sharma, A. and Sivapullaiah, P.V., Improvement of strength of expansive soil with waste granulated blast furnace slag, in: *GeoCongress 2012: State of the Art and Practice in Geotechnical Engineering*, pp. 3920–3928, 2012.
29. Dayalan, J. and Dayalan, J., Comparative study on stabilization of soil with ground granulated blast furnace slag (GGBS) and fly ash. *Int. Res. J. Eng. Technol.*, 3, 5, 5, 2016.
30. Butt, W.A., Gupta, K., Jha, J.N., Strength behavior of clayey soil stabilized with saw dust ash. *Int. J. Geo-Eng.*, 7, 1–9, 2016.
31. Rawat, A., Jain, R., Parte, S.S., Improvement of CBR and compaction characteristics of black cotton soil using lime and blast furnace slag. *Int. J. Res. Appl. Sci. Eng. Technol.*, 5, 523–528, 2017.

32. Murmu, A.L., Dhole, N., Patel, A., Stabilisation of black cotton soil for subgrade application using fly ash geopolymer. *Road Mater. Pavement Des.*, 21, 3, 867–885, 2020.
33. IRC: 37-2018, Guidelines for the design of flexible pavements, 2018.
34. Al-Refaie, A. and Al-Tahat, M.D., Solving the multi-response problem in Taguchi method by benevolent formulation in DEA. *J. Intell. Manuf.*, 22, 505–521, 2011.

8

Comparative Study on Strength and Compaction Characteristics of Sand and Waste Materials Blended Expansive Soil

Manisha Khuntia[1]*, Sourav Mahanand[2], Harpreet Singh[2] and Raj Kishor Gonjhu[2]

[1]*Department of Civil Engineering, Odisha University of Technology and Research, Bhubaneswar, Odisha, India*
[2]*Department of Civil Engineering, SIT West, Sambalpur, Odisha, India*

Abstract

Foundation is the most important part of a structure as it shoulders the entire weight of the structure, transferring it to the underlying soil or rock. It is required to have knowledge about different properties of soil and factors affecting its behavior before working with such problematic soil. Expansive soil undergoes significant volume change upon wetting and drying which makes it unsuitable for direct engineering application in its original form. The rapid growth of coal and marble industries leads to the production of hazardous waste materials to a large extent which affects living-being and the environment. A large percentage of these wastes remain unutilized giving rise to the need for effective utilization by exploiting their inherent properties so as to solve the above stated problem. To make such soil suitable for construction purposes, we need to enhance their properties by using various ground improvement method; sand soil stabilization is one of them. This chapter demonstrates the efficacy of combining fly ash, sand, and tile waste with soil by observing the strength and compaction properties that are being assessed, taking into account the outcomes of the laboratory-based experimental program. To obtain the appropriate mixes, admixtures are added to the soil in a suitable proportion that is determined with the aid of a proctor compaction test. Expansive soil specimens combined with different proportions of sand-between 10% and 50%-with a 10% sand increment were used in a series of experimental

**Corresponding author*: manisha.khuntia8@gmail.com

Sandeep Samantaray, Subhashree Samantasinghar, Soumendra K. Mohanty, Vivek, Abinash Sahoo and Deba Prakash Satapathy (eds.) Waste Utilization in Geotechnical Practice, (209–238) © 2025 Scrivener Publishing LLC

projects. The preferred proportion of clay-sand composition is then combined with 10% fly ash, increasing the amount of fly ash by 5% until 25%. The developed composite of locally available soil-fly ash is further incorporated with 5%tile-waste up to 20%with an incremental of 5%. When compared to natural expansive soil, California bearing ratio values of these optimum mixes showed a significant improvement. The best combinations for compaction characteristics on the basis of this experimental study are 60:40, i.e., clay: sand, 54:36:10, i.e., clay: sand: fly ash and 48.6:32.4:9:10,i.e., clay: sand: fly ash: tile waste. The California bearing ratio value of each optimal mixes exhibits an upward trend. The key objective of this study is to incorporate waste resources (river sand, fly ash, and tile fragments) as an additive to lessen the disposal issue and create a more economical construction material.

Keywords: California bearing ratio, sand, compaction, fly ash, expansive soil, tile-waste

8.1 Introduction

The primary concern of a civil engineer is soil stability which can be boosted by improving the engineering properties of soil. This process of improving the engineering characteristics and general stability of soil is referred to as soil stabilization [1–7]. The purpose of this process is to increase the resilience of soil to settlement and failure as well as its strength, durability, and load-bearing capacity. To change the properties of the soil and make it appropriate for use in construction or other applications where improved soil performance is crucial, various materials or techniques are added. When the soil that is available for construction is inappropriate for the intended use, it becomes necessary. Pre-consolidation, compaction, drainage, and many other processes are all included in the broad category of stabilization. Nonetheless, the term stabilization is typically limited to the procedures that modify the soil material itself to enhance its characteristics [8]. For stabilization purpose, a chemical or cementing material is added to naturally occurring soil. This procedure is utilized to improve the shear strength of the soil while lowering its permeability and compressibility in earth structures. One of the techniques for improving the ground is soil stabilization, which makes a soil material more stable and capable of supporting more weight, strengthening and resisting against stress and adverse moisture conditions [9]. There are numerous proven soil stabilization methods that are in use all over the world, improving the soil with proportionate waste materials like fly ash, tile waste, cement, lime, etc. Such materials are added to the soil, which not only greatly improves its qualities at a lower cost but also eliminates industrial and other wastes in a very environmentally friendly manner.

The occurrence of significant volume changes in expansive soils due to variations in moisture content and it poses a significant challenge [10]. The phrase "expansive soil" is typically used to describe any rock or soil material that has the capability to expand or contract in response to variations in moisture levels [11]. The swelling and shrinkage behavior of clayey soils makes construction work very challenging. When exposed to moisture, these soils expand; when left dry, they begin to contract. Because of their property for swelling and shrinking, volume of clayey soils changes significantly, which causes substantial variations in water content. Their high susceptibility to volume changes—they swell in presence of moisture and shrink in absence of it—may have an unfavorable effect on stability and structural integrity of soil [12, 13]. Clay mineral content, void ratio and water content all affect volumetric changes in clayey soils. But the main cause of the seal iterations is thought to be the minerals called montmorillonite that are present in the clayey soil which has a high potential for swelling. This mineral has a substantial impact on volume change of the soil and is a major factor in its expansive behavior. In fact, when dry, clayey soils have significant strength. But when they get wet, their behavior changes significantly, which causes instability. These soils can experience large volume changes when they get wet, which causes swelling. This tendency to swell puts structures built on top of these types of soils at significant risk of damage or instability because of the altered and less predictable properties of the wet soil [14]. Particularly, light-weight constructions built on clayey soils have suffered damage in several nations, including the United States of America, India, and China. Given all of these problems, clayey soils are usually perceived to be subpar materials for slopes, pavement subgrade and foundations [15]. As a result, building on these difficult soils is a challenging task that causes great concern among geotechnical engineering professionals [16]. Therefore, stabilization of clayey soils is necessary prior to any agreed-upon construction activity to lessen the impact of water on the clayey soil [17]. Due to industrialization, population growth is happening very quickly which leads to the issue of availability of land. This sort of issue compelled researchers to brainstorm on how to alter the engineering characteristics of existing problematic clayey soils to enable construction even on these soils [18, 19]. To resolve the issues related to clayey soils, many researchers have put forth a variety of strategies for altering their geotechnical characteristics. There are various methods by which we can improve the overall performance, strength, and stability of clayey soil for different engineering and construction applications. One of the methods which can overcome these issues is soil stabilization by improving the geotechnical characteristics of these type of problematic soils [20].

Finely broken rock and mineral particles combine to form sand, a naturally occurring granular material. It is finer than gravel but coarser than silt. It can be used as a filler material because of its strong ability to support loads in confined spaces. As a result, different amounts of sand could be added to cohesive soils to modify the plasticity properties, compaction characteristics and strength behavior of soil additive mix [21]. Finer soil particles are substituted with coarser sand admixture particles in accordance with stabilization techniques used to improve weak soil. This gives the soil a consistent particle gradation and produces a composite mix that is cohesive and frictional. Additionally, the soil has a higher load carrying capacity when it is mixed, placed, and compacted correctly on site. If the materials are appropriate, subgrade is crucial to the safe and economical construction of pavements. Generally, local soils need to have both their plasticity and their bearing capacity improved. One common way to accomplish such goals is through stabilization. Granular materials like sand are found in nature. It has a high load bearing capacity in confined spaces, making it a potential filler material. Cohesive soils could benefit from the addition of sand as an admixture to alter their strength, compaction, and plasticity [22].

Fly ash, sometimes known as flue ash, is a light-weight fine derivative formed by means of coal combustion in thermal power plants. The amount of fly ash produced globally is increasing annually [23]. Fly ash is a noncohesive, silt-sized substance with a specific gravity that is lower than that of typical soils. Fly ash disposal unveils a significant risk to the environment and also its disposal can cost millions of rupees. In several nations, including India, fly ash has been utilized in a range of purposes, including the creation of embankments and compacted fills, concrete, bricks and liners. Although fly ash doesn't have much of its own cementitious value, it integrates chemically with moisture for generating cementitious compounds, which improves strength and compressibility. Fly ash is divided into two groups:

- Fly ash of Class – C type
- Fly ash of Class – F type

The ash's amounts of iron, calcium, silica, and alumina primarily distinguish these groups from one another. The chemical characteristics of fly ash that are produced are significantly influenced by bituminous, lignite, and anthracite chemicals found in the burned coal. The word "ceramic" has its root in the Greek term "keramos" which means "potter's clay." The earliest ceramics were made of fire-hardened clay either pure or combined with other materials. Later ceramics underwent firing and glazing to produce a smooth, colorful surface. The term "ceramic tiles" refers to the glazed tiles that were once made by potters using clay. An enormous quantity of waste is generated during production, installation and shipping of ceramic tiles. Thirty percent of the output of ceramics is lost in the manufacturing, handling, and shipping stages, according to numerous studies [24]. With a 5% annual growth rate, the output of ceramic tiles was recorded as 13056 million square meters in 2016. With a predicted Compound Annual Growth Rate (CAGR) of 9%, the ceramics market is expected to grow, by 2022, to a value of 7.5 billion euros with the ceramic tiles market in India accounting for the lion's share of the 4.9 billion euros in 2017. This waste or scrap material is a dangerous inorganic material. These days, there are many ways to successfully modify the geotechnical characteristics of clayey or weak soils by making use of leftover ceramic materials. Enhancing the stability and properties of clayey soils can be done more affordably and sustainably by using ceramic waste as a soil stabilizer [15, 18, 25].

Using fly ash and tile waste has many benefits, some of which include:

- Saving disposal space
- Saving natural resource scarcity
- Saving energy, as the material is generated automatically as a by-product and requires no energy to produce; and secondly, it can replace materials that would otherwise require energy consumption
- Protecting the environment, as it can partially replace cement in construction, whose production requires energy consumption and CO_2 emissions.

Studies on waste utilization and applications for soil stabilization have a long history. Various efforts have been done in the geotechnical characterization of clay and sand that has fly ash and tile waste as an additive. The identification and remediation of problematic subgrade soils is therefore, a major goal. Replacing poor soil is one of the most popular solutions. However, given the amount of labor required for such projects, it is both expensive and impracticable.

Numerous research works have concentrated on the geotechnical characterization of clay and sand with additives like fly ash and tile waste. A brief overview of a few of these papers is provided in this section. Investigations into the application of fly ash [26], shows that in US during 1996,

101.8 million short tons(mst) of by products from coal were produced of which 59.4 mst were fly ash and out of this total fly ash produced, 16.2 mst or 27.4% was utilized and the remaining 43.2 mst was disposed of by means of 'landfilling'. He [27], investigated the characterization of fly ash in relation to geotechnical applications to have a bulk usage of the fly ash. The compaction behavior of coal ashes and soil is nearly the same rather than having a variation in specific gravity. Based on the investigation, the fly ash from Vindyanagar had more cohesiveness when black cotton soil was added. Increased California bearing ratio (CBR) value can be ascribed to the increased cohesiveness that arises from the combination of fly ash and soil, signifying an improved ability of the mixture to support loads. As analyzed in [28], how effectively fly ash integrates into expansive soil is determined by considering a CH type soil based on its grain size distribution and performing a number of tests by mixing soil with various percentages of fly ash, with increments of 5% to 20% depending on dry weight. The increase in the amount of fly ash was found to be correlated with the rise in MDD (maximum dry density) and reduction in OMC (optimum moisture content).The aforementioned alteration in properties suggests that the soil with a greater fly ash content can withstand greater loads in comparison to virgin soil. Fly ash treated soil and its effectiveness was investigated [29]. Three different types of soil with various fly ash content were taken for this study. On the basis of test results, an addition of fly ash was found to lower the DD and unit weight of soil. It is interesting to note that although the dry density and unit weight decreased, the CBR value increased significantly. Better resistance to deformation under applied loads is indicated by a higher CBR value, which may be advantageous in a variety of engineering applications. Based on a variety of soil compositions, the outcomes demonstrate that fly ash can improve structural stability and load-bearing capacity, among other advantages of soil properties. The efficacy of stabilizing fine-grained soft soils by self-cementing fly ash was studied [30]. The outcome for fly ash treated soil, CBR illustrates an inverse relationship with a rise in water content and a direct correlation with a rise of fly ash content. Additionally, it was observed that adding fly ash significantly improved CBR of wetter fine-grained soil. Furthermore, it was revealed that soil-fly ash mixtures show significantly lower CBR and M_r compared to organic matter when soil contains 10% organic matter. He [31], examined the effects of incorporating fly ash for stabilization of expansive soil and its implications in the geo-engineering. Notable trends were observed in their study as the quantity of fly ash varies in the soil mixtures. The MDD showed an increase up to 20% fly ash content, after which it declined. In contrast, the OMC and CBR (unsoaked and soaked) values consistently reduced with an increase in fly ash proportion. Furthermore, a rise in the cohesive component brought about by the addition of soil (black cotton soil) to the mixtures was responsible for the higher CBR values that were noted. Notably, the CBR values exhibited peaks at 20% and 80% fly ash content in tests where clay-fly ash blends were evaluated without prior soaking. He [32], found that combining waste ceramic dust with the local clayey soil by gradually increasing the amount from 0% to 30% enhanced the compaction properties of the soil. MDD increased to 18.1 KN/m^3 from 15.6 KN/m^3while OMC decreased to 17.6% from 20.4%. The improved soaked CBR is indicative of this improvement, rising from 1.6% to 4%. The use of up to 30% ceramic dust for reinforcing flexible pavement subgrades was recommended by economic analysis, offering significant construction cost savings. To stabilize dune sand, he [33], added residual ceramic tile. The results of their investigation showed a correlation between the particle size of additive and the CBR value of additive mixed soil. Additionally, an increase in the admixture percentage or quantity had a positive influence on CBR value. Interestingly, the CBR values demonstrated more significant increases in unsoaked conditions than in soaked conditions, across a range of admixture proportions, indicating a greater improvement in load-bearing capacity without moisture saturation. Fly ash mixtures, comprising both fine and coarse particles and comprising five to thirty percent of the sample compositions, were utilized in an experiment [34], to stabilize expansive soil. The OMC and MDD were ascertained by curing three identical samples for 1, 7, 8, and 28 days. The lowest dry density range in their results was roughly 0.6g/cc which is for a mix of 70% soil and 30% fly ash, while the highest dry density range was 1.35g/cc for a mix of 95% soil and 5% fly ash. Further, the study found that an increase in water content caused a 20–30% decrease in dry density. The DD gradually decreased as the water content increased. He [35], mixed sand, fly ash, tile waste, jute fibers, and clayey soil available locally. In their study, they found that the optimal mix proportion for jute fiber, tile waste, fly ash, sand and clay was 0.5: 9:10:27:63.They noticed that the MDD dropped and the OMC went up when there is a rise in the fly ash content. In the clay-sand-fly ash mixture, the addition of tile waste was found to increase the maximum dry density initially. However, as more tile waste was added, the MDD gradually decreased. A similar pattern in MDD was observed with an addition of jute fiber. The inclusion of jute fiber had no noticeable effect on the OMC. This ideal mixes with the local clayey soil revealed significant improvements for both soaked and unsoaked CBR. The final combination that was optimized proved to be a great building material that offered significant cost savings in applications involving the construction of flexible pavements.

The foundry sand stabilized clayey soil and the tile waste treated expansive soil has the potential to be used as subgrade materials, especially for building flexible pavements on rural roads with low traffic volumes. As a consequence of the test results being analyzed [36], it was discovered that while MDD and CBR (both soaked and unsoaked) increased with an addition of CWD, liquid limit (LL), plastic limit (PL), plasticity index, and OMC consistently decreased up to a 30% level. Between unmixed laterite and 30% CWD addition, LL dropped from 59.62% to 35.61%. After mixing laterite with 30% CWD, PL dropped linearly from 40.11% (unmixed) to 23.31%. When adding 5% more CWD up to 30%, the percentages for the soaked and unsoaked CBR increased from 6.82% to 21.97% and 4.55% to 8.39%, respectively. The study came to the conclusion that, due to its benefits to the environment, economy, and durability, CWD should be used to stabilize lateritic soil. 21 highly generated but underutilized waste materials from the agricultural, industrial, construction and demolition, and mining sectors have been selected for review, out of the many waste materials generated from these sectors. The current study has evaluated their use as additives in soil stabilization, either in full or in part replacement of conventional/traditional additives. With an ideal dosage of 10% to 30% for soil stabilization, palm oil fuel ash (POFA), one of the many agricultural waste products taken into consideration, has been found to be the most effective. It was discovered that the best dosages for iron ore tailings (IOT) and cement kiln dust (CKD) with 50% lime were 16% and 20%, respectively. In the given data, C&D waste was applicable in the 10%–24% range. There was an impact of marble dust (MD) and granulated blast furnace slag (GBFS) additives on enhancing the geotechnical characteristics of low and high plasticity clays as investigated in this study [37]. Different proportions of MD and GBFS (5-10-15–20%) were combined with clay-containing soils. Numerous tests of unconfined compression and compaction were conducted on soil and mixes of soil treated with additives.

When MD or GBFS are mixed with clayey soils, OMC decreases; however, with an increase in curing time, there was a rise in unconfined compressive strength, MDD and bulk unit weight. For clay with high plasticity (CH) soil, the ideal addition amount was 10% MD or 15% GBFS, moreover; for clay with low plasticity (CL) soil it was 15% MD or 10% GBFS. Additionally, 15% GBFS on CH soils and 15% MD on CL soils produced the highest soil improvement ratio.

8.2 Material and Methodology

8.2.1 Materials Used

Expansive soil (C) was gathered from Kuchinda in Odisha state. The earth was initially dug up to a depth of three meters, after which it was collected. The soil has a significant clay percentage and is dark grey to black in color. After the soil was collected, it was air dried and pulverized manually; thereafter, the virgin soil properties were determined. The sand used in this experiment was grouped as poorly graded sand (SP) and was collected from the IB Valley area in Jharsuguda, Odisha. Figure 8.1 shows the specific sand and the expansive soil used in this study. These materials underwent a rigorous preparation process that included thorough washing and cleaning, oven drying, and removal of organic matter, roots, and other impurities. Tables 8.1 and 8.2 depict physical characteristics of expansive soil and sand utilized in this study respectively.

The soil was categorized using standard methods that adhered to the Indian classification system. Sieve analysis, hydrometer analysis, and Atterberg's limits assessment were some of these methods. The sieve analysis has been conducted following IS: 2720 (Part 4)- 1980 [38]. Figure 8.2 displays the particle size distribution curves for expansive soil (C), fly ash (FA), tile waste (TW), and sand (S). Using the plasticity index and particle size distribution curve, it was found that the soil is clay with high plasticity according to the IS classification criteria. Multiple analyses and evaluations were used to develop this classification system, which provided crucial information about engineering properties of soil and its behavior under various conditions. It made it easier to understand the characteristics and performance of the soil, particularly with regard to plasticity. Graphical representation of different grain size distributions among the designated soil constituents in Figure 8.2 made it easier to conduct comparative analyses and brought attention to the

(a) (b)

Figure 8.1 Materials used: (a) Expansive soil (b) Sand.

Table 8.1 Index and engineering properties of expansive soil.

Properties	**Soil**
IS Soil Classification	CH
LL (%)	57.11
PL (%)	28.67
Plasticity Index (%)	28.44
G (Specific Gravity)	2.65
OMC (%)	23.86
MDD (g/cm³)	1.545
c(Cohesion)	45.5kPA
Φ (Angle of Internal Friction)	15.05°
CBR value Unsoaked (%)	2.923
Soaked (%)	1.705
Free Swell Index (%)	68.18

Table 8.2 Physical characteristics of sand.

Properties	**SA**
IS Classification	SP
G (Specific Gravity)	2.62
Cohesion (kPa)	-
Φ (Angle of Internal Friction)	35.44°
Cu (Coefficient of Uniformity)	1.6
Cc (Coefficient of Curvature)	6.42

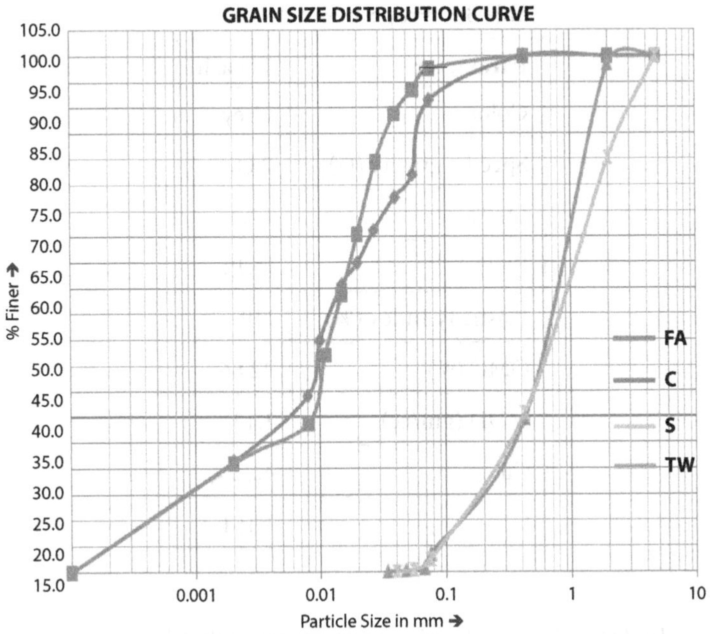

Figure 8.2 Particle size distribution curve of different raw materials.

distinct particle proportions in each material. With an emphasis on engineering applications, building projects, and soil modification initiatives, this methodical approach produced crucial data about the classification of soil and enabled a deeper understanding about its behavior.

Fly ash for this study was procured from the Hirakud, Sambalpur, Odisha-based HINDALCO power plant. Based on data extracted from the power plant's records, the fly ash sample was categorized as class F. To facilitate analysis, the obtained sample was carefully dried in an oven set between 105°C and 110°C. The dried samples were then carefully placed in airtight containers to maintain their integrity for future studies and experiments. Enumerated below are the physical and chemical properties of fly ash sample. Along with an extensive number of other details that provide insight into the elemental and chemical composition of fly ash as well as its physical properties, these tables provide a detailed breakdown of its composition. Table 8.3 focuses on the physical characteristics and clarifies important factors like specific gravity, particle size distribution, and other relevant features of fly ash. Contrariwise, the elemental composition is explored in the chemical analysis shown in Table 8.4, which also highlights the presence and concentration of important components in the fly ash sample.

Table 8.3 Properties of fly ash.

Parameter	Fly ash
Color	Grey
Specific gravity (G)	2.08
LL (%)	57.2
PL (%)	(NP)
Plasticity Index (%)	NP
OMC (%)	42.86
MDD (gm/cc)	1.13
Cu	4.22
Cc	1.151

Table 8.4 List of Chemical components of fly ash.

Chemical components	Available % of components
SiO_2(Silica)	60.84
Al_2O_3(Alumina)	29.15
Fe_2O_3 (Oxide of Iron)	3.02
TiO_2 (Titanium Oxide)	0.67
CaO (Lime)	1.87
MgO (Magnesia)	0.36
K_2O (Potash)	0.55
Na_2O (Soda)	0.46
SO_3 (Sulphate)	0.06
P_2O_5 (Phosphate)	0.69
Loss of ignition	2.06
Others	0.27

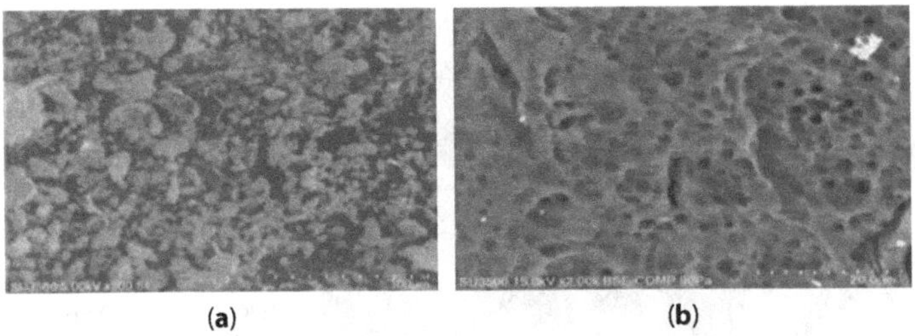

Figure 8.3 (a) SEM of fly ash (b) SEM of fly ash showing hollow structure.

Scanning Electron Microscope (SEM) was used to investigate surface morphology of fly ash providing a thorough study about the surface texture and individual particle morphology from both external and internal viewpoints. Micrographs with the best resolution were obtained at 5kV×500, 10kV×500, and 15kV×2 magnification factors when the voltage was increased. To make the analysis process easier, fly ash samples were divided into small portions and placed on double-sided carbon tape that was attached to a stub of a scanning electron microscope as a supporting material. To characterize the fly ash particles, a field of view was chosen and each fly ash particle inside that field was carefully examined and represented, as shown in Figure 8.3.

EDS or energy-dispersive X-ray spectroscopy, is a method for analyzing chemical or elemental composition of a sample. It depends on how the sample and an X-ray excitation source interact. The fundamental idea behind this method is based on the fact that every element has a different atomic structure, which produces different peaks for each element X-ray emission spectrum. The EDS spectrum of a fly ash sample is shown in Figure 8.4, indicating that the main components of the fly ash specimen are silicon and aluminum.

The tile waste was collected from a manufacturing factory in Sambalpur, Odisha. The details of properties are given in the Table 8.5.

8.2.2 Methodology

Three separate phases made up the structure of the experimental program. Firstly, different percentages of locally available sand from 10% to 50% with an increment of 10% was mixed with the expansive soil. Making use of a 2.6 kg rammer with a compactive effort of 595 kJ/m^3, light compaction test was carried out as part of the mechanical performance assessment to determine the OMC and MDD of various treated soil mixtures by taking

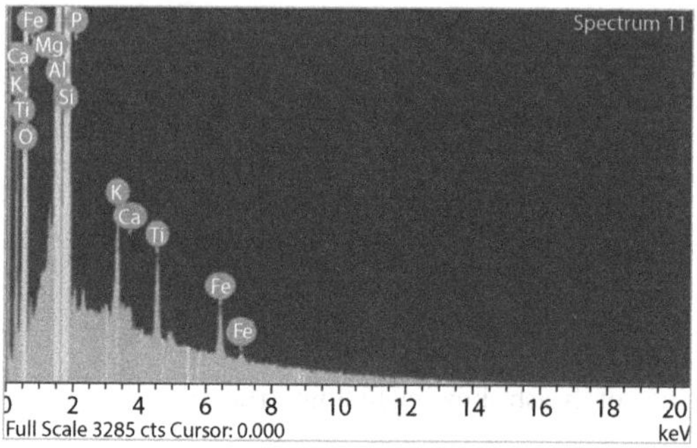

Elem...	Weight%	Atomic%
O K	57.57	70.47
Mg K	0.17	0.13
Al K	16.89	12.26
Si K	23.33	16.27
P K	0.17	0.11
K K	0.40	0.20
Ca K	0.10	0.05
Ti K	0.66	0.27
Fe K	0.71	0.25
Totals	100.00	

Figure 8.4 EDS of fly ash sample.

Table 8.5 Details of tile waste properties.

Parameter	Values
Specific gravity (G)	2.64
Cu (Coefficient of uniformity)	7.33
Cc (Coefficient of curvature)	1.07

into consideration the respective compaction curves. The established OMC was used to calculate the soaked and unsoaked CBR values. In the second stage, various proportions of fly ash (10%, 15%, 20%, and 25%) were mixed with the ideal mixture of sand and expansive soil. In the last phase, tile waste was added in 5%, 10%, 15%, 20%, 25%, and 30% percentages with the goal of figuring out the best combination of sand, fly ash, and expansive soil to improve engineering qualities. CBR features and compaction were the main topics of the investigation report. The tests were carried out in an OMC condition that resembled the characteristics of undisturbed soil for assessing the strength properties of sand, fly ash, and tile waste incorporated soil. In this study, geotechnical characteristics of expansive soil stabilized with fly ash, sand, and tile waste were assessed as the main objective. An overview of the overall strategy used to accomplish various goals and reach the final target is provided by the flowchart shown in Figure 8.5.

The objective of the study has been accomplished by undergoing a well-designed methodology, and the different parameters and materials considered for this study are listed as shown in Table 8.6.

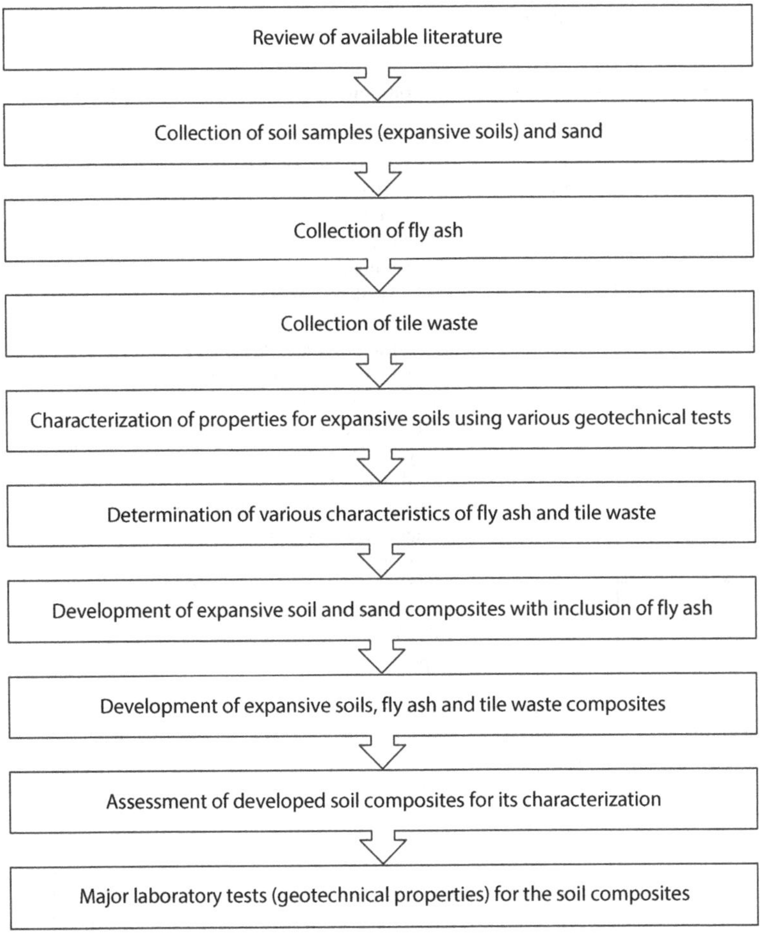

Figure 8.5 Flowchart of methodology.

8.3 Results and Discussion

For the comparative study, we have undergone a number of experimental programs with the developed composite material which is given in the Table 8.7.

8.3.1 Compaction Characteristics

This is an in-depth analysis of the experiment that blended virgin expansive soil with different proportions of sand, thereby noting out the results

Table 8.6 Parametric variation.

Laboratory investigation	Parameters	Variations	Remarks
Characterization of constituents	Expansive soil Sand Fly ash	Physical Properties: Specific Gravity, Particle Size Distribution, Free Swell Index, Atterberg's Limit Engineering Properties: Compaction, soaked and unsoaked CBR, Direct Shear Test Physical Properties: Specific Gravity, Grain Size Distribution Engineering Properties: Direct Shear Test Physical Properties: Particle size distribution, Atterberg's Limit, Specific Gravity Engineering Properties: Compaction	
	Tile waste	Physical Properties: Particle Size Distribution, Specific Gravity	
Development of Composite material	Expansive soil (Clay)+sand (S)	90% C +10%S 80% C +20%S 70% C +30%S 60% C +40%S 50% C +50% S	The proportion which will show highest MDD and OMC will be considered for Mixed proportion with fly ash and tile waste X=Clay: sand Proportion which will show highest OMC and MDD Y=Clay: sand: fly Ash proportion which will show highest OMC and MDD
	Expansive soil (Clay)+Sand+ Fly ash	X+10%FA X +20% FA X +30%FA	
	Expansive soil (C)+Sand+ Fly ash+ Tile waste	Y +5%TW Y+10%TW Y+15%TW Y+20%TW	
Geotechnical characterization		Compaction characteristics, California bearing ratio (unsoaked and soaked)	

Table 8.7 Development of composite material.

Soil used	Stabilizing agent	Mix designation	Variation of composite material
Expansive soil(C)	Sand(SA)	Sample 1	100% C
		Sample 2	90% C + 10% S
		Sample 3	80% C + 20% S
		Sample 4	70% C + 30% S
		Sample 5	60% C + 40% S
		Sample 6	50% C + 50% S
Expansive soil(C) + Sand(S)	Fly ash(FA)	Sample 7	54% C + 36% S + 10% FA
		Sample 8	48% C + 32% S + 20% FA
		Sample 9	42% C + 28% S + 30% FA
Expansive soil(C) +Sand(S) + Fly ash(FA)	Tile waste(TW)	Sample 10	51.3% C + 34.2% S + 9.5% FA + 5% TW
		Sample 11	48.6% C + 32.4% S + 9% FA + 10% TW
		Sample 12	45.9% C + 30.6% S + 8.5% FA + 15% TW
		Sample 13	43.2% C + 28.8% S + 8% FA + 20% TW

in terms of maximum dry density and optimum moisture content. The study explores the fascinating dynamics that exist between mixtures of sand and expansive soil, examining what effects that changing the sand proportions has on the composite material's physical characteristics. The correlation between dry density and moisture content for various soil-sand mix compositions is illustrated by the moisture content-dry density curves, shown in Figure 8.6. IS:2720 (Part 7)-1980 [39] was followed to conduct the proctor compaction test in the laboratory.

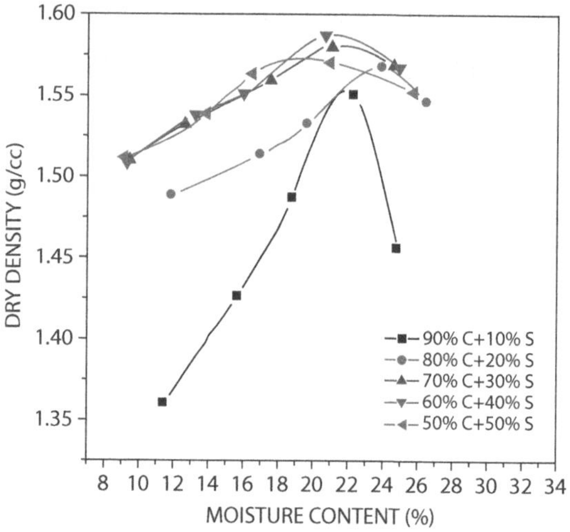

Figure 8.6 Compaction curve of different samples of clay–sand mix.

At the same time, Figure 8.7 shows how the addition of sand in different amounts alters MDD. A key finding of the research is the apparent pattern in MDD with respect to sand addition. At first, the MDD shows a steady increase as the mixture's sand content rises from 10% to 40%. This is explained by the way the sand and clay particles interact inside the

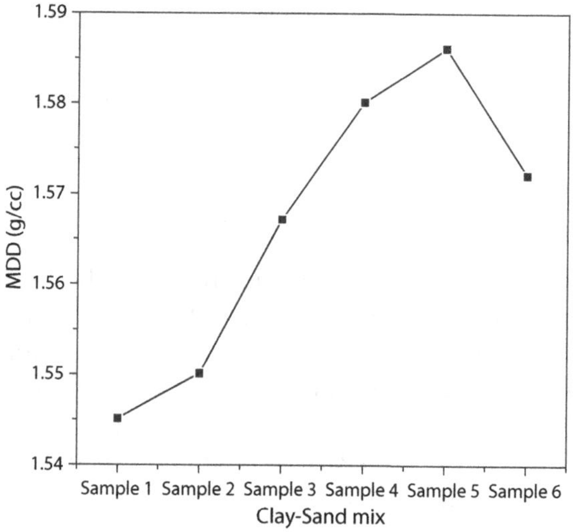

Figure 8.7 MDD of different samples of clay-sand mix.

composite. Higher MDD values are obtained at lower sand percentages because the clay particles efficiently fill the spaces left by the sand particles; hence, improving compaction. However, when the sand content rises above 40%, the trend reverses and the MDD starts to decline. This reversal is a sign of possible particle segregation because the clay particles may separate due to a higher sand content, which would result in a less effective packing arrangement and lower MDD.

Moreover, the examination of OMC fluctuation in the clay-sand composite, as illustrated in Figure 8.8, reveals a subtle pattern. In comparison to the virgin expansive soil, the overall trend shows a decrease in OMC. This is because of the substitution of cohesive clay particles by cohesion less sand particles. The cohesive quality of the soil decreases with increasing sand content because there are fewer clay minerals present. Thus, a lower OMC for the composite soil as a whole, result from the substitution of cohesive particles with cohesion less particles.

Interestingly, there is a significant anomaly in the OMC trend within the clay-sand mixture. The OMC shows a rise in sand content up to 20%, then a fall in sand content with additional inclusion. Up to 20 percent sand content, it is possible that the sand addition will cause a specific configuration or interaction between the sand and clay particles that will temporarily increase the composite's ability to retain moisture. Above this ideal ratio, though, additional sand additions may upset the advantageous configuration and cause the OMC to drop.

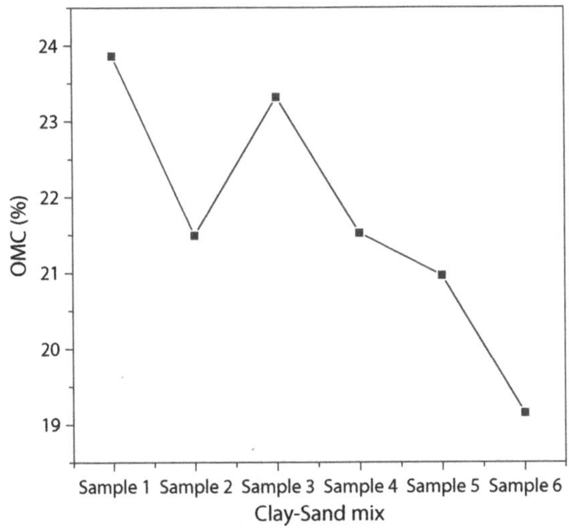

Figure 8.8 OMC of different samples of clay-sand mix.

These findings have significant ramifications, especially in the fields of geotechnical engineering and construction methods. Making educated decisions in soil stabilization, foundation design, and earthwork construction projects requires an understanding of how the addition of sand affects the compaction properties and behavior of expansive soils. By optimizing soil mixtures, engineers and construction professionals can ensure better stability and performance in a range of structural applications. The intricate relationship between sand mixtures and expansive soil is highlighted in the conclusion of this study, which also shows how various sand proportions impact the MDD and OMC. The observed trends in OMC and MDD variations provide valuable insights into the moisture characteristics and compaction behavior of these composite soils, facilitating improved decision-making for geotechnical engineering projects.

The figures, particularly the compaction curves shown in Figure 8.9, shed light on how the addition of fly ash, which can vary from 10% to 30%, impacts compaction behavior of the clay-sand composite. These results provide important new information about how the addition of fly ash modifies compaction characteristics of the composite. The compaction curves show a distinct pattern that suggests an intriguing finding that the maximum dry density of the composite mixture decreases with the addition of fly ash in comparison to the blend of clay and sand without fly ash.

This remarkable result is caused by the complex interaction between the fly ash and clay particles in the composite matrix. Specifically, when

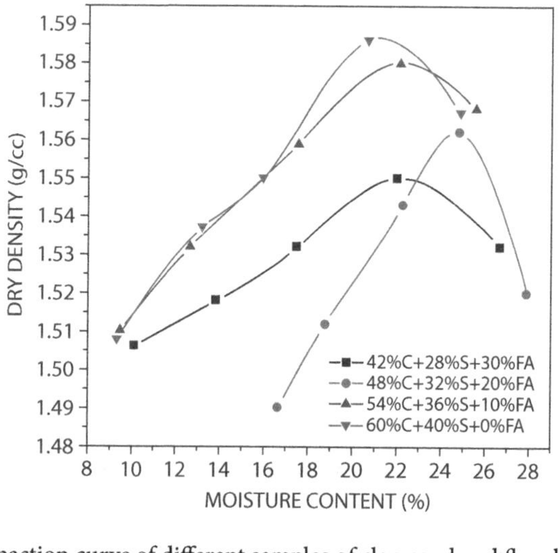

Figure 8.9 Compaction curve of different samples of clay, sand and fly ash mix.

combined in a 60:40 ratio (i.e., 60% clay to 40% sand), the clay particles exhibit a remarkable ability to absorb a significant portion of void spaces present in the sand. Consequently, the MDD of the composite material is lowered when fly ash is added to the mixture due to effective filling action by the clay particles. Furthermore, variation of OMC and MDD from Figures 8.10 and 8.11 respectively helps us to understand how the presence of fly ash affects the features of compaction of composite materials in a better way. These graphic representations unambiguously demonstrate an inverse relationship between MDD and moisture content as fly ash content increases. Increase in fly ash content causes a simultaneous rise in moisture content and fall in MDD. This fascinating finding sheds light on the inherent characteristics of the composite material at various concentrations of fly ash. Fly ash contributes to a higher OMC, which affects the compaction characteristics of the material. The overall increase in moisture content is mostly due to this OMC.

A specific blend of clay, fly ash, and sand exhibits potential in the pursuit of optimizing compaction and engineering qualities. Careful testing and analysis reveal that a blend ratio of 54:36:10 which represents the percentages of clay, sand, and fly ash respectively; is a favorable combination with the required compaction qualities. This specific ratio maximizes the unique properties that each component material contributes by establishing a balance between moisture content, MDD and effective compaction.

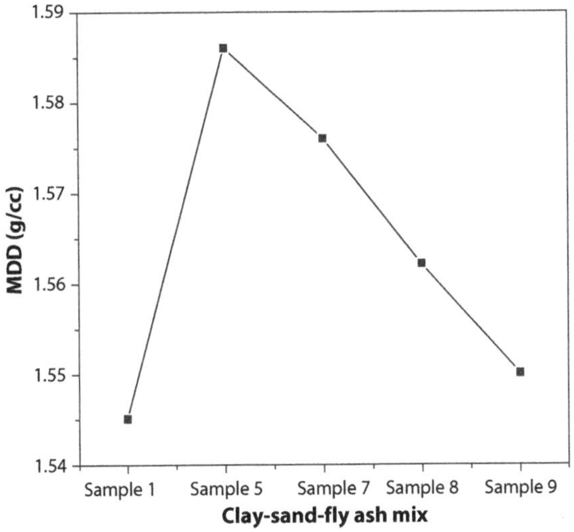

Figure 8.10 Variation of MDD in the clay and sand mixture with varying fly ash proportions.

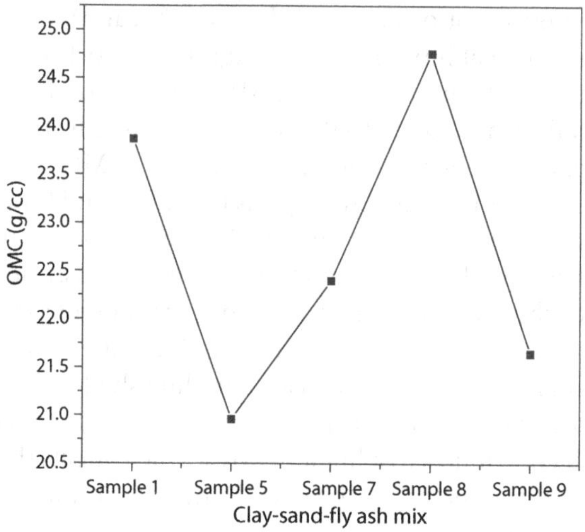

Figure 8.11 Variation of OMC in the clay and sand mixture with varying fly ash proportions.

The compaction curves in Figure 8.12 provide a thorough understanding of how adding tile waste in different proportions affects the compaction characteristics of the optimal fly ash treated expansive soil and sand. This study explains how adding tile waste to the mix impacts the maximum dry density by examining the dynamics of compaction. The pattern of the

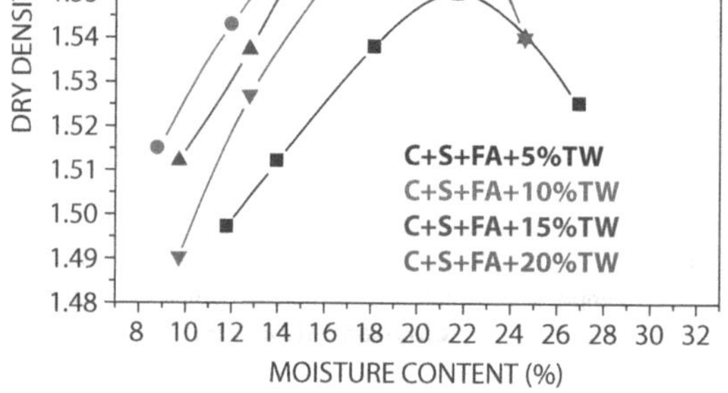

Figure 8.12 Compaction characteristics of tile waste incorporated fly ash treated soil.

curves shows that a composite blend containing tile waste performs better than a blend without tile waste in terms of maximum dry density. The intricate interactions between the various components of the composite lead to these interesting findings. More precisely, when mixed in a particular proportion of 48.6% clay, 32.4% sand, 9% fly ash, and 10% tile waste, the clay particles demonstrate an amazing ability to occupy the spaces left by the addition of crushed tile waste and sand. A larger maximum dry density is attained within the composite material as a result of this efficient filling action.

On the other hand, an intriguing finding emerges about how different tile waste contents affect the composite's maximum dry density from Figure 8.13. The MDD of the tile waste blended clay, sand and fly ash composite is found to rise initially before declining as the tile waste content progressively increases up to 10%.This fascinating behavior clarifies the intricate dynamics between the component materials. When the proportion of waste tile is still relatively low, the finer clay particles effectively occupy the voids between these sand and waste tile particles. This results in an ideal packing arrangement that raises the maximum dry density. However, a critical point is reached where the particles tend to segregate as the tile waste content rises above a given threshold. The packing arrangement is upset by this segregation, which lowers the maximum dry density.

The data taken out of Figure 8.14 shows a strong correlation that the optimum moisture content significantly decreases when tile waste is added.

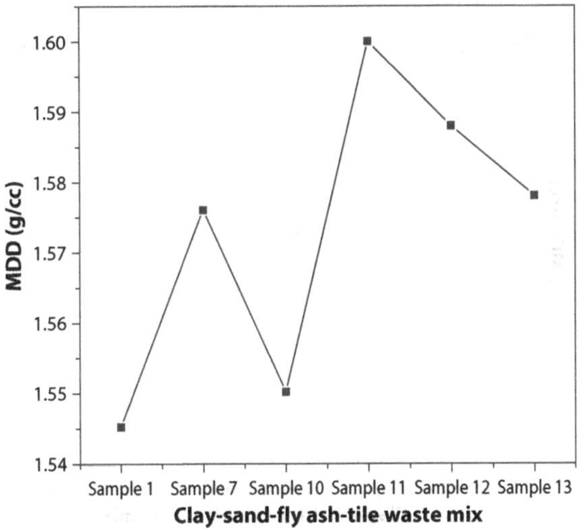

Figure 8.13 MDD of different samples of tile waste incorporated fly ash treated soil.

More specifically, there is a noticeable drop in OMC from an initial value of 23.86% to a final measurement of 19.76% as the percentage of ceramic dust increases from 0% to 20%. This interesting change in OMC can be explained by the way the soil particles in the composite mixture interact with the ceramic dust particles. This behavior is essentially explained by the replacement mechanism that takes place between the soil particles and the ceramic dust particles. Part of the soil particles is gradually replaced by these particles as the amount of ceramic dust in the mixture rises. The overall affinity or attraction of the composite material towards water molecules is subsequently decreased as a result of this substitution. In other words, the ceramic dust particles may have a diminished propensity to interact with water molecules in comparison to the soil particles they are replacing. When this happens, the composite material exhibits a reduced ability to absorb water, which causes the OMC to actually drop.

8.3.2 California Bearing Ratio

The strength and load-bearing capacity were to be assessed in laboratory-based California bearing ratio (CBR) tests conducted in accordance with IS 2720-16 (1987) [40] standard. The fact that the unsoaked CBR values were constantly higher than the soaked CBR values was an interesting test result. This pattern is consistent with the previous research conducted by Murthy and Sharma *et al.*, (1996), who also found that the addition of sand,

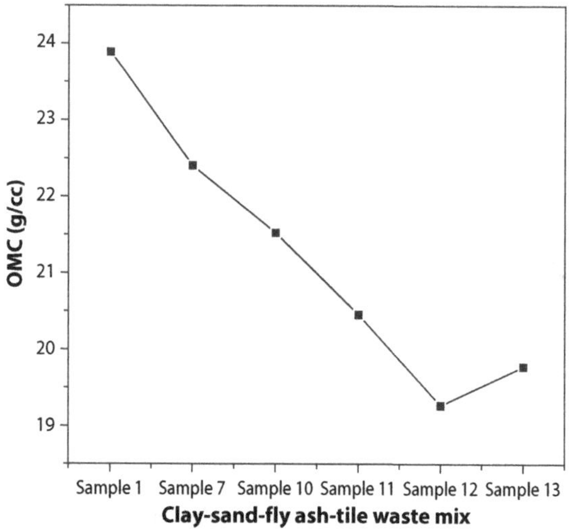

Figure 8.14 OMC of different samples of tile waste incorporated fly ash treated soil.

different soils and additives like cement and lime, raised the CBR values of fly ash. CBR values often result in more conservative design values when measured under soaked conditions. This indicates that these values are typically lower, indicating a more cautious assessment of the strength of the soil in the presence of moisture or saturation. Soaking the soil simulates actual situations where soil strength is impacted by moisture, which increases the applicability of the findings in those circumstances. However, CBR values that are obtained under "unsoaked" or "undisturbed" conditions-that is, without soaking-can also demonstrate conservatism, i.e., these tends to be higher than those under soaked conditions, but in situations where the soil stays relatively dry, they may nevertheless represent a conservative estimate of the strength of the soil. Load-penetration curves of soaked and unsoaked mix are shown in Figures 8.15 and Figure 8.16 respectively by providing a thorough comparison between the expansive soil in its virgin state and the same soil modified with different admixtures. In particular, both soaked and unsoaked CBR tests were performed on the optimized blends, such as clay: sand: fly ash (54%: 36%: 10%) and clay: sand: fly ash: tile waste (48.6%: 32.4%: 9%: 10%).

The CBR values for these optimized mixtures were significantly higher than those of pure soil, according to the results. This improvement implies that the incorporation of fly ash, sand, and tile waste may have had a beneficial impact on the compaction and particle packing of the mix, raising the CBR values. Flexible pavement sub-grade design places a strong emphasis on the importance of CBR values. In general, a CBR value greater than 5.0 is

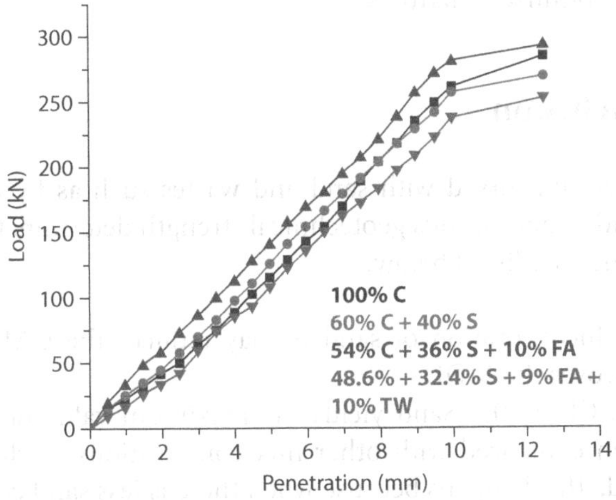

Figure 8.15 Comparison of soaked CBR of different sample mix.

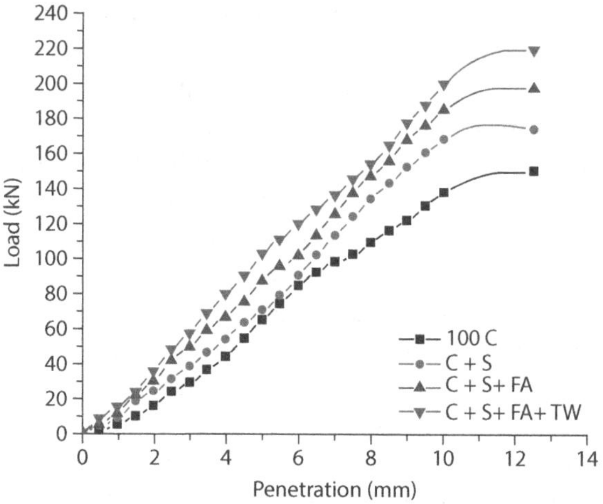

Figure 8.16 Comparison of unsoaked CBR of different sample mix.

considered adequate for flexible pavement design for traffic loads between one and ten million standard axles (msa).Thus, constructing sub-grades on roads intended for lower traffic volumes using clayey soil along with sand, fly ash and tile waste is a workable strategy. By combining these materials, the subgrade's structural integrity and stability may be improved, fulfilling the design specifications for roads with lower traffic volumes.

The variations in the soaked and unsoaked CBR values for each of the optimized mixtures are shown in Figure 8.17. This graphic depiction sheds light on how the CBR values vary with soaking and without soaking among the different optimized mixtures.

8.4 Conclusion

The expansive soil mixed with sand and wastes such as fly ash and tile waste has undergone various geotechnical strength determination and the results obtained are listed below:

1. The incorporation of sand in clay reduces the OMC and improves the MDD.
2. 60% Clay: 40% Sand yields the maximum value of MDD when compared with other mix compositions of clay and sand. This happens because when there is less sand content, the void spaces within the grains of sand are filled by clay

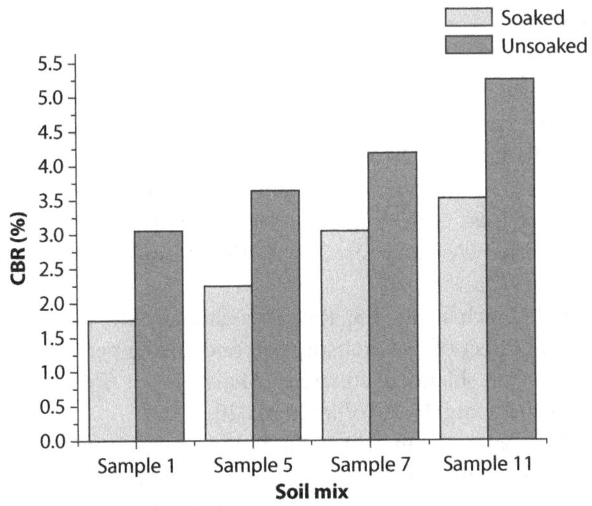

Figure 8.17 CBR for various optimum soil mixes (unsoaked and soaked).

particles, but when there is a greater percentage of sand content, the particles segregate thereby lowering the MDD.

3. The inclusion of fly ash, a material that is lighter than clay and sand, reduces the MDD of the clay-sand (60% C: 40% S) mix. This is primarily explained by the flocculated structure that is created when fly ash with a lower specific gravity is added.
4. The sand, fly ash, and tile waste mix expansive soil which attains the highest possible MDD is 48.6% C: 32.4% S: 9% FA: 10% TW, followed by other proportions. This is because, after a certain percentage, the tile particles fill up the voids in the mix, causing segregation within the mix and eventually lowering the MDD.
5. The addition of sand to expansive soil considerably raised the CBR value for both soaked and unsoaked conditions. When fly ash and tile waste are added to the soil mixture, a similar pattern is observed in the CBR values.
6. The higher CBR value indicates increased load-carrying capacity, thereby improving soil stability, which makes this material a viable and reliable option for building roads in areas with low traffic volumes.

References

1. Singh, P., Dash, H.M., Samantaray, S., Effect of silica fume on engineering properties of expansive soil. *Mater. Today Proc.*, 33, 5035–5040, 2020, https://doi.org/10.1016/j.matpr.2020.02.839.
2. Kumar, M.H., Mohanta, N., Samantaray, S., Kumar, N.M., Combined effect of waste glass powder and recycled steel fibers on mechanical behavior of concrete. *SN Appl. Sci.*, 3, 1–18, 2021a, https://doi.org/10.1007/s42452-021-04422-w.
3. Kumar, M.H., Macharyulu, I.S., Ray, T., Mohanta, N.R., Jain, M., Samantaray, S., Sahoo, A., Effect of water absorption and curing period on strength and porosity of triple blended concrete. *Mater. Today Proc.*, *43*, 2162–2169, 2021b, https://doi.org/10.1016/j.matpr.2020.12.092.
4. Kumar, M.H., Saikrishnamacharyulu, I., Kumar, U., Jena, S., Mohanta, N.R., Samantaray, S., Coupling effect of fly ash, metakaoline and different types of steel fibers on mechanical performance of concrete. *AIP Conf. Proc.*, 2417, 1, 2021c, https://doi.org/10.1063/5.0072573.
5. Ray, T., Mohanta, N.R., Kumar, M.H., Samantaray, S., Study of effect of temperature on behavior of alkali activated slag concrete. *Mater. Today Proc.*, *43*, 1352–1357, 2021, https://doi.org/10.1016/j.matpr.2020.09.169.
6. Kumar, M.H., Saikrishnamacharyulu, I., Mohanta, N.R., Ashutosh, A., Mishra, P., Samantaray, S., Mechanical behaviour of high strength concrete modified with triple blend of fly ash, silica fume and steel fibers. *Mater. Today Proc.*, *65*, 933–942, 2022a, https://doi.org/10.1016/j.matpr.2022.03.528.
7. Kumar, M.H., Mohanta, N.R., Patel, N., Samantaray, S., Reddy, S.V.B., Influence of Fly ash and Metakaoline on the crack resistance and shrinkage property of concrete. *Iran. J. Sci. Technol. Trans. Civ. Eng.*, 46, 2011–2026, 2022b, https://doi.org/10.1007/s40996-021-00748-y.
8. Mohanta, N.R. and Samantaray, S., Study of Combined Effect of Metakaolin and Steel Fiber on Mechanical Properties of Concrete. *Pertanika J. Sci. Technol.*, 27, 3, 1381–1396, 2019, http://journals-jd.upm.edu.my/resources/files/Pertanika%20PAPERS/JST%.
9. Joel, M. and Agbede, I.O., Mechanical-Cement Stabilization of Laterite for Use as Flexible Pavement Material. *J. Mater. Civ. Eng.*, 23, 86–152, 2011.
10. Wheeler, S.J., Sharma, R.S., Buisson, M.S.R., Coupling of hydraulic hysteresis and stress–strain behaviour in unsaturated soils. *Géotechnique*, *53*, 1, 41–54, 2003.
11. Nelson, J. and Miller, D.J., *Expansive soils: problems and practice in foundation and pavement engineering*, John Wiley & Sons, USA, 1997.
12. Sabtan, A.A., Geotechnical properties of expansive clay shale in Tabuk, Saudi Arabia. *J. Asian Earth Sci.*, *25*, 5, 747–757, 2005.
13. Yadav, J.S. and Tiwari, S.K., Evaluation of the strength characteristics of cement-stabilized clay–crumb rubber mixtures for its sustainable use in geotechnical applications. *Environ. Dev. Sustain.*, *20*, 1961–1985, 2018.

14. Chen, J.A. and Uloko, J.O., Comparative Study of Effect of Waste Ceramic and Marble Dusts on Geotechnical Properties of Expansive Soil. *Electron. J. Geotech. Eng.*, *22*, 5073–5086, 2017.
15. Sharma, R.K., Utilization of fly ash and waste ceramic in improving characteristics of clayey soil: a laboratory study. *Geotech. Geol. Eng.*, *38*, 5, 5327–5340, 2020.
16. Pourakbar, S., Asadi, A., Huat, B.B., Fasihnikoutalab, M.H., Stabilization of clayey soil using ultrafine palm oil fuel ash (POFA) and cement. *Transp. Geotech.*, *3*, 24–35, 2015.
17. Al-Baidhani, A.F. and Al-Taie, A.J., Recycled crushed ceramic rubble for improving highly expansive soil. *Transp. Infrastruct. Geotechnol.*, *7*, 3, 426–444, 2020.
18. Al-Bared, M.A.M., Harahap, I.S.H., Marto, A., Sustainable strength improvement of soft clay stabilized with two sizes of recycled additive. *GEOMATE J.*, *15*, 51, 39–46, 2018.
19. Samantasinghar, S. and Singh, S.P., Strength and durability of granular soil stabilized with FA-GGBS geopolymer. *J. Mater. Civ. Eng.*, *33*, 6, 06021003, 2021.
20. Harichane, K., Ghrici, M., Kenai, S., Grine, K., Use of natural pozzolana and lime for stabilization of cohesive soils. *Geotech. Geol. Eng.*, *29*, 759–769, 2011.
21. Louafi, B. and Bahar, R., SAND: An Additive for Stabilzation of Swelling Clay Soils. *Int. J. Geosci.*, 03, 719–725, 2012.
22. Roy, T.K., Influence of Sand on Strength Characteristics of Cohesive Soil for Using as Subgrade of Road. *Procedia Soc. Behav. Sci.*, 104, 218–224, 2013.
23. Samantasinghar, S., Singh, S.P., Palsule, P., Synthesis and Characterization of Geo-polymer Grout Material. *Indian Geotechnical Conference 2017 GeoNEst*, IIT Guwahati, India, 2017.
24. Senthamarai, R.M. and Manoharan, P.D., Concrete with ceramic waste aggregate. *Cem. Concr. Compos.*, *27*, 9-10, 910–913, 2005.
25. Cabalar, A.F., Hassan, D.I., Abdulnafaa, M.D., Use of waste ceramic tiles for road pavement subgrade. *Road Mater. Pavement Des.*, *18*, 4, 882–896, 2017.
26. Scheetz, B.E. and Earle, R., Utilization of fly ash. *Curr. Opin. Solid State Mater. Sci.*, *3*, 5, 510–520, 1998.
27. Pandian, N.S., Fly ash characterization with reference to geotechnical applications. *J. Indian Inst. Sci.*, 84, 189–216, 2004.
28. Phanikumar, B.R. and Sharma, R.S., Effect of Fly Ash on Engineering Properties of Expansive Soils. *J. Geotech. Geoenviron. Eng.*, 130, 764–767, 2004.
29. Prabakar, J., Dendorkar, N., Morchhale, R.K., Influence of fly ash on strength behavior of typical soils. *Constr. Build. Mater.*, 18, 263–267, 2004.
30. Edil, T.B., Acosta, H.A., Benson, C.H., Stabilizing Soft Fine-Grained Soils with Fly Ash. *J. Mater. Civ. Eng.*, 18, 283–294, 2006.
31. Bose, B., Geo-engineering properties of expansive soil stabilized with fly ash. *Electron. J. Geotech. Eng.*, 17 J, 1339–1353, 2012.

32. Sabat, A.K., Stabilization of expansive soil using waste ceramic dust. *Electron. J. Geotech. Eng.*, 17 Z, 3915–3926, 2012.
33. Ameta, N.K., Wayal, A.S., Hiranandani, P., Stabilization of dune sand with ceramic tile waste as admixture. *Am. J. Eng. Res.*, 2, 9, 133–139, 2013.
34. Takhelmayum, G. and S.A.L.K.G., Soil Stabilization By Fly Ash. *Int. J. Eng. Sci. Innov. Technol.*, 2, 1, 1–6, January 2013.
35. Singh, B. and Sharma, R.K., Evaluation of geotechnical properties of local clayey soil blended with waste materials. *Jordan J. Civ. Eng.*, 8, 135–151, 2008.
36. Onakunle, O., Omole, D.O., Ogbiye, A.S., Stabilization of lateritic soil from Agbara Nigeria with ceramic waste dust. *Cogent Eng.*, 6, 1, 1710087, 2019.
37. Soğancı, A.S., Yenginar, Y., Orman, A., Geotechnical Properties of Clayey Soils Stabilized with Marble Dust and Granulated Blast Furnace Slag. *KSCE J. Civ. Eng.*, 27, 4622–4634, 2023.
38. IS:2720 (Part 4)-1985, Indian Standard, Methods of Test for Soils, Part 4: Grain Size Analysis, Bur. Indian Stand, New Delhi, India, vol. Reaffirmed, 2006, pp. 1–38, 1985.
39. IS:2720 (Part 7)-1980, Method of test for soils, Part 7: Determination of water content- dry density relation using light compaction, Bur. Indian Stand., vol. Reaffirmed, 2011, pp. 1–10, 1980.
40. IS:2720 (Part 16)-1987, Method of test for soils, Part 16: Laboratory determination of CBR, Bur. Indian Stand., vol. Reaffirmed, 2002, pp. 1–10, 1987.

9

Investigation of the Strength Parameters for Red Soil with Coir and Pet Fiber Reinforcement

Nandita Das[1], Chhabirani Tudu[1]*, Biswajit Majhi[1] and Prateek Kumar Dhir[2]

[1]School of Infrastructure and Planning, Odisha University of Technology and Research, Bhubaneswar, India
[2]University of Naples Federico II, Durham University, England, UK, Italy

Abstract

The topmost surface of the earth or the soil is the most exploited material in civil engineering field. Whenever it is utilized in construction as a foundation or pavement, it provides enough strength for structure's stability. Some construction in buildings or pavement is very difficult in red soil, therefore, soil stabilization method is adopted to increase the engineering properties of soil, shear strength of soil, and bearing capacity of the soil. By increasing the strength of soil, construction will be strong and it will be very effective for ages. There are numerous ways to stabilize the soil; the methods are cement stabilization, lime stabilization, bitumen stabilization, and the method using electrochemical, thermal action, and artificial freezing. These methods, however, can have a detrimental effect on the environment. Researchers have found many effective and low-cost materials for improving shear strength and bearing capacity of soil. In this direction, a study has been conducted to examine the values of California bearing ratio (CBR) and modified proctor of soil stabilized with polyethylene terephthalate (PET) bottles and coir fibers (CF). Initially, PET fiber and coir fiber were individually blended with virgin soil at varying percentages by weight of soil; and then, a modified proctor test and CBR test were done to find the optimum percentages to be mixed to the soil that increases the engineering property of soil. Also, coir fiber is reinforced into the soil to know how much strength is increased; and then, coir fiber at its optimum percentage and PET bottle strips are mixed with virgin soil at different percentages

**Corresponding author*: chhabiranice@outr.ac.in

Sandeep Samantaray, Subhashree Samantasinghar, Soumendra K. Mohanty, Vivek, Abinash Sahoo and Deba Prakash Satapathy (eds.) Waste Utilization in Geotechnical Practice, (239–260) © 2025 Scrivener Publishing LLC

with the dry weight of soil, so that we can know the influence of strength on virgin soil with coir fiber individually or when two materials are mixed.

Keywords: Polyethylene terephthalate (PET), coir fibers, modified proctor compaction test, shear strength, CBR

9.1 Introduction

Soil stabilization has the most important role to make the foundation strong. Enhancing the soil property is the primary goal of soil stabilization which is necessary in the situation where the local soil is poor and construction of buildings or pavement is extremely challenging [1]. The primary purpose of stabilization is to decrease the permeability, compressibility and settlement which increases the bearing capacity and shearing strength of soil. Many buildings are constructed but they are not effective for a long time. The foundation of the building should be strong enough to endure external forces and ensure structural integrity. Several methods are considered to improve the quality of soil but it depends upon the type of soil, the availability of materials, its cost etc. So, the soil stabilizing technique is adopted for convenience. The stabilizing method improves the geotechnical and important engineering properties of soil so that the soil can be replaced by better quality of soil which is feasible for construction. There are many methods of stabilizing the soil for enhancing bearing capacity of soil, such as lime, cement, sawdust, stone dust, tire crumble, marble dust, etc. have been used and a mixture of two additives like plastic with tire crumble, brick kiln dust with coir fiber, flash with marble dust, etc. are used in increasing the geotechnical properties of soil. Red soil has been stabilized; it was locally available, and in many parts of Odisha, it contains red soil. Stabilizing red soil will be very effective for building construction, embankments, pavement, etc. Hence, in many places the soil quality can be improved. There are many methods of stabilization. Nowadays, the stabilization of soil by new approaches has been experimented and it is showing good results.

Many different types of waste material are leading to environmental pollution [2–7]. Of which, one of the environmentally polluted wastes is polyethylene terephthalate. India has a high population, therefore, in recent decades roads and buildings have also been built over less stable ground. In India, usage of plastic bottles is increasing day by day and plastic waste is also rapidly increasing. As plastic is non-biodegradable, it cannot be disposed of properly. One of the modern, widely utilized, affordable and efficient method of stabilizing soil is by the use of ET bottles. Many researchers have investigated that soil stabilization is an adequate method for enhancing geotechnical and index properties of soil. Several studies have been conducted on PET (polyethylene terephthalate) waste plastic bottles as an admixture mixed with soil. They found that the strength parameter of soil significantly changes when plastic waste is used as stabilizing agent [8]. It has been demonstrated that adding PET waste plastic bottles to soil increases it strength and bearing capacity. Again another waste material, i.e., coir fibers is also being examined by researchers. An increase in CBR was noticed when CF was added to the soil [9].

On the other hand, there are not enough experimental studies using coir fiber and PET plastic bottles in red soil even though these two wastes are easily available. However, some researchers have experimented individually on the effect of these wastes and they got good results. This paper mainly emphasizes the study of several properties, i.e., the maximum dry density (MDD), CBR, and optimum moisture content (OMC) of soil with plastic PET waste bottle and coir (coconut) fiber; and also, individual experiments are done only with coir fiber to know the optimum dosages of coir fiber to be added to the soil with PET bottle so that the strength is increased.

9.2 Background of the Study

9.2.1 Polyethylene Terephthalate

They [10], conducted a series of triaxial test using various percentages of plastic strip (0.5 to1.5) % by dry weight of soil. The result shows that (25% to 125%) improved, and the shear strength of soil increased by 0.5 % of plastic strip addition to soil, and the ultimate load carrying capacity of soil is improved. Finally, he concluded that the increase in PET strips enhances the shear strength and bearing capacity of soil. They [11], used a new technique of soil stabilization by using coir fiber and waste PET bottle. They used these materials to enhance the safe bearing capacity of soil to increase the density and CBR of soil and additionally to look for another way to dispose of plastic waste. After the comparison according to the test result, it was discovered that by increasing plastic (PET) bottle strip into the soil, it improves the property of the soil, CBR strength is increased but coir fiber stabilized the soil gives lesser strength as compared to plastic strips.

They [12], took expansive soil, which was locally available, and treated them with brick waste powder at different percentages of 20%, 30%, and 40% plastic waste PET bottle strips with (0.25, 0.5, and 0.75) % by dry weight of soil. They conclude that by mixing the soil with brick powder and plastic PET bottle strip, a noteworthy change in compaction properties, and California bearing ratio was observed. Moreover, adding more plastic strip to the soil improves its subgrade strength, and, and by adding brick powder content contributes to slight increases in MDD and OMC (Amena 2021). They [13], utilized plastic waste material with lime as subgrade material in the pavement which can be eco-friendly, cost-effective, and environmentally friendly. Their observation suggests that plastic waste pollution can be decreased by stabilization of black cotton soil utilizing lime and plastic PET bottle strip, which has a significant impact on plasticity and strength parameters of soil.

9.2.2 Coconut Coir Fiber

They [14], studied the combined effect of coconut fiber with pond ash and stone dust as a stabilizing material. The proportion of fiber (0.5 to 1.5%) is taken. After the experimental laboratory test, the stone dust strength is slightly increased with coir fiber as compared to the stone dust with pond ash. The CBR value is increased when the stone dust is added to silty soil. Hence, the strength parameter of the coir fiber and stone dust soil sample shows better results than that of the coir fiber and pond ash sample. They [15], studied the effect of coir fiber reinforcement in clayey soil on different properties. They observed that the addition of coir fiber resulted in a decrease in MDD and more water absorption capacity enhancing optimum water content. Also, the CBR value was reported to be enhanced with the addition of coir fiber. They [16], concluded that coir fiber is a beneficial biodegradable waste that increases engineering properties of expansive soil. They also suggested a rigorous experimental evaluation before its application in the field. They [17], evaluated the effect of the coir fiber (0.5%, 1%, and 1.5%) along with lime (5%) in the stabilized soil. They reported an increase in CBR and shear strength in modified soil and also suggested an optimum combined dosage of 1% coir fiber and 5% lime. They, undertook an experimental study on the use of coir and PET fiber for soil stabilization. According to their observation, the maximum MDD and optimum unconfined compressive strength (UCS) were found for 0.75 % of coir fiber.

Literature reveals that various tests have been conducted on expansive and clay soil which was mixed with different additives, including fly ash, lime, stone dust, marble dust, cement, and others. However, a combined effect of coir fiber and PET waste as a stabilizing agent has not been explored yet. Observing these gaps, several laboratory experiments using red soil stabilized with PET bottle strips and coir fiber with varying content levels are needed to be investigated. The main purpose of the project is that by adding plastic PET bottle strips and locally available coir fiber to the nearby available red soils, it intensifies the engineering properties of red soil. In this project, shear strength, MDD, CBR values, and UCS are calculated by executing different types of tests in virgin soil. Also, these tests are done by adding plastic PET bottle strips that are cut into different sizes, and natural coir fibers are cut into pieces to a particular length, and then they are mixed with the undisturbed soil. A series of geotechnical laboratory tests is carried out on both virgin soil and soil mixed with plastic PET bottles and natural coir fiber to evaluate soil improvement and increase strength and bearing capacity of soil.

9.3 Materials

9.3.1 Red Soil

Red soil is very common and it is available largely in India. It is locally available to stabilize red soil; different methods of stabilization are used. Many researchers utilized various types of admixtures with red soil; they are either chemicals or organic additives. For this current work, the red soil was collected from the nearby institute of Gastro and Kidney Care Hospital, Kalinga Nagar, Bhubaneswar, 751003 (20°16'24.4"N 85°45'54.7" E). The soil at the site can be seen in Figure 9.1a. The soil was dug out by machinery tools up to a depth of 2.5m from the earth's surface. After collecting the soil sample first, it is oven-dried for 24 hours at (105 to 110°c), and then wet sieving is conducted on the soil sample.

9.3.2 Plastic Waste Pet Bottles

The plastic bottles used in this study were waste PET beverage bottles in fiber form to stabilize the red soil. The empty PET waste water bottles were collected from the nearby area of Bhubaneswar. The average weights of the transparent bottles were 18 gm. First, the bottles were washed properly, and then they were cut manually into strips of (15mm x 3mm) sizes, as shown in Figure 9.1b.

9.3.3 Coir Fiber

The naturally occurring coconut coir fiber was used for this study. It is a fibrous material from coconut husk and was collected from a nearby fruit shop in Bhubaneswar. The coir fiber was air-dried for 48 hours to remove moisture from it. Figure 9.1c shows the sample of coir fiber. The coir fiber used in this paper has a diameter of mm. The coir fiber, (70mm-90mm) long is taken from the literature review. The coir fiber is cut into this length using scissors and scale and utilized as an additive material for stabilizing the virgin soil. The main advantage of using coir fiber is that it is

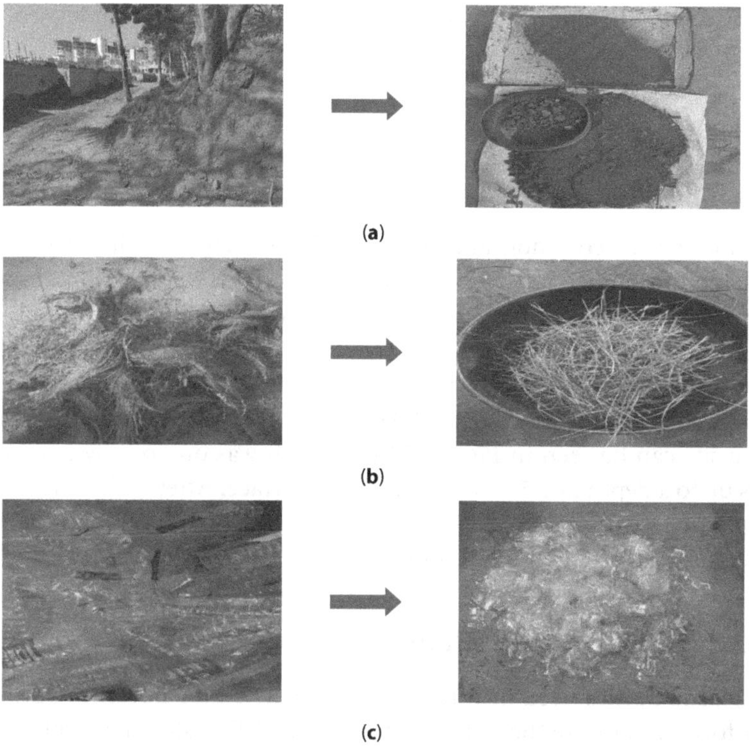

Figure 9.1 (a) Red soil sample at site and laboratory. (b) Coconut husk to coir fiber form. (c) PET bottle to strip form.

eco-friendly. improves the strength of soil subgrade. and is low-cost, available in nearby markets, and eco-friendly. Fiber mixed with soil has been used in many countries in the recent past.

9.4 Methodology

The sample of red soil was gathered from the not so far distance. From a depth of (2 to 2.5 m) from ground surface and the sample was collected. The chosen additives namely, plastic PET bottles and coir fiber were also collected locally. An experimental study was done to analyze the different properties of collected red soil sample stabilized with PET plastic strips and coir fibers with varying proportions of dry weight of soil. To accomplish this, the following process has been adopted.

1. The *in-situ* and engineering property of soil sample were experimentally evaluated.
2. Initially, coir fiber was cut into (70mm-90mm) long and added to soil with different percentages of (0.1%, 0.2%, 0.3%, 0.4%, 0.5%, 0.6%); the optimum dosage of coir fiber was assessed by performing various geotechnical test on soil mixed with coir fiber. Then, different tests were done to know the engineering properties of soil with the collected soil sample. After conducting all these tests, it will give the safe bearing capacity of soil.

Table 9.1 List of experiments conducted.

Tests conducted	**IS code**
Grain size distribution	IS 2720-Part 4 (1985) [18]
Specific gravity	IS:2720-Part 3 (1980) [19]
Liquid limit	IS :2720-Part 5 (1985) [20]
Plastic limit	IS :2720-Part 5 (1985) [20]
Light compaction test	IS: 2720-Part 7 (1983) [21]
Direct shear test	IS: 2720-Part 13 (1986) [22]
CBR	IS: 2720-Part 16 (1985) [23]

3. The PET waste plastic bottles were cut into the size of (15mm x 3mm) manually and these plastic strips were added to the soil sample with optimal coir fiber content in varying percentages of 0.1%, 0.2%, 0.3%, 0.4%, 0.5%, 0.6 %, 0.7%, and 0.8%. Subsequently, the geotechnical properties (OMC, MDD, CBR, DST) of stabilized soil were evaluated to obtain the optimum dosages of plastic strips.
4. The list of experiments conducted in laboratory with IS code is shown in Table 9.1.

9.5 Experimental Program

9.5.1 Grain Size Distribution

Grain size distribution curve of red soil has been determined by conducting a sieve analysis test in accordance with IS 2720-Part 4 (1985) [18].

9.5.2 Specific Gravity Test

Specific gravity test has been conducted according to IS:2720-Part 3 (1980) [19]. The standard range of specific gravity lies between (2.60 to 2.80) for red soil.

9.5.3 Plastic Limit (PL) Test

This test was done based on IS :2720-Part 5 (1985) [20].

9.5.4 Liquid Limit (LL) Test

This test was done based on IS :2720-Part 5 (1985) [20].

9.5.5 Standard Protector Test

Through the standard proctor test, the OMC and MDD can be calculated. The standard proctor test was conducted on red soil mixed with varying proportions of coir fibers and PET strips according to the specifications of IS: 2720-Part 7 (1983) [21].

9.5.6 Direct Shear Test

The cohesion and angle of internal friction of soil and modified soils are determined by conducting direct shear test following IS: 2720-Part 13 (1986) [22].

9.5.7 CBR Test

To determine the relationship between force and penetration, CBR experiments were performed in the laboratory. CBR test was done on red soil and mixed with varying proportions of coir fibers and PET strips according to the specifications of IS: 2720-Part 16 (1985) [23].

9.6 Results and Discussion

9.6.1 Properties of Red Soil

The geotechnical and index properties of collected red soil locally were investigated experimentally. The results are shown in Table 9.2.

Table 9.2 Index and geotechnical properties of the soil sample.

Description	Parameters	Values
Soil classification	Grain size distribution	Cu=4.2, Cc=1.15
In-situ properties	Specific gravity	2.68
Atterberg limits	LL (%) PL (%) Plasticity index (%)	25.605 14.18 11.425
Compaction characteristics	MDD (g/cm^2) OMC (%)	2.121 9.05
CBR test	CBR (2.5mm penetration)	12.463
	CBR (5mm penetration)	15.69
Soil classification (USCS)		ML (Silty soil with low LL)
UCS test		119.507

9.6.2 Standard Proctor Test

9.6.2.1 *Impact of Coir Fiber on OMC and MDD of Stabilized Soil*

The outcome of common standard proctor compaction test is discussed here. For both natural soil and soil blended with different coir percentages, the OMC begins to fall and the MDD rises when compared to soil without admixture. When the admixture of varying percentages is mixed with red soil, OMC starts decreasing up to 0.3% by dry weight of soil and MDD rises to 0.3% of dry weight of soil. The OMC of virgin soil is 16.92% and the MDD is 1.721(g/cc). The variation in a sample of OMC compacted

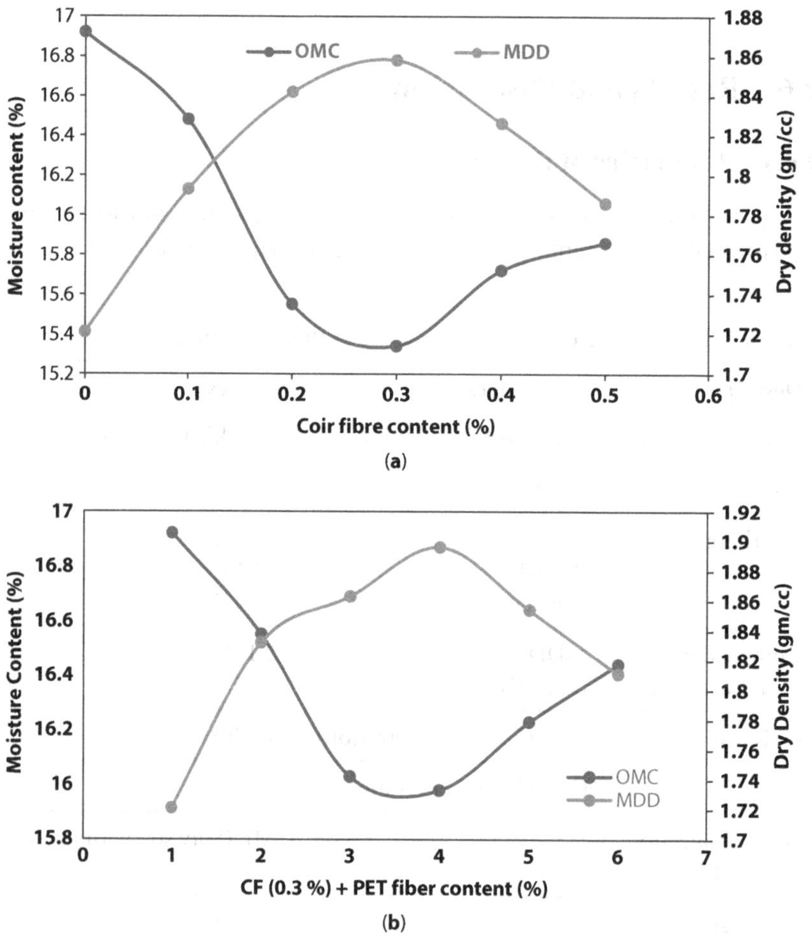

Figure 9.2 (a) OMC and dry density with varying content of coir fibers. (b) OMC and dry density with varying content of PET fibers with constant coir fiber content of 0.3%.

with different percentages of coir fiber shows that OMC decreases as fiber content increases. Figure 9.2a shows the effect of coir fibers of different percentages in MC and DD. The OMC values decreased. Figure 9.2b shows the impact of varying content of PET fibers in MC and DD. OMC values were seen to have decreased; however, the MDD value rose with the rise in coir fiber percentage.

9.6.2.2 *Impact of Plastic (PET) Strips on OMC and MDD*

The optimum coir fiber content was 0.3% in terms of OMC. Keeping the coir fiber constant at 0.3% by weight of virgin soil, further PET strips were added from 0.2%, 0.4%, 0.6%, 0.8%, and 1% to the soil. Stabilization of soil was done with coir fiber and PET strips at different percentages and for each percentage of admixture, different OMC and MDD values were obtained, which are shown in Table 9.3. As the percentage of coir fiber increases the OMC starts decreasing and MDD increases as compared to virgin soil. The OMC starts decreasing up to (15.98%) at 0.6% PET fibers and 0.3% coir fibers.

Figure 9.3 illustrates the relationship between OMC and MDD of soil samples stabilized with coir fibers. This gives us a clear idea of how the coir content from 0.1% up to 0.5% behaves in terms of OMC and MDD. In the same manner, the relationship between OMC and MDD of the soil sample stabilized with PET fibers with optimum coir fibers in it which is 0.3% is shown. Furthermore, Figure 9.4 shows the relation between OMC and MDD by keeping the optimum 0.3% of coir fiber and increasing the dosages of plastic fibers from 0.2% to 1% of plastic fibers. The combination

Table 9.3 Illustration of MDD and OMC of stabilized soil sample.

Coir fiber %	OMC	MDD	PET fibers + coir fibers (PF+CF) %	OMC	MDD
0	16.92	1.721	0+0	16.92	1.721
0.1	16.48	1.793	0.2+0.3	16.55	1.832
0.2	15.55	1.842	0.4+0.3	16.03	1.863
0.3	15.34	1.858	0.6+0.3	15.98	1.896
0.4	15.72	1.826	0.8+0.3	16.23	1.854
0.5	15.86	1.786	1+0.3	16.44	1.811

of 0.6% plastic fibers and 0.3% coir fibers was found to have a MDD of 1.896 gm/cm^3.

9.6.3 Direct Shear Test (DST)

In this section, experimental results of a DST on both unreinforced soil and soil that had been blended with various admixture percentages are discussed. In comparison to virgin soil, the cohesiveness (c) and angle of internal friction (ϕ) increase as the percentage of coir fiber increases up to 0.3%, after which the values steadily drop.

9.6.3.1 Cohesion and Angle of Internal Friction

Direct shear test procedure was performed by IS 2720-part 13-1986, and the results are shown in Table 9.4. In comparison to virgin soil, the angle of internal friction (ϕ) and cohesiveness increase as the percentage of coir fiber increases up to 0.3%, after which the values steadily drop. At 0.3% coir fibers concentration, the DST value increases up to 0.6337 for 0.5 kg, 1.076 for 1 kg, and 1.4694 for 1.5 kg; further increase in coir fiber percentage increases, and the DST value gradually falls. 0.3% of coir fiber is therefore chosen as the optimum percentage. The ϕ and cohesion (*c*), show that all values of blended soil samples are greater than the values of natural soil samples by 0.2434, 0.2676, 0.2891 for cohesion at 0.1%, 0.2%, and 0.3%, respectively: and angle of internal friction is increased to 32.3, 35.8, 38.2 at

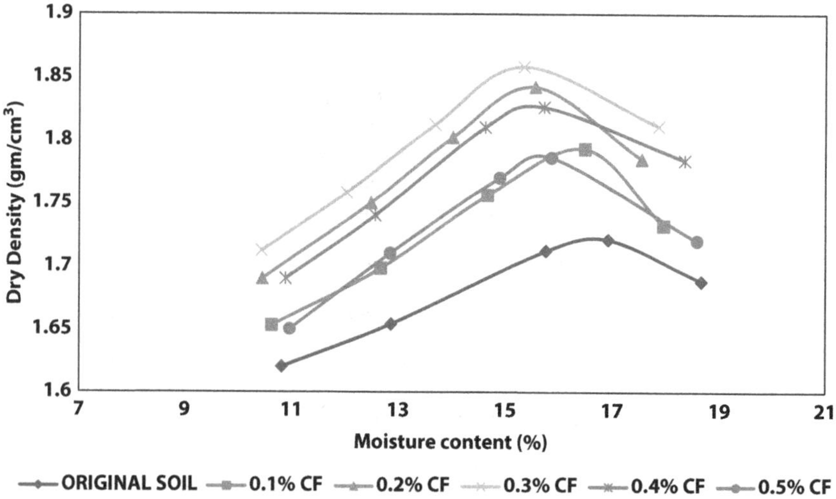

Figure 9.3 Relation between MDD and OMC of virgin soil stabilized with coir fibers.

0.1%, 0.2%, 0.3% when compared to natural soil and after that, by increasing the percentage of CF fiber, it gradually decreases. Figure 9.5 and Figure 9.6 show the relation between ϕ and *c* of stabilized soil with different content of only coir fiber and with mixed fiber of PET and coir respectively.

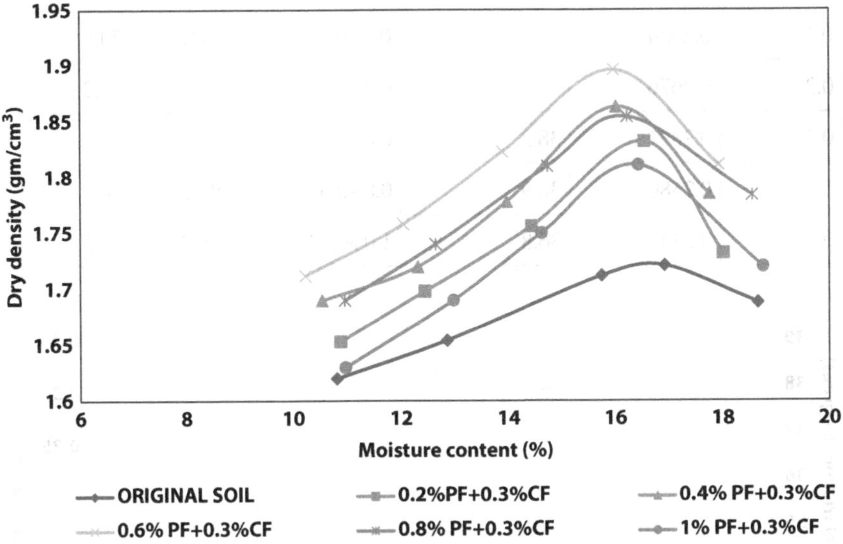

Figure 9.4 Relation between OMC and MDD of soil stabilized with coir fibers (0.3%) and PET fibers.

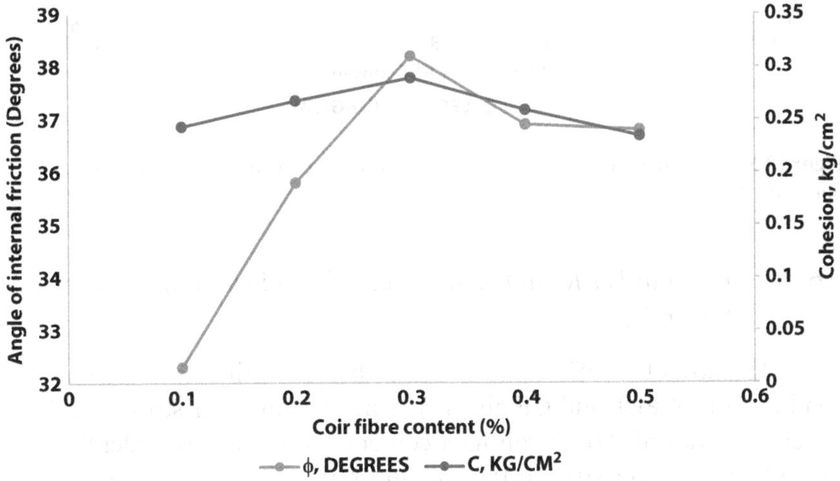

Figure 9.5 Relation between angle of internal friction (ϕ) and cohesion (c) for various percentage of coir fibers.

Table 9.4 C and ϕ values for various admixtures at different percentage.

Coir fiber content in %	Cohesion(C) in kg /cm²	Angle of internal friction (Ø) (Degree)	PET content (%) + CF (0.3)%	Cohesion (C) in kg /cm²	Angle of internal friction (Ø)
0.1	0.2434	32.3	0.2+0.3	0.2691	34.5
0.2	0.2676	35.8	0.4+0.3	0.3021	37.2
0.3	0.2891	38.2	0.6+0.3	0.3211	38.5
0.4	0.2580	36.9	0.8+0.3	0.2923	36.2
0.5	0.2343	36.8	1.0+0.3	0.2613	32.3

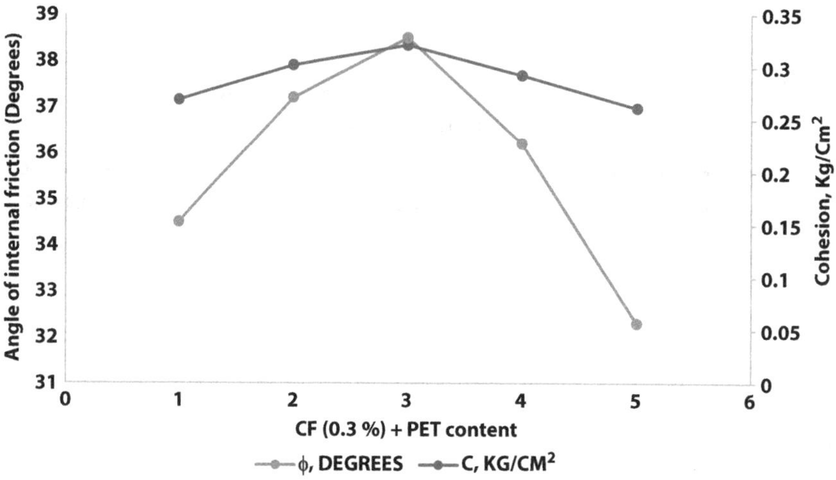

Figure 9.6 Relation between cohesion and ϕ for various percentages of PET and coir fibers (0.3 %).

9.6.3.2 *Effect of Pet Reinforcement and Coir Fiber on the Direct Shear Test*

Table 9.5 shows the DST value of soil stabilized with coir fibers and the combination of PET and CF fibers. The maximum shear stress of 1.4694 Kg/cm² is obtained at 0.3% coir fiber content in the soil. It is evidently clear that adding the coir strip to the soil and mixing it can increase the soil shear strength. The strength of the sample without the strip is lower than the sample with the strip. It is also evident that the DST value is enhanced by the use of recycled plastic waste and coir fiber. The combination of 0.6% PET fibers and 0.3% coir fiber gives the optimum shear stress of 1.5142 Kg/cm^2.

A linear relation can be seen between the shear stress and the normal stress values for the soil stabilized with coir fiber only from 0% to 0.5% content, which is shown in Figure 9.7. The relation between both normal and shear stress of soil reinforced with various percentage of PET (0.2 % to 1 %) and coir fibers (0.3 %) are presented in Figure 9.8.

Table 9.5 Maximum values for various admixtures at different percentage.

Coir fiber content (%)	Maximum shear stress (Kg/cm^2)	PET + coir fiber content (%)	Maximum shear stress (Kg/cm^2)
0.1 %	1.191	0.2+0.3	1.2999
0.2 %	1.3494	0.4+0.3	1.406
0.3 %	1.4694	0.6+0.3	1.5142
0.4 %	1.3845	0.8+0.3	1.39
0..5 %	1.3563	1.0+0.3	1.2094

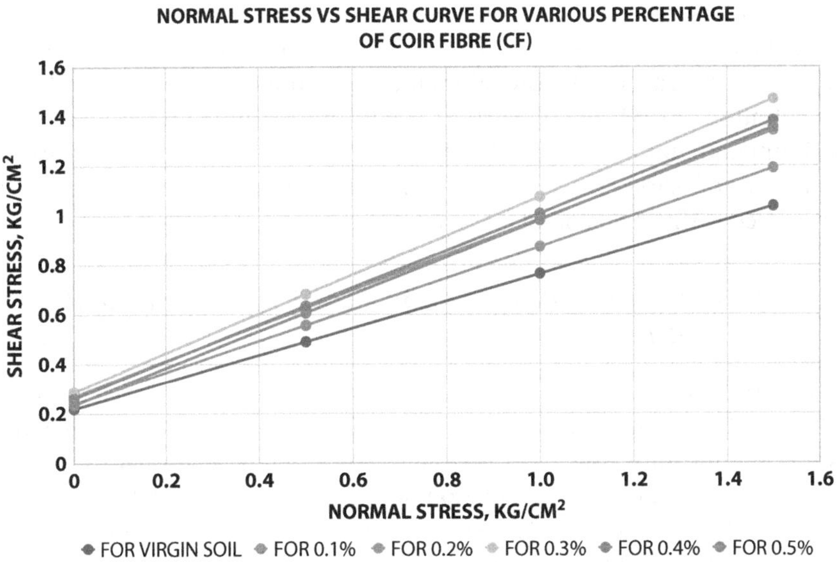

Figure 9.7 Relation among normal and shear stress of coir fiber reinforced soil.

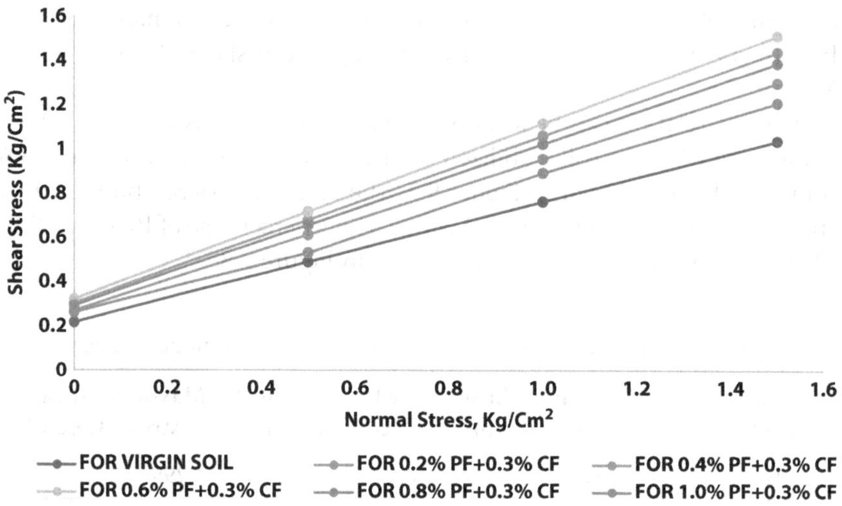

Figure 9.8 Relation among normal stress and shear stress of reinforced soil with various percentage of PET and coir fibers (0.3 %).

9.6.4 California Bearing Ratio Test-Disturbed Soil

9.6.4.1 Effect of CF on Unsoaked CBR Test

The California bearing ratios in unsoaked conditions for coir stabilized soil were evaluated and the load versus penetration curve was plotted in Figure 9.9. It is observed that for coir fiber at different contents, the unsoaked CBR values increase from 7.21 to 11.42 at 2.5mm and then decrease to 10.78 %. The unsoaked CBR values increase from 6.90 to 12.21% at 5mm penetration at 0.3% of CF. Further, the California bearing ratio values at 2.5 mm and 5mm are reduced to 10.78 % and 10.02 % with an increase of coir fiber of 0.5%. Hence, the optimum percentage for which strength is increased is the inclusion of 0.3% of CF.

9.6.4.2 Effect of PET Reinforcement and CF Mix on Unsoaked CBR Test

It is also evident that the CBR value is boosted by the use of recycled plastic PET bottle. When a soil sample is used with coir and PET fiber, the CBR value significantly rises. This is evident from Figure 9.10, that the maximum CBR value of the combination of 0.6% PET and 0.3% CF is found to be 14.89 % at 2.5 mm and 12.89 % at 5 mm. A similar result is observed in increasing the CBR values when coir fiber is added to the soil as reported by Kaushik and Singh (2021) [24].

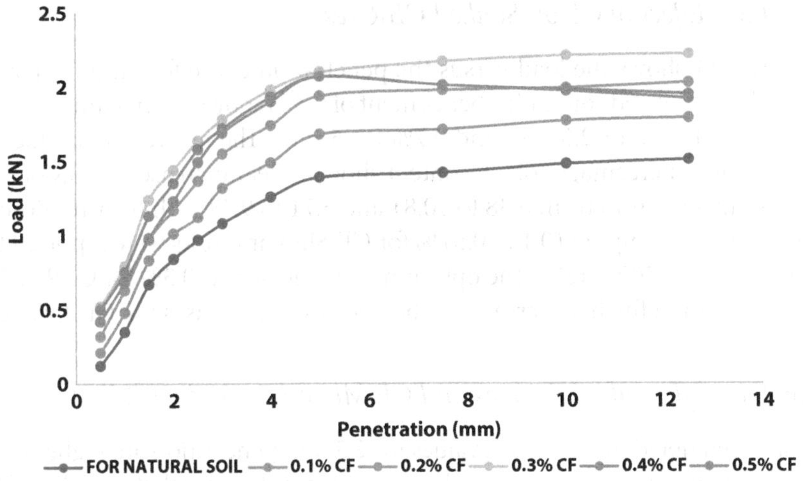

Figure 9.9 Load-Penetration relationship of soil stabilized with various (%) of coir fiber stabilized for unsoaked condition.

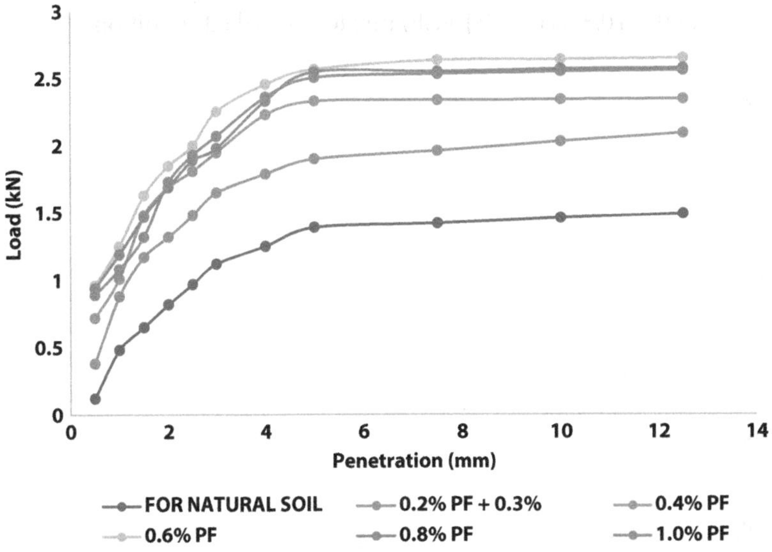

Figure 9.10 Load-Penetration relationship of soil stabilized with various (%) of coir fiber for unsoaked condition.

9.6.4.3 *Effect of CF on Soaked CBR Test*

Figure 9.11 shows the load versus the penetration curve for virgin soil and coir fiber stabilized soil. Coir fiber content of 0.3 % shows the maximum CBR value of 10.81% for 2.5 mm and 9.7% for 5 mm. The test result conducted with various percentages of CF content shows a clear increase in CBR of soil at 2.5mm and 5mm from 6.38 to 10.81 and 5.71 to 9.78 with fiber reinforcement within a range of (0.1 to 0.5) % for CF. Similar results were reported by Sridhar *et al.*, (2023) [25]. The optimum percentage was 0.3% for CBR at 2.5 mm and 5mm; further increase in fiber, however, decreases the CBR value.

9.6.4.4 *Effect of PET Strips and CF Mix on Soaked CBR Test*

It has been noted that the CBR values for 2.5 mm penetration are higher than those for 5 mm penetration. From Figure 9.12, it is observed that for (PET+CF) content at different percentages, the CBR values increase. The California bearing ratio (CBR) at 2.5 mm is increased from 6.38 to 15.23% and the CBR at 5mm is increased from 5.71 to 13.24% at (0.6+0.3)% of PET and CF mix. Further, the California bearing ratio (CBR)at 5mm and 2.5mm is reduced to 12.77% and 14.41% by increasing admixture mix i.e., 1% PET and 0.3% CF. Hence, the optimum percentage for which the strength is increased is the inclusion of (0.6+0.3)% of (PET+CF) mix for unsoaked conditions.

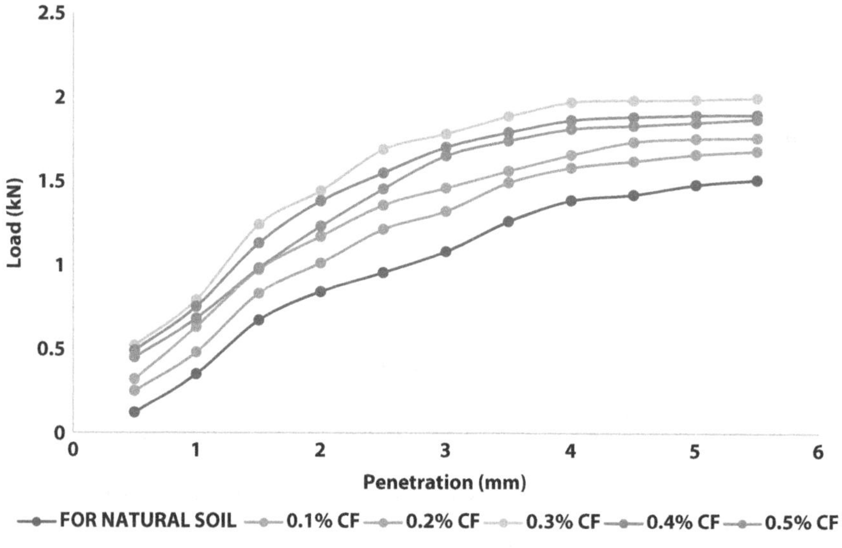

Figure 9.11 Load-Penetration relationship of soil stabilized with various percentages of coir fibers for soaked condition.

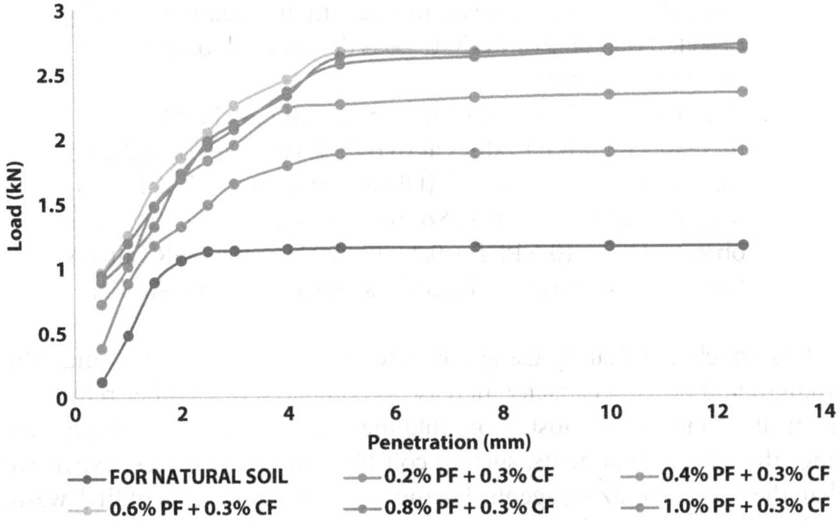

Figure 9.12 Load-Penetration relationship of soil stabilized with different percentage of coir fibers for soaked condition.

9.7 Conclusion

In this study, the combined effect of both plastic PET bottle strips and coir fibers was studied on the stabilized red soil. Different experiments were conducted on the red soil stabilized individually only with coir fiber and by combining the PET strips and 0.3% of coir fiber. The following are the conclusions:

1. The introduction of coir fiber to the red soil causes a decrement in optimum moisture content of 15.55% at 0.3% and an increase of MDD 1.721 to 1.826 (g/cc) at 0.3% fiber content. After that, there is a slight increase of OMC to 15.72% and decrease of MDD values to 1.826(g/cc).
2. After the inclusion of coir fiber, the DST value and California bearing ratio value increased. However, up to the addition of 0.3% fiber content, the DST value and CBR value increased, and with further addition of fiber content, there was a decline in strength characteristics. The CBR value is increased from 7.21 to 11.43 and the DST value is increased from 0.4904 to 0.6825 at 0.5kg, 0.7641 to 1.076 at 1 kg, and 1.0378 to 1.4694 at 1.5 kg., i.e., at 0.3% fiber content. Further increase of fiber

content, causes a decrease in strength and bearing capacity of soil. Based on this study, 0.3% is determined to be the optimum fiber content.

3. The use of PET and CF fiber mix reduces the OMC and increases the MDD values at (0.6 PET +0.3 CF) %, and after that, the OMC decreases. MDD increases at (PET+CF) content at (0.6+0.3) % mix. So, the optimum percentage was observed to be (0.6 PET + 0.3 CF) %. Here, the CBR value is higher as compared to the soil stabilized only by coir fiber.

It is concluded that by using coir fiber and plastic PET strip mix, the strength and bearing capacity increase as compared to CF fiber reinforcement. It is one of the most affordable methods of stabilizing clayey soil, since the basic components, such as coir fiber and PET, are less expensive than the other stabilizing agents. Furthermore, the utilization of PET waste and agro-waste in the form of coir fiber as soil stabilizing agent can reduce the mismanagement of this waste. This stabilizing technique can be considered for minimizing waste disposal to the environment. The coir fiber length can be increased and many tests can be conducted to improve the properties of soil. The PET plastic bottle strip can increase the strength by increasing strip size and length. In this paper, plastic strips are placed randomly but the plastic can be laid vertically and horizontally. In this paper, tri axial test is not done; this text can be done by increasing the plastic strip size to increase more strength of the soil, and different tests can be done to study the effect of strip size and length on soil.

References

1. Mohanta, N.R. and Samantaray, S., Study of Combined Effect of Metakaolin and Steel Fiber on Mechanical Properties of Concrete. *Pertanika J. Sci. Technol.*, 27, 3, 1381–1396, 2019, http://journals-jd.upm.edu.my/resources/files/ Pertanika%20PAPERS/JST%.
2. Singh, P., Dash, H.M., Samantaray, S., Effect of silica fume on engineering properties of expansive soil. *Mater. Today Proc.*, 33, 5035–5040, 2020, https://doi.org/10.1016/j.matpr.2020.02.839.
3. Kumar, M.H., Mohanta, N., Samantaray, S., Kumar, N.M., Combined effect of waste glass powder and recycled steel fibers on mechanical behavior of concrete. *SN Appl. Sci.*, 3, 1–18, 2021a, https://doi.org/10.1007/s42452-021-04422-w.
4. Kumar, M.H., Macharyulu, I.S., Ray, T., Mohanta, N.R., Jain, M., Samantaray, S., Sahoo, A., Effect of water absorption and curing period on strength and

porosity of triple blended concrete. *Mater. Today Proc., 43*, 2162–2169, 2021b, https://doi.org/10.1016/j.matpr.2020.12.092.

5. Ray, T., Mohanta, N.R., Kumar, M.H., Samantaray, S., Study of effect of temperature on behavior of alkali activated slag concrete. *Mater. Today Proc., 43*, 1352–1357, 2021, https://doi.org/10.1016/j.matpr.2020.09.169.
6. Kumar, M.H., Saikrishnamacharyulu, I., Mohanta, N.R., Ashutosh, A., Mishra, P., Samantaray, S., Mechanical behaviour of high strength concrete modified with triple blend of fly ash, silica fume and steel fibers. *Mater. Today Proc., 65*, 933–942, 2022a, https://doi.org/10.1016/j.matpr.2022.03.528.
7. Kumar, M.H., Mohanta, N.R., Patel, N., Samantaray, S., Reddy, S.V.B., Influence of Fly ash and Metakaoline on the crack resistance and shrinkage property of concrete. *Iran. J. Sci. Technol. Trans. Civ. Eng.*, 46, 2011–2026, 2022b, https://doi.org/10.1007/s40996-021-00748-y.
8. Acharyya, R., Lahiri, A., Mukherjee, S.P., Raghu, P.V., Acharyya, R., Lahiri, A., Mukherjee, S.P., Raghu, P.V., Improvement of undrained shear strength of clayey soil with PET bottle strips. *Proceedings of Indian Geotechnical Conference*, December, 22-24, 2013.
9. Kodicherla, S.P.K. and Nandyala, D.K., Influence of randomly mixed coir fibres and fly ash on stabilization of clayey subgrade. *Int. J. Geo-Eng., 10*, 1, 3, 2019.
10. Karmacharya, R. and Acharya, I.P., Reinforcement of soil using recycled polyethylene terephthalate (PET) bottle strips, in: *Proceedings of IOE Graduate Conference*, vol. 5, pp. 153–156, December 2017.
11. Vinoth, M., Ayyappan, A., Keerthana, P., Pandiselvi, D., Dharshini, K.P., Radhika, S., Experimental Investigation on Comparison of Soil Stabilization by Using Waste Plastic and Coirfibre. *Int. J. Emerg. Technol. Eng. Res. (IJETER), 6*, 3, 156–161, 2018.
12. Amena, S. and Chakeri, D., A study on the effects of plastic waste strips and lime on strength characteristics of expansive soil. *Adv. Civ. Eng.*, 2022, 1–6, 2022.
13. Amena, S., Experimental study on the effect of plastic waste strips and waste brick powder on strength parameters of expansive soils. *Heliyon, 7*, 11, e08278, 2021.
14. Nitish, S.S.S., Sangita, D., Ramya, A.V.S.L., Kumar, G.S., Comparative Study on Soil Stabilization using Industrial by Products and Coconut Coir. *J. Phys. Conf. Ser.*, 2040, 1, 012014, October 2021.
15. Hubballi, R.M. and Rahman, S.K., Soil improvement using coir fiber: a case study. *Technol. (IJARET), 12*, 2, 593–598, 2021.
16. Qureshi, N., Jain, S.I., Garg, C., A Review on Stabilization of Soil Using Coconut Coir Fibre, Lime and PSC. *Int. J. Recent Adv. Multi. Top., 3*, 5, 88–90, 2022.
17. Boobalan, S.C. and Devi, M.S., Investigational study on the influence of lime and coir fiber in the stabilization of expansive soil. *Mater. Today Proc., 60*, 311–314, 2022.
18. IS: 2720 (Part-4): 1985, Methods of test for soils: Grain size analysis, Bureau of Indian Standards, 1985.

19. IS: 2720 (Part-3): 1980, Methods of test for soils: Determination of specific gravity, Section 1: Fine grained soils, Bureau of Indian Standards, 1980.
20. IS: 2720 (Part-5): 1985, Methods of test for soils: Determination of liquid and plastic limit, Bureau of Indian Standards, 1985.
21. IS: 2720 (Part-7):1980, Determination of water content-dry density relation using light compaction, Bureau of Indian Standards, 1980.
22. IS: 2720 (Part-13): 1986 Methods of test for soils, Direct Shear Test.
23. IS: 2720 (Part-16): 1965, Methods of test for soils, Determination of California Bearing Ratio, Bureau of Indian Standards, 1965.
24. Kaushik, D. and Singh, S.K., Use of coir fiber and analysis of geotechnical properties of soil. *Mater. Today Proc.*, *47*, 4418–4422, 2021.
25. Sridhar, R., Guruprasad, H.C., Naveenkumar, D.T., Sangeetha, D.M., Influence of treated coir fiber on durability properties of black cotton soil. *Mater. Today Proc.*, *80*, 1611–1616, 2023.

10

Study on Geotechnical Properties of Red Soil by Stabilizing with LDPE Plastic

Rosalin Dalai[1]*, Debasish Kar[1], Abinash Sahoo[1], Pramodini Sahu[1] and Haran Pragalath[2]

[1]*Department of Civil Engineering, Odisha University of Technology and Research, Bhubaneswar, Odisha, India*

[2]*Haran Pragalath, British Applied College, Umm Al Quwain, UAE*

Abstract

Global technological progress is leading to a rise in the usage of plastics, including polyethylene bottles and bags. Due to the fact that the majority of trash made of plastic are non-biodegradable and must not be burned since they release hazardous gasses, disposing them presents a significant difficulty. Red soil is one of the largest soil groups in India covering an area of around10.6%.Red soil needs to be stabilized for minimizing the possibility of settlement, cracks, and other damage over time, and to guarantee a strong foundation for buildings/embankments. Every year, hundreds of tons of LDPE are produced in India; this plastic is difficult to dispose of and bad for the environment. This study aims to show how low-density polyethylene strips (LDPE) obtained from waste plastic can be used as soil reinforcement to enhance the red soil's engineering performance. LDPE strips made from leftover plastics were mixed with the soil in a horizontal layer-by-layer fashion. In reinforced soil, a sequence of California Bearing Ratio (CBR) tests was conducted at six different percentages of total soil weight 0%, 0.25%, 0.5%, 0.75%, 1%, and 1.25% and size (0.5 cm x 0.5 cm). The results of CBR testing showed that adding 0.75% waste LDPE strips to the soil increased its strength and bearing capacity. Thus, building roads and embankments can benefit from the suggested process.

***Keywords*:** Stabilization, red soil, LDPE plastic strips, horizontal layer-by-layer mixed method, CBR value

**Corresponding author*: rosalin.iitk@gmail.com

Sandeep Samantaray, Subhashree Samantasinghar, Soumendra K. Mohanty, Vivek, Abinash Sahoo and Deba Prakash Satapathy (eds.) Waste Utilization in Geotechnical Practice, (261–280) © 2025 Scrivener Publishing LLC

10.1 Introduction

Environmental contamination is a big worry for the entire world nowadays, with traditional plastic being a major contributing component. However, the desire for and use of industrially produced products made of plastic compounds, rapid population expansion, cultural shifts in consumption patterns, are all contributing to an exponential increase in the amount of non-biodegradable garbage being produced. Conventional plastic is posing a significant problem for the world because it is not biodegradable and cannot be disposed of by burning because it releases toxic gasses into the atmosphere [1]. Therefore, it's critical to create new technologies and applications for making good use of this plastic trash to lessen the negative effects of the garbage on the geo-environment. One of the most significant factors revealed in a report is that consumers' careless disposal of plastic waste materials choked sewers, which contributed to the 2005 Mumbai city flood. This is a perfect illustration of the problems that non-biodegradable plastic waste causes in our daily lives. Therefore, it is necessary to appropriately dispose of these waste products. One relevant, practical, and economical option to recycle these materials is to use them as reinforcing materials to strengthen poor soils. These materials can be utilized for a range of civil engineering applications, such as for base stabilization, sub-base course of pavements, reinforcement for earthen embankment, and reducing soil settlement in foundation.

To achive sustainability various agencies around the globe have emphasized on utilization of different solid waste material in construction industry [2–5]. The growing rate of plastic waste which is non biodegradable raises concern on waste management as plastic pollutes the soil, ocean, river, ground water, emits harmful gas, litters landscape; overall, damaging the ecosystem. Table 10.1 compares India's per capita plastic use to that of the rest of the globe.

Table 10.1 Plastic waste consumption (*ICPE, 2007*).

Description	India	World
Plastic in solid waste (%)	9	7
Per capta per year plastic consumption (kg)	12-16	24-28
Recycling (%)	60	25

According to Central Pollution Control Board (CPCB) report for 2017–18 fiscal year, India annually produces 9.4 million tons of plastic waste, of which 3.8 million tons are left uncontrolled or littered, and 5.6 million tons are recycled. Almost similar situation has been found in other countries also. Hence, effective management of plastic waste is the need of the hour. Plastic can be classified in different ways, but most commonly classified according to their physical properties, chemical sources and source of waste. According to physical properties plastics are classified as thermoplastic and thermosetting. Details of these materials are tabulated in Table 10.2. According to the source of waste plastic generation, plastics are classified into different types presented in Table 10.3.

However, only plastic that complies with the recommendations for pavement construction in India is LDPE, HDPE and PET (IRC: SP:98-2013). Many researchers reported that performance properties such as drain down, marshal stability, indirect tensile strength, moisture induced resistance, rutting and fatigue properties of bituminous mixes used in various layers of pavement, i.e., base layer (DBM) and surface layer (BC) enhanced by utilizing waste plastics in the form of LDPE, HDPE and PET fiber. Also mixing process such as dry mixing and wet mixing process influence the property of bituminous mixes [6–8]. Similarly, the sub grade soil strength can be increased by utilizing various waste fibers which act as a reinforcement to the soil. Overall, the utilization of waste fibers for soil stabilization offers a sustainable, cost effective and environmentally friendly solution for improving soil properties by increasing its geotechnical properties like stability cohesion, load bearing capacity and reduced

Table 10.2 Typical thermoplastic and thermosetting, (IRC: SP:98-2013).

Thermoplastic	Thermosetting
HDPE	Alkyd
LDPE	Ultra – Formaldehyde
Poly Vinyl Acetate (PVA)	Polyurethane
Polystyrene (PS)	Melamine
Polypropylene (PP)	Polyester
Poly Vinyl Chloride (PVC)	Epoxy
Polyethylene Terephthalate (PET)	Bakelite

Table 10.3 Waste plastic and its source, (IRC: SP:98-2013).

Waste plastic	Origin
Poly Vinyl Chloride (PVC)	Credit card, toys, Mineral water bottles, medical disposal, pens, pipes, electrical fittings and, furniture folders
PP	microwave trays, wrappers of detergent & biscuits and bottle cap
HDPE	household articles, carry bags, bottle caps
PS	protective packaging, disposable cups, Bottle caps, food tray and egg packs
PET	Drinking water bottles
LDPE	detergent and cosmetic bottles, carry bags and milk pouches

Table 10.4 List of waste fibers used in soil stabilization.

Paper fibers	Waste paper fibers obtained from newspapers, cardboard and office package can be shredded and mixed with soil to improve its stability and structure.
Textile fiber	Waste textile fiber such as cotton, wool, or synthetic fibers from discarded clothing or manufacturing scraps, can be used to reinforce soil and prevent erosion.
Natural fiber	Natural fibers such as coconut fiber, jute fiber and sisal fiber is used for soil stabilization due to their high strength and bio degradability.
Agricultural residues	Residues from agricultural crops, such as rice husks, sugarcane bagasse, or corn stalks, can be repurposed as fibers for soil stabilization, providing a sustainable and cost-effective solution.
Recycled non-biodegradable waste	Waste plastic, polypropylene fibers are obtained from recycled plastic packaging.

erosion and settlement. The various waste fiber used in soil stabilization are given in Table 10.4.

Researchers looked into and reported enhanced geotechnical properties of subgrade soil altered with various types of recycled plastic waste (HDPE, PP, LDPE and PET) in the form (pellets, flakes, and ground) such as improved compaction characteristics, CBR, unconfined shear and compressive strengths, resilient modulus and permeability [9, 10].

On the other hand, stabilization is a common process to increase strength, stiffness and durability of soil. Stabilization with commonly used stabilizer (such as cement and lime) offer advantages such as providing high strength and stability for soil, improving resistance to water, chemicals and weathering as well as allowing for faster construction. However, production of cement is costly and availability of lime is rare [11, 12]. An alternative technique to enhance stability and subgrade of earthen embankments is soil stabilization utilizing raw waste plastic bottle strips. Using this inventive soil stabilizing method, society's problems may be successfully addressed and the amount of plastic waste can be decreased, making a safe and friendly environment. PET, PVC, HDPE, LDPE, PS, and PP are among the common plastic wastes. In this work, engineering properties of soil are enhanced through the use of PET plastic bottle strips.

Therefore, given its abundance, it makes sense to investigate wide application of waste plastic especially LDPE, in stabilization of soil.

10.1.1 Background

Many researchers reported that plastic strips, plastic flakes or fibers obtained from HDPE or PET bottles have improved the geotechnical property of soil when used as stabilizer [13, 14]. They [15], performed CBR test on clay sand stabilized with fiber obtained from plastic waste and scrap tire and concluded that the optimum percentage of 0.5-1.5% of plastic waste increased the bearing capacity of soil. However, soil with plastic showed higher CBR value than soil with scrap tire fiber because of poor dispersion of fiber in soil matrix. According to them [16], the geotechnical property of clay soil was significantly improved in terms of CBR test, UCS test, and resilient modulus (Mr) test by adding 1-2% of polyethylene (PE) and PP as fiber measuring 1 cm and 2 cm long. Additionally, they reported that plastic fiber decreased the optimum moisture content (OMC) and maximum dry density (MDD) of stabilized soils. Many researchers examined plastic as stabilizer for sandy soil and found that reinforced sandy soil showed better geotechnical property than unstabilized soil. They [17], investigated the friction properties between sand particles and LDPE, sand particles and sand particles, and LDPE and LDPE, ascribed strength fluctuation and stiffness decline of sand mixed with LDPE to microscopic observations of the specimens. In comparison to unreinforced specimens, they observed that the strength of sand mixed with LDPE improved by 40% while the shear stiffness of sand mixed with LDPE reduced by 25–35% for LDPE contents ranging from 1.5% to 2.0%. Thus, 1.5% to 2.0% LDPE content was recommended as the optimum dose for stabilization. They (2016) [18], evaluated the properties of dune sand with 0%, 0.05%, 0.06%, 0.7%, 0.8%, 0.9% and 1% LDPE stripes of size 5mm×5mm and reported that LDPE reinforced sand showed an increase in strength and decrease in settlement than unreinforced sand. They [19], performed laboratory tests on uniform fine sand with plastic fiber content (up to 0.9 wt.%), cement content (from 0 to 7 wt.%), fiber length (up to 36 mm), and concluded that fiber reinforced cement stabilized sand showed better results when subjected to splitting tensile test, saturated drained triaxial compression test, and UCS test. He [1], found that the engineering properties of silty sand improved with the addition of optimum percentage of 0.4% plastic content with 15 mm × 15 mm strip size. They, compared the properties of black cotton soil by stabilizing with plastic sheets at different depth, i.e., H, H/2 and H/3 and found that soil stabilized at H/2 depth showed higher bearing capacity when compared to other depth. They [20], stabilized black cotton soils in proportions of 0.5%, 1% and 1.5% by mass of soil with various aspect ratios (8×8mm, 8×16mm, and 8×24mm). They found that adding plastic strip significantly reduced the swelling potential index, and increased other properties like compressive strength and CBR value. They [21], also obtained similar conclusion on black cotton soil. Many researchers evaluated the geotechnical properties of various soil like clayey soil and black cotton soil stabilized with various industrial solid waste like fly ash, red mud, waste marble dust and rice husk ash further stabilized with plastic strips and reported that plastic reinforced added with industrial solid waste showed better results [22–25].

Numerous studies have found that the type, proportion, and size of fiber used to reinforce the soil affect CBR value regardless of type of soil [26–31]. They [32], discovered that when 1%, 2%, and 3% of polypropylene fiber were added to red soil as a stabilizer, the addition of polypropylene fiber increased the attributes of red soil, including the UCS and CBR value. They proposed that 3% of polypropylene fiber is the optimum dose to stabilize soil. They [33], conducted a number of tests on red soil before and after stabilizing with 2, 4, 6, 8, and 10% cement by weight of dry soil for varying curing times. These tests included durability, water absorption, ultrasonic pulse velocity (UPV), and UCS tests, in addition to pH and electrical conductivity (EC) tests, energy-dispersive X-ray analysis (EDX), scanning electron microscopy (SEM) imaging, and X-Ray diffraction (XRD). They discovered that at an ideal dosage of 6% cement, adding cement to red soil that is hematite-rich encourages the production of ilavite (C-F-S-H) and calcium ferrite hydroxide (C-F-H), which in turn increases thecompressive strength and stability of red soil.They also reported that the formation of Fe-containing nanostructures and the presence of hematite, helped red soil specimen improve compressive strength which is 17 times that of untreated soil in seven days. They [34], concluded that 15% of hypo sludge with molasses and cement as additives gives optimum stabilization to red soil in terms of unconfined compression test. The red soil's strength characteristics and compaction were studied [35], at various dosages of terra-zyme, a bio-enzyme used as a soil stabilizer. When compared to untreated soil, red soil treated with terra-zyme exhibits a notable improvement in strength. It is also said to be less expensive than using conventional soil stabilizers like cement, fly ash, wood ash, and lime. They considered [36], F1 ionic soil stabilizer to have considerably enhanced water sensitivity, compaction properties, and shear strength of red clay at an ideal percentage of 0.5 L/m^3. Also reported was that the stabilized red clay showed increase in plastic limit (PL), OMC and MDD. However, liquid limit (LL) and plasticity index (PI) reduced. According to reports, thermal stabilization, i.e., the process of exposing soil to high temperatures as a result of particle fusion improves the index, engineering, and swelling qualities of red and black cotton soil. The majority of treated soil outcomes are comparable to those of soils stabilized by chemical or mechanical stabilization [37, 38]. From the review of literature, it has been concluded that very few studies have been done on red soil, particularly with plastic waste, where LDPE plastic strips by varying percentage, were mixed horizontally layer by layer rather than randomly following the mixed method.

The main objectives of this research work are to evaluate the optimum plastic content to stabilize soil and compare geotechnical properties of untreated red soil with soil treated with plastic waste in the form of strips of size 0.5cm×0.5cm with six different percentage (0%, 0.25%, 0.5%, 0.75%, 1%, 1.25% of total soil weight) by in horizontally layer by layer mixed methods.

10.2 Materials and Methodology

10.2.1 Soil Sample

Index properties of soil is provided in the following Table 10.5. Soil in present study is identified as red soil that is collected from Odisha University of Technology and Research (OUTR) campus, Bhubaneswar. Red soil is India's 3rd largest soil group spreading over an area of around 3.5 lakh sq. km. (10.6% of India's area). Soil sample for laboratory study is taken from 2mt depth from ground surface, shown in Figure 10.1. On virgin soil (red soil), sieve analysis is conducted for determining particle size distribution of granular material in soil. Figure 10.2 represents the curve showing grain size distribution for given virgin soil sample.

Table 10.5 Index properties of virgin soil.

Properties	Test results	Test procedure
Coefficient Of Curvature (CC)	0.94	
Coefficient Of Uniformity (CU)	4.83	
LL	29.86%	IS:2720 (Part 5,1985) [40]
PL	16.20%	
PI	13.66	
Specific Gravity (G)	2.676	IS:2720 (Part 3,1980) [39]

Figure 10.1 Soil sample collected from a depth of 2mt.

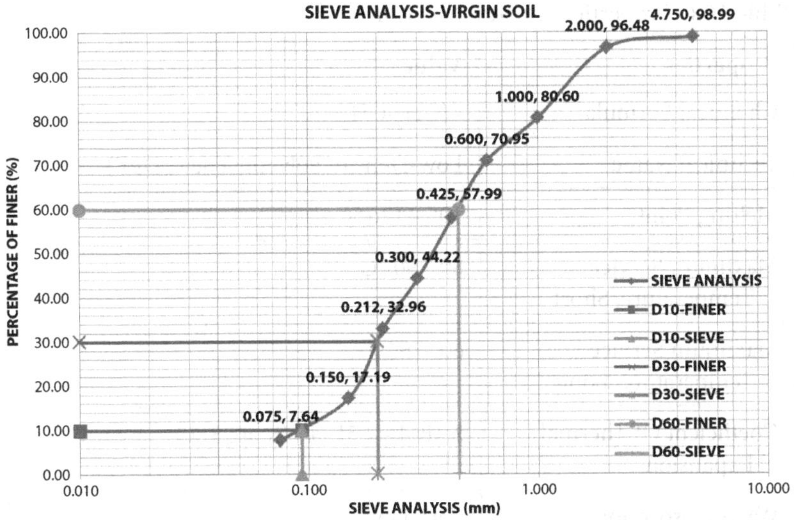

Figure 10.2 Sieve analysis curve of virgin soil.

10.2.2 Waste Plastic as Soil Stabilizer

In this research LDPE has been used as a stabilizer to enhance the soil loading capacity. The waste plastic bags cut into 0.5cm*0.5cm strips (shown in Figure 10.3) and further experiment has been performed on the soil mixed with the plastic strips of varying percentage of 0%, 0.25%, 0.50%, 0.75%, 1%, 1.25% of dry weight of soil. Properties of LDPE are given in Table 10.6.

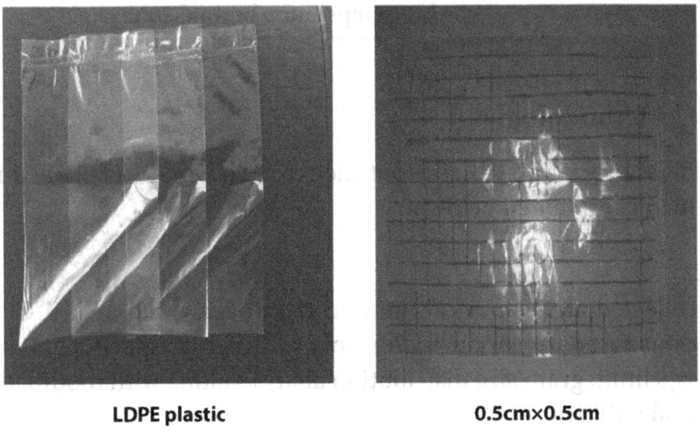

LDPE plastic **0.5cm×0.5cm**

Figure 10.3 LDPE plastic & size of plastic strips.

Table 10.6 Properties of LDPE plastic used.

Properties	Value
Chemical Formula	$-(CH_2\text{-}CH_2)_n-$
Technical Name	Low-Density Polyethylene (LDPE)
Melting Point	105 - 120 °C
Maximum Service Temperature, Short	80 - 90 °C
Maximum Service Temperature, Long	60 -75 °C
Coefficient Of Thermal Expansion	2.3E-4 - 2.5E-4 1/K at 23 °C
Water Absorption	Very Low
Tensile Strength	6 - 17 MPa at 23 °C
Elastic Modulus	0.2 - 0.48 GPa at 23 °C
Density	0.92 g/cm^3 at 23 °C
Strength	Better impact strength in cold conditions
Chemical resistance	Good resilient to alcohols, alkalis, and acids Less resilient to hydrocarbons and oxidizing agents
Heat resistance	Density decreases when exposed to temperature beyond 20°C

10.2.3 Methodologies

The detailed procedure to carry out the research work is explained in two steps which is explained below.

Step – I
Soil is collected at a depth of 2mt and then oven dried at 110°C for 24 hrs. Various tests are conducted on this soil such as specific gravity test, Atterberg's limit, grain size distribution analysis along with modified proctor test and CBR test.

Step-II

In this laboratory study LDPE household plastic bags were collected and cut into 0.5cm*0.5cm (Aspect Ratio-1) strips and mixed with virgin soil with varying percentage of 0%, 0.25%, 0.50%, 0.75%, 1%, 1.25% by the dry weight of soil. LDPE plastic strips mixed with soil by oriented fiber reinforcement method, i.e., horizontally layer by layer mixed method. Then modified proctor test, and CBR test (unsoaked and soaked) are conducted on soil containing 0%, 0.25%, 0.50%, 0.75%, 1%, 1.25% LDPE by dry weight of soil.

For experimental program, LDPE plastic strips of size of 0.5cm*0.5cm has been placed at 1 cm interval between them (Figure 10.4). For example, at 1 cm interval there are 90 numbers of plastic strips placed on one layer, which have a weight of 3.0gm and according to the requirement, the experiment has been conducted with five different percentages of total soil weight, i.e., 0.25%, 0.50%, 0.75%, 1.00%, 1.25% (Figure 10.5). For conducting a series of tests, around 5 kg of soil sample was taken. By the 5 kg soil sample, the calculated weight of different percentages of plastic strips, the number of layers & distance between two layers is given below in Table 10.7: The flow chart of the detailed experimental program is shown in Figure 10.6.

Compaction with a light rammer like the one employed to enhance the compaction properties of an expansive soil using plastic bottle strips, is the first kind of mechanical loading [1]. For investigating the compaction properties of soil and determining how plastic strip reinforcement affects MDD, light compaction tests were assessed. In accordance with IS:2720 (Part 8–1983) [41], light compaction tests were conducted on specimens with various aspects ratio and content combinations to determine their ideal moisture content and maximum dry unit weight.

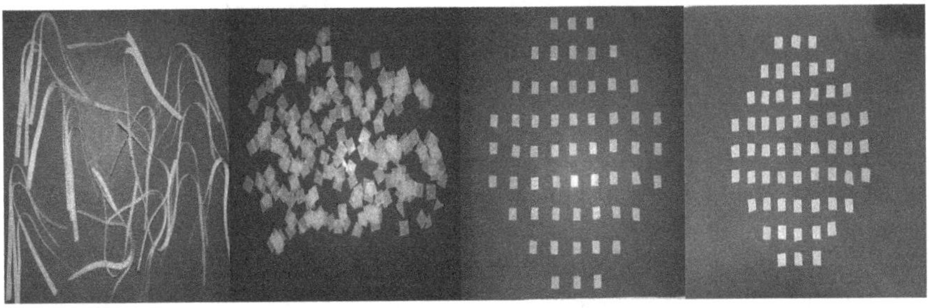

Figure 10.4 Cutting and placing of LDPE plastic strips.

Figure 10.5 CBR test of soil sample.

Table 10.7 Detailed calculation of LDPE plastic strips.

% of LDPE strips added (%)	Weight of LDPE strips added (gm)	No. of layer as per weight of LDPE strips
0.25%	12.5 gm	5
0.50%	25 gm	9
0.75%	37.5 gm	13
1.00%	50 gm	17
1.25%	62.5 gm	21

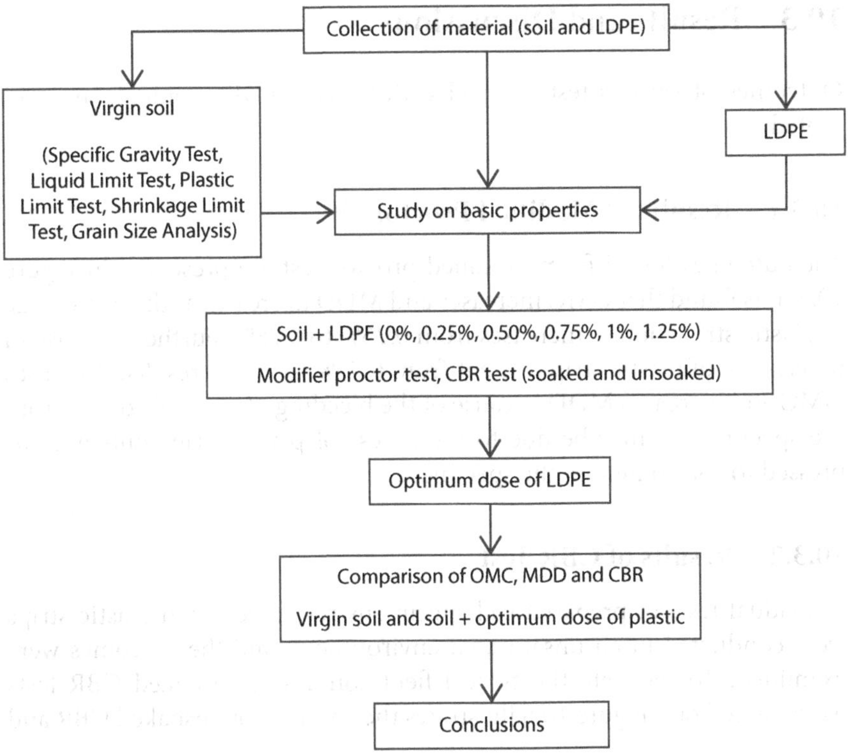

Figure 10.6 Flow chart of detailed experimental program.

The CBR properties were tested in a second kind of experiment using sample soil that had been strengthened with plastic strips. The potential strength of subgrade soil, or CBR, is a crucial component for rapid performance. It has been decided to evaluate CBR values of plastic strip and soil mixes according to IS:2720 (Part 16–1987) [42]. In a similar vein, HDPE's capacity for fortifying soil has been observed to enhance subgrade soil's design qualities [33]. Freshly made specimens were tested in strain-controlled CBR machines while they were not moistened. Before being placed in the testing equipment, the CBR samples were submerged in water for 96 hours in order to assess them. The results have a discussion on the impact of strip size. CBR values were computed for 2.5 mm of penetration, and the following mathematical method can be used for determining the desired CBR value:

10.3 Results and Discussions

Outcomes of various tests on soil with varying LDPE contents are presented below.

10.3.1 Results of Modified Proctor Test

The outcomes found from modified proctor test are presented in Figure 10.7. It is found that OMC increases and MDD decreases as the percentage of plastic strip content increases from 0.25% to 0.75%. Further increase in percentage of plastic strip content from 0.75% to 1.25% results, decreases OMC and increases MDD because of the bleeding of water absorbed from the specimen. It may be due to the excess of plastic strip content compressed to soil content of the specimen.

10.3.2 Results of CBR Test

Individual tests of prepared soil specimens reinforced with plastic strips were conducted in an unsaturated environment, and the outcomes were examined. To replicate the typical field condition, unsoaked CBR tests were carried out. Figure 10.8 illustrates the variance of unsoaked CBR and

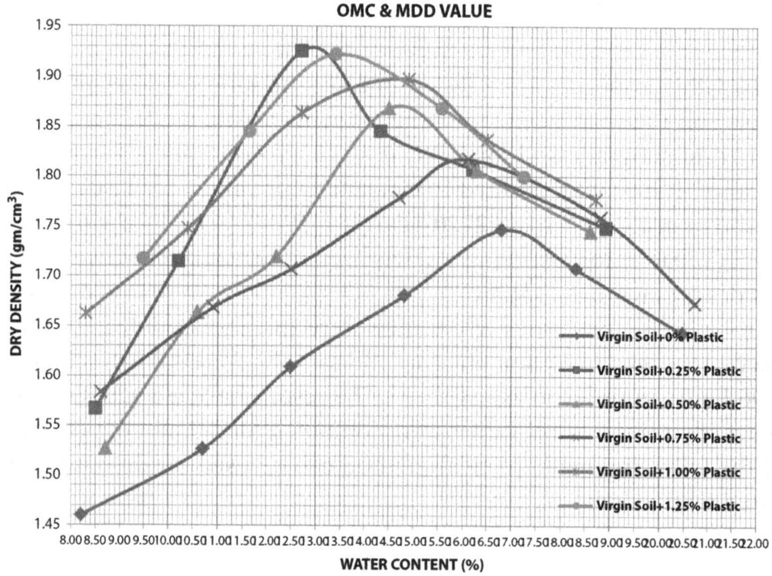

Figure 10.7 Comparison of compaction curve with different LDPE contents.

demonstrates the CBR variation of red soil of various sizes under unsoaked conditions. In the lab, sopped CBR specimens were also examined. When exposed to rain, red soil has a tendency to enlarge and lose its firmness. To replicate the worst field conditions, prepared soil samples were therefore immersed in water for 96 hours. This allowed for the conducting of soaked CBR testing. The reinforced soil sample's soaking CBR tests were also assessed. The results obtained from CBR test i.e., unsoaked CBR value and soaked CBR value are presented in Figures 10.8 and 10.9 respectively.

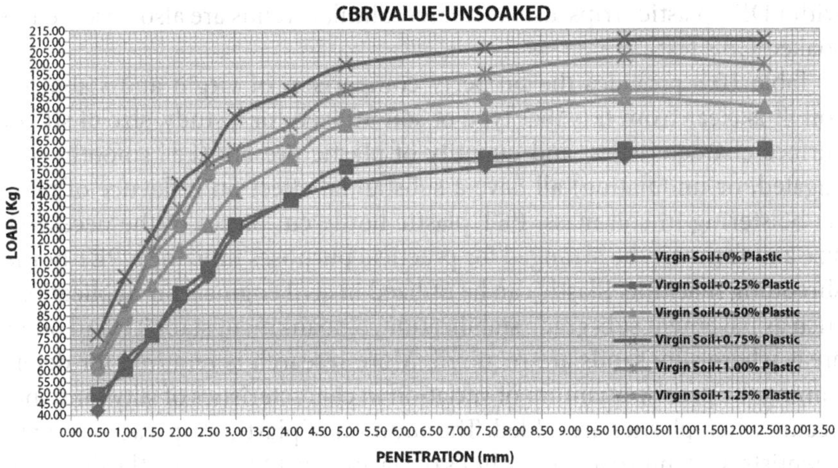

Figure 10.8 Comparison of unsoaked CBR with different LDPE contents.

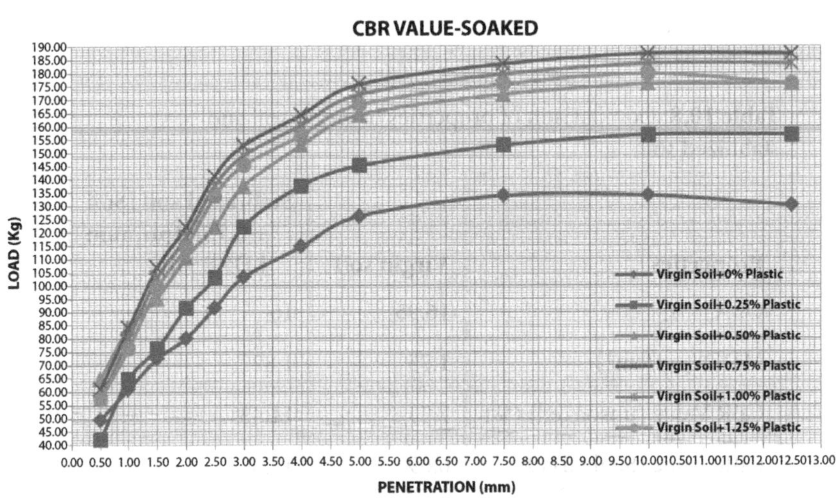

Figure 10.9 Comparison of soaked CBR with different LDPE contents.

It is found that unsoaked CBR increases with increases in percentage of plastic strips content up-to a certain limit. Subsequently, it decreases, reaching a maximum at 0.75% of the dry weight of soil. Unsoaked CBR value depends upon the OMC, MDD and the amount of plastic strips present in the soil. Because further addition of percentage of plastic strips, i.e., from 0.75% to 1.25% the OMC decreases and the MDD increases, which means that the soil became dry. So, the binding between soil particles is not properly done or the binding capacity of soil has diminished. Therefore, the unsoaked CBR value as well as the strength of soil specimen with LDPE plastic strips also decreases. Similar trends are also observed in soaked CBR test.

Table 10.8 provides the property comparison of virgin and stabilized soil. It is clear from this that type of soil utilized for a study, size of strips, surface characteristics, and quantity of plastic present (i.e., smooth, corrugated, or undulating) all have a substantial effect on influence of plastic bolstering in soil mass. PET plastic bottle cut pieces of the necessary size and form can be arranged for practical purposes utilizing a PET bottle shredding machine. Plastic can be utilized in civil engineering applications such as pavement sub-grade stabilization, embankment stability, and other areas where silty sands are required. More research is required for determining the precise quantum of variation in characteristics of silty sand that occurs when plastics are added. The addition of plastic strips, according to scientists, will improve the qualities of clayey soil. Once more, the degree of change in soil qualities would differ whether the plastic type was switched to HDPE or LDPE.

Table 10.8 Comparison of properties of virgin soil and stabilized soil.

Properties	**Virgin soil**	**Stabilized soil [Soil + 0.75(Optimum LDPE %)]**
OMC (%)	16.95	16
MDD (gm/cm^3)	1.75	1.822
CBR Value (unsoaked) (%)	7.53	11.43
CBR Value (Soaked) (%)	6.69	10.32

Cost of soil stabilization can be reduced if plastic strips made from discarded plastic bottles are utilized in place of more costly admixtures like cement, lime, etc. Additionally, the problem of how to dispose of plastic garbage would also be successfully resolved. Therefore, plastic reinforcement can take the place of pricey stabilization techniques, saving money on construction while also assisting in the recycling of non-biodegradable plastic trash.

10.4 Conclusions

In this research work, a sequence of CBR test and Modified Proctor test was performed and the conclusions drawn on that basis are summarized below:

- MDD decreases and OMC increases as the percentage of plastic strip content increases from 0.25% to 0.75%. Further increase in percentage of plastic strip content from 0.75% to 1.25%, results in the decrease in OMC and increase in MDD because of bleeding of water absorbed from the specimen. It may be due to the excess of plastic strip content compressed to soil content of the specimen.
- It was found that soil with 0.75% LDPE strips of size 0.5cm×0.5cm showed 16% OMC, i.e., 5.60% decrease in contrast to virgin soil. Similarly, MDD increased by 4.11% in contrast to virgin soil.
- So, from the above study, it is determined that the ideal quantity of plastic strips to be utilized effectively with red soil is 0.75% with LDPE addition.
- It was observed that unsoaked and soaked CBR value increased by 51.79% and 54.26% respectively, with regards to soil with 0.75% LDPE as compared to CBR value of virgin soil.
- CBR value of red soil taken increases after plastic stabilization (with 0.75%); it typically indicates the soil's improved strength and load bearing capacity.

LDPE plastic stabilization can help enhance the soils such as the red soil mechanical properties, making it more suitable for construction purposes such as roadways and embankment foundations.

References

1. Peddaiah, S., Burman, A., Sreedeep, S., Experimental study on effect of waste plastic bottle strips in soil improvement. *Geotech. Geol. Eng.*, *36*, 2907–2920, 2018.
2. Singh, P., Dash, H.M., Samantaray, S., Effect of silica fume on engineering properties of expansive soil. *Mater. Today Proc.*, 33, 5035–5040, 2020, https://doi.org/10.1016/j.matpr.2020.02.839.
3. Ray, T., Mohanta, N.R., Kumar, M.H., Samantaray, S., Study of effect of temperature on behavior of alkali activated slag concrete. *Mater. Today Proc.*, *43*, 1352–1357, 2021, https://doi.org/10.1016/j.matpr.2020.09.169.
4. Kumar, M.H., Saikrishnamacharyulu, I., Mohanta, N.R., Ashutosh, A., Mishra, P., Samantaray, S., Mechanical behaviour of high strength concrete modified with triple blend of fly ash, silica fume & steel fibers. *Mater. Today Proc.*, *65*, 933–942, 2022a, https://doi.org/10.1016/j.matpr.2022.03.528.
5. Kumar, M.H., Mohanta, N.R., Patel, N., Samantaray, S., Reddy, S.V.B., Influence of Fly ash & Metakaoline on the crack resistance & shrinkage property of concrete. *Iran. J. Sci. Technol. Trans. Civ. Eng.*, 46, 2011–2026, 2022b, https://doi.org/10.1007/s40996-021-00748-y.
6. Kar, D., Panda, M., Jena, S., Utilization of Waste Polyethylene in Open Graded Friction Course, in: *International Conference on Transportation Planning & Implementation Methodologies for Developing Countries*, pp. 121–132, Springer Nature Singapore, Singapore, 2022.
7. Giri, J.P., Panda, M., Sahoo, U.C., Performance of bituminous mixes containing treated recycled concrete aggregates & modified by waste polyethylene. *J. Mater. Civ. Eng.*, *30*, 8, 04018184, 2018.
8. Punith, V.S. and Veeraragavan, A., Characterization of OGFC mixtures containing reclaimed polyethylene fibers. *J. Mater. Civ. Eng.*, *23*, 3, 335–341, 2011.
9. Gangwar, P. and Tiwari, S., Stabilization of soil with waste plastic bottles. *Mater. Today Proc.*, 47, 3802–3806, 2021.
10. Abukhettala, M. and Fall, M., Geotechnical characterization of plastic waste materials in pavement subgrade applications. *Transp. Geotech.*, 27, 100472, 2021.
11. Behnood, A., Soil & clay stabilization with calcium-& non-calcium-based additives: A state-of-the-art review of challenges, approaches & techniques. *Transp. Geotech.*, 17, 14–32, 2018.
12. James, J. and Pandian, P.K., Soil stabilization as an avenue for reuse of solid wastes: a review. *Acta Technica Napocensis: Civ. Eng. Archit.*, 58, 1, 50–76, 2015.
13. Rawat, P. and Kumar, A., Study of CBR Behaviour of Soil Reinforced with HDPE Strips. *Indian Geotechnical Conference IGC2016*, IIT Madras, Chennai, India, 15-17 December, 2016.

14. Babu, G.S. and Chouksey, S.K., Stress–strain response of plastic waste mixed soil. *Waste Manage. (Oxford)*, 31, 3, 481–488, 2011.
15. Gardete, D., Luzia, R., Sousa, M., Carronda, S., Simão, A., Soil stabilization with waste plastic & waste tyre fibres, in: *Proceedings of the XVII European Conference on Soil Mechanics & Geotechnical Engineering (ECSMGE)*, 2019, ISBN (978-9935).
16. Hassan, H.J.A., Rasul, J., Samin, M., Effects of plastic waste materials on geotechnical properties of clayey soil. *Transp. Infrastruct. Geotechnol.*, 8, 3, 390–413, 2021.
17. Balani, D.M.P., Kikumoto, M., Cui, Y., Enhanced s& strength through low-density polyethylene reinforcement. *Constr. Build. Mater.*, 409, 133928, 2023.
18. Kumar, P., Das, V., Gupta, A., Ameta, N.K., Stabilization of dune s & mixed with plastic (LDPE) waste strips for design of flexible pavement in construction of roads. *Am. J. Eng. Res. AJER*, 5, 315–320, 2016.
19. Consoli, N.C., Montardo, J.P., Prietto, D.M., Pasa, G.S., Engineering behavior of a sand reinforced with plastic waste. *J. Geotech. Geoenviron. Eng.*, 128, 6, 462–472, 2002.
20. Saravanan, R., Murthi, P., Poongodi, K., Raju, A., A study on the effect of waste plastic strips in the stabilization of clay soil. *IOP Conf. Ser.: Mater. Sci. Eng.*, 981, 3, 032062, IOP Publishing, 2020.
21. Kumar, M., Azhar, M., Mondal, S., Singh, R.P., Stabilization of expansive soil subgrade by waste plastic. *Arabian J. Geosci.*, 15, 10, 936, 2022.
22. Amena, S. and Kabeta, W.F., Mechanical behavior of plastic strips-reinforced expansive soils stabilized with waste marble dust. *Adv. Civ. Eng.*, 2022, 9807449, 2022.
23. Paramkusam, B.R., Prasad, A., Arya, C.S., A study on CBR behavior of waste plastic (PET) on stabilized red mud & fly ash. *Int. J. Struct. Civ. Eng. Res.*, 2, 232–240, 2013.
24. Muntohar, A.S., Widianti, A., Hartono, E., Diana, W., Engineering properties of silty soil stabilized with lime & rice husk ash & reinforced with waste plastic fiber. *J. Mater. Civ. Eng.*, 25, 9, 1260–1270, 2013.
25. Sobhan, K. and Mashnad, M., Mechanical stabilization of cemented soil–fly ash mixtures with recycled plastic strips. *J. Environ. Eng.*, 129, 10, 943–947, 2003.
26. Patel, S.K. and Singh, B., Investigation of glass fiber reinforcement effect on the CBR strength of cohesive soil, in: *Ground Improvement Techniques & Geosynthetics: IGC 2016 Vol. 2*, pp. 67–75, Springer Singapore, 2019.
27. Moghal, A.A.B., Chittoori, B.C., Basha, B.M., Effect of fibre reinforcement on CBR behaviour of lime-blended expansive soils: reliability approach. *Road Mater. Pavement Des.*, 19, 3, 690–709, 2018.
28. Sarbaz, H., Ghiassian, H., Heshmati, A.A., CBR strength of reinforced soil with natural fibres & considering environmental conditions. *Int. J. Pavement Eng.*, 15, 7, 577–583, 2014.

29. Kumar, D., Nigam, S., Nangia, A., Tiwari, S., Improvement in CBR values of soil reinforced with jute fibre. *Int. J. Eng. Tech. Res.*, 3, 5, 290–294, 2015.
30. Choudhary, A.K., Jha, J.N., Gill, K.S., A study on CBR behavior of waste plastic strip reinforced soil. *Emir. J. Eng. Res.*, 15, 1, 51–57, 2010.
31. Tiwari, S.K. and Sharma, J.P., Influence of Fiber-Reinforcement on CBR-Value of S&. *Electron. J. Geotech. Eng.*, 18, 4303–4311, 2013.
32. Suriya, A., Sangeetha, S.P., Abirami, R., Subathra, P., Stabilization of red soil using polypropylene. *Mater. Today Proc.*, 46, 5881–5884, 2021.
33. Amiri, M., Sanjari, M., Porhonar, F., Microstructural evaluation of the cement stabilization of hematite-rich red soil. *Case Stud. Constr. Mater.*, 16, e00935, 2022.
34. Mahendran, K. and Vignesh, N., Utilization of hypo sludge for the stabilization of red soils along with cement and molasses. *Indian J. Sci. Technol.*, 9, 2, 1–8, 2016.
35. Nandini, D., Amate, V., Kumar, M.P., Compaction and strength characteristics of terra-zyme stabilized red soil. *Int. J. Res. Publ. Eng. Technol. Manage.*, 1, 1, 1–3, 2015.
36. Wang, X., Li, J.D., Wang, X., Zhang, Y., Jiang, D., Zhao, G., Study on Strength & Microstructure of Red Clay Reinforced by F1 Ionic Soil Stabilizer. *Appl. Sci.*, 12, 19, 9831, 2022.
37. Hanuma, K.N.V.K. and Prasad, D.S.V., A study on stabilization of black cotton soil & red soil by using heat treatment method. *Int. Res. J. Eng. Technol.*, 4, 6, 779–782, 2017.
38. Amiri, M., Kalantari, B., Porhonar, F., The effect of thermal stabilization process on mineralogy, morphology & engineering properties of red soil in southern Iran. *Case Stud. Constr. Mater.*, 19, e02454, 2023.
39. IS Code, Determination of specific gravity, BIS (Bureau of Indian Standards, 1980, 2720: (P-3).
40. IS Code, Determination of Liquid & Plastic limit, BIS (Bureau of Indian Standards, 1985, 2720:(P-5).
41. IS Code, Determination of water content-dry density relation using heavy compaction, BIS (Bureau of Indian Standards, 1983, 2720(P-8).
42. IS Code, Determination of California bearing ratio, BIS (Bureau of Indian Standards, 1980, 2720:(P-16).

11

Effect of Marble Dust on the Properties of Black Cotton Soil

Subhadipta Chaudhury* **and Abinash Sahoo**

Department of Civil Engineering, Odisha University of Technology and Research, Bhubaneswar, Odisha, India

Abstract

Stabilization of soil is performed to modify soil characteristics to improve its engineering performances. This research is based on an intention to probe the impact of marble dust (MD) upon stabilization of expansive soil such as black cotton soil. In this analysis black cotton soil taken as the soil specimen, is characterized by high compressibility, high shrinkage, high moisture content and high expansion on drying including deep cracks and excessive swelling on wetting. Black cotton soil exists in various parts of Odisha. In Odisha province, a site is selected and soil samples are collected for determining engineering and index properties. While selecting site, the soil is visually inspected and soil samples are collected from areas in which the soil during its dry condition has deep and wide cracks. The index properties of soil samples taken were classified as expansive soil (black cotton soil) and the liquid limit (LL) is above 35% and plasticity index (PI) is above 20%. Various techniques used for improving the black cotton soil are mostly uneconomical nowadays. For the improvement of black cotton soil, we use marble dust which is oppressing the natural environment. Therefore, utilizing these wastes in order to stabilize the black cotton soil would be the best alternative. The marble dust used in the present study was obtained by cutting of the Makrana marble. A radical increase was attained in shrinkage limit (SL) from 18.35% to 34.72%. The LL and PI were decreased gradually from 52.20% to 38.32% and 34.42% to 20.62% respectively. The inclusion of marble dust improved its maximum dry density (MDD) from 1.77 to 1.804.Results divulged that an inclusion of marble dust pronounces a favorable outcome on California bearing ratio (CBR) values for both soaked and unsoaked conditions. CBR value increased from 6%

**Corresponding author*: csubhadipta@gmail.com

Sandeep Samantaray, Subhashree Samantasinghar, Soumendra K. Mohanty, Vivek, Abinash Sahoo and Deba Prakash Satapathy (eds.) Waste Utilization in Geotechnical Practice, (281–316) © 2025 Scrivener Publishing LLC

to 8.58% on inclusion of 20% marble dust. It also revealed that the inclusion of marble dust in the expansive soil gradually decreases the hydraulic conductivity from 1.38x10^{-2} cm/sec to 8.73x10^{-3} m/sec. A descending trend of settlement was found from the consolidation test from 0.309cm to 0.187cm. Therefore, from the above results, it was proven that marble dust is economically an alternative to the use of ordinary Portland cement and chemical stabilizers and is a fruitful way for the treatment of expansive soil.

Keywords: Expansive soil, marble dust, maximum dry density, hydraulic conductivity, consolidation, CBR

11.1 Introduction

Expansive soil, regarded as shrink or swell soil, tends to shrink and swell with differing moisture content [1]. Due to this, significant disturbances in the soil takes place causing tremendous damage to the structures lying above it. At the time of monsoon season, the water is imbibed by the expansive soil; thus, it swells, becomes soft and the load carrying ability of soil decreases, whereas in dry/arid season the above said soil shrinks and hardens by evaporation of water. Expansive soil, often found in semi-arid and arid areas is considered as a budding natural peril, and if not treated well, it will cause relentless damage to structures constructed over it and to the human life and property too [2]. Hydrated lime/slaked lime is considered quite efficient in the treatment of plastic clay soil while utilized alone or may be combined with bitumen, fly ash or cement. lime, which reduces the plasticity index of highly plastic soils, making it more apt to handle and pulverize. Marble is primarily composed of mineral calcite ($CaCO_3$)/dolomite. Therefore, in the present study MD is utilized for stabilization of black cotton soil and is also used to reduce the potential natural hazard which arises due to MD. Stabilization is a method to treat the expansive soil, i.e., black cotton soil to make it favorable for any construction. The stabilizers are categorized in three categories: by-product stabilizers (quarry dust, slag, fly ash, silica fumes, etc.), conventional stabilizers (lime, cement, etc.) and non-conventional stabilizers (polymer, ammonium chlorides, enzymes, potassium compounds, sulfonated oils, etc.). For the removal of such huge capacities of industrial by-products (fills) on disposal sites adjoining the industries will not only necessitate huge spaces but will also generate a lot of environmental nuisance. Thereby, various efforts are being put forward by different organizations as well as researchers for using them at convenient places. Stabilizing of expansive soil is one of the methods of utilizing generated by-products, thus reducing environmental pollution [3–8].

They [9], elaborated the geotechnical features of expansive soil by adding pyroclastic rock dust. Various parameters such as linear shrinkage, compaction, plasticity, shear strength and CBR values of soil were found out by mixing varying proportions of pyroclastic rock dust. They [10], assessed the impact of coir fiber's inclusion on shear strength of marine clay and the outcomes indicated that there is a rise in shear strength with enlarging fiber content up to 0.8% by weight. They [11], evaluated the durability and strength of rice husk ash stabilized expansive soil by adding marble dust as a stabilizer. The materials used were rice husk ash, MD and expansive soil. Results revealed that MDD of soil decreases and the optimum moisture content (OMC) increases with the preeminent stabilization, which was observed to be 70:10:20. They [12], dealt with geotechnical properties of pond ashes which were taken from Rajghat Power Plant, Badarpur and Dadri located in NCR, Delhi. The study depicted the properties of pond ashes when mixed with waste - marble dust, and the properties were also found out. They [13], probed the impact of marble dust and bagasse ash on stabilization of expansive soils. 4%, 8% and 12% marble dust and bagasse ash were added to the soil sample which led to reduce the plastic limit (PL), expansive index, PI and LL. They [14], investigated the effect of sand, fly ash and marble dust on sub-grade features of expansive soil. They (2014) [15], presented a laboratory study result which was undertaken to explore the index properties of black cotton soil by adding MD powder. The SL augmented from 8.06% to 18.39%, PI reduced from 28.35% to 16.67%, the LL reduced from 57.67% to 33.9% by adding marble dust. They (2014) [16], analyzed the shrinkage and swelling aspects of expansive soil by utilizing brick dust as stabilizer. Various tests like Atterberg's limit test, swelling test, compaction test, and linear shrinkage test were conducted on both normal and stabilized soil and observed a great decline in shrinkage and swelling of soil.

He [17], stabilized the problematic soils, especially swelling clays by utilizing waste marble dust. The results obtained revealed that the marble dust on mixing with clay samples reduced construction cost of structures on challenging soil. They [18], determined the chemical composition of soil and waste marble fines (WMF) by using atomic absorption spectrometry and X-ray fluorescence, respectively. Atterberg limits, specific gravity, permeability, compaction and strength characteristics of soil were found by utilizing varying ratios of soil-WMF mixture. Bansal and Sidhu (2016) [19], mixed marble waste in soil and related properties of parent soil with that of stabilized soil. The authors evaluated the index properties like plastic limit, LL and engineering properties like permeability, compressibility and CBR and a comparison was done. They [20], carried out the stabilization of clayey soil by utilizing marble dust using sodium silicate as the binder due to which the plastic limit and LL decreased up-to 15% and then increased beyond this limit. They [21], utilized waste marble powder and performed UCS tests and CBR for treated and untreated soils with optimum dosages of waste marble powder, whereby the readings were recorded. They [22], collected expansive soil locally which was stabilized with marble dust. Proctor test was performed on black cotton soil diversified with different percentage of MD. Test results revealed a rise in MDD and a decline in the OMC values with the increment in marble dust percentages. They [23], evaluated the potentiality of MD and coconut shell powder used as supplement to expansive soil. Unconfined compression check was conducted by creating specimens of black cotton soil by adding variable percentages of coconut shell powder and marble mud. They [24], stabilized the soil with Alccofine and marble dust. Both Alccofine and marble dust were blended in various percentages to soil sample from which elementary properties of soil such as specific gravity, Atterberg limit, gradation analysis, CBR and UCS were conducted on the samples. They [25], focused on the ability of the MD to upgrade physical characteristics of the soil. Various physico-chemical and experimental studies were carried out to identify the materials used and for understanding the potentiality of the marble dust. Results depicted that LL of soil reduced considerably with the rise in marble dust.

They [26], studied the utilization of marble dust and waste foundry sand as a soil stabilizing supplement for the soil. By the utilization of these waste materials, not only the pollution was reduced but also the human reliability on natural expedient. They [27], evaluated the result of the inclusion of paddy straw and marble dust on soil strength parameters for clay soil. Soil is replaced by marble dust in varied percentages and paddy straw fiber has been blended with optimum soil marble dust in various percentages and various tests such as MDD, OMC, CBR and UCS were performed. They [28], focused on geotechnical behavior and performed micro-analysis of expansive soil altered with marble dust. Wide-ranging geotechnical tests were conducted in varying percentages of marble dust. They [29], focused on behavior of the soil due to the impact of granite powder subjected to loading. Therefore, the soil samples were examined mixing it with different percentages of granite powder of the total weight. Direct shear, compaction and CBR tests were conducted for evaluating the performance of treated and untreated soils.

They [30], explored the impact of calcitic marble dust and dolomitic marble dust on geotechnical properties of fine-grained soil. Expansion index, linear shrinkage, consistency limits and 1-D consolidation tests were conducted on stabilized as well as non-stabilized samples and determined the fruitfulness of waste marble dust in soil stabilization. They [31], examined effects of cement, magnesium phosphate, and waste marble powder on soil properties. Different tests were performed on soil samples in changing percentages of MP and MPC. They [32], made an effort to identify a variety of solid wastes which was utilized in stabilizing the soil mixed without or with cement and lime, so as to find various possible ways for implementing solid waste for diversified geotechnical applications. They [33], conducted a study on granular soil samples. The tests conducted included unconfined compression strength, CBR and freezing-thawing. They [34], improved the sub grade properties of clayey soil by incorporating incorporation ash, marble dust, and municipal solid waste. The purpose was to use waste materials for stabilizing the soil and create less traffic volume on roads; hence, decreasing the cost for the construction of roads along with a solution to the disposal problems of waste materials, thereby providing sustainable management for the environment. They [35], used Kota stone slurry powder as a stabilizing agent for black cotton soil and then predicted and verified the behavior of the mixture by conducting several geological tests. He [36], performed the stabilization with the help of crushed waste glass powder (CWG) with waste foundry sand (WFS) and the values of MDD, OMC and CBR were recorded.

The objective of this research is to perform various tests (LL, PL, SL, Proctor compaction, CBR, consolidation and permeability) for finding the engineering properties as well as the index properties of soil, and also to evaluate the suitability of the expansive black cotton soil for stabilization by utilizing marble dust.

11.2 Methodology

In this experiment, expansive soil is blended with marble dust based on different percentages, i.e., 0%, 5%, 10%, 15% and 20% by weight of dry soil. Tests were performed on marble dust mixes and expansive soil in accordance with IS Code 2720:

i) LL Test
ii) PL Test
iii) SL Test
iv) Specific Gravity
v) Compaction Test

vi) CBR
vii) Consolidation Test
viii) Permeability Test

11.2.1 Soil Sample

In the present study, distributed soil sample was collected from Nayagarh district, Odisha at a depth 1.5m between latitude 20.116°N and longitude 85.01°E. The appearance of collected sample is mainly black in color.

11.2.2 Materials Used

The expansive soil considered for the study was black cotton soil and the same has been collected from Nayagarh, Odisha, India. The waste marble dust was used as stabilizer, which was procured from India MART. The test value of liquid limit of marble dust was found to be 26.15% and the specific gravity was 2.373 respectively. Several tests were performed on original soil sample for determining different properties. Table 11.1 presents the obtained test results. For determining its strength characteristics of soil sample, marble dust was added with varying percentages in conjunction with varying curing periods of 0 day and 4 days.

Table 11.1 Virgin soil properties.

Sl. no.	Description	Test value
1	LL	52.20%
2	PL	17.78%
3	PI	34.42%
4	SL	18.35%
5	Specific gravity	2.8
6	Maximum dry density	1.77
7	Optimum moisture content	17.02%
8	CBR a. Soaked b. Unsoaked	2.74% 6%

11.2.3 Various Tests Conducted

11.2.3.1 Particle Size Distribution of Soil

Soil sample (oven dried) was taken for dry sieve analysis. IS sieves are selected and arranged as 4.75mm, 2mm, 1mm, 0.6mm, 0.425mm, 0.3mm, 0.212mm, 0.150mm, 0.075mm respectively from top to bottom. At the bottom, one collecting pan was attached. On the top, i.e., above the 4.75mm sieve 1000grams of oven dried soil was poured. After pouring the soil, a lid was fixed above the 4.75mm sieve. It was placed in a mechanical sieve shaker for ten minutes to allow the soil to pass through the arranged sieves. The weight of the soil retained in each sieve was noted. The percentage finer was calculated and a graph between the percentage finer and the particle size was plotted.

11.2.3.2 Specific Gravity of Soil

Empty density bottles along with stopper were dried in an oven at 105°C - 110°C, cooled in a desiccator and was weighed (W_1). About 20-25 grams of soil sample (oven dried) having size smaller than 425-micron sieve, was transferred into the density bottles and was weighed along with the stoppers (W_2). Distilled water was poured in some quantities so as to soak the soil for at least two to three hours. Then, the distilled water was poured along, gently stirred with a clean glass rod until the density bottle was filled. Then, the stopper was fixed into the density bottle. Then, density bottle was wiped to remove the excess water adhering outside the density bottle and was weighed (W_3). Next, the density bottle was deflated and thoroughly washed. After that, the distilled water was used to fill the density bottle. Then, the stopper was inserted, and the density bottle was wiped and dried from outside and weighed (W_4).

$$\text{Specific gravity, G of soil} = (W_2 - W_1)/[(W_4 - W_1) - (W_3 - W_2)] \tag{11.1}$$

11.2.3.3 Liquid Limit of Soil

A Casagrande's apparatus was adjusted in such a way that the drop between point of contact on base and screw which had tightened after rearranging should be exactly 1cm. About 120 grams of soil sample (oven dried) sieved through 425-micron sieve was taken. To make a uniform paste, the test sample was thoroughly blended with clean water for fifteen minutes. To mature for sufficient time, this soil water mixture was kept under humid conditions. A part of the matured paste was taken and again mixed systematically. Then, the uniform paste was placed in a cup of Casagrande's apparatus by a spatula and was levelled to a depth of 1 cm at maximum thickness point. Surplus soil was trimmed out and transferred to a dish. After that, a grooving tool was held in perpendicular direction to the cup and taken through the paste in the cup in symmetrical axis along the diameter of center line of cup. To provide the blow, the device's handle was rotated at a rate of two revolutions per sec. Only when the two halves of specimen came into contact along a distance of 12mm at bottom of groove because of flow rather than sliding, the number of blows was counted. Moving the spatula from one edge to another edge of the soil cake in width wise direction, a representative specimen of the soil was collected. The collected specimen was kept in an air tight tin container and placed in a hot air oven for 24 hours and at 105°C-110°C in order to find moisture content. Then, the left-over soils on the cup were removed and collected in the dish and were mixed thoroughly. Subsequently, water was added while mixing in the dish to change the water content. The above steps were repeated to find water content and different number of blows. After that, a graph was plotted between water content and number of blows. The water content at 25 numbers of blows was determined and was taken as LL of the soil.

11.2.3.4 Plastic Limit of Soil

The soil sample (oven dried) which was finer than 425 micron IS sieve was taken. A whole of 30 grams of soil sample and clean water was taken in a dish and mixed thoroughly until it became plastic and could easily be molded with the. fingers. To make a soil thread of uniform diameter, a sum of 10 grams of plastic soil was taken and a hand ball was formed. Then, it was rolled with the help of fingers. The soil thread which was made should not have diameter more than 3mm and that should be rolled crack free. Then, the soil was kneaded further and was rolled repeatedly again and again until the thread crumbled. The moisture content, while the soil can be rolled into a thread having diameter of 3mm without crumbling was taken as the plastic limit of soil. The test was done each time with fresh soil sample and was repeated for 3 times. The mean of the three values was considered as PL of the soil.

11.2.3.5 Shrinkage Limit of Soil

Soil sample (oven dried) finer than 425 micron IS sieved about 100grams was taken. About 30 grams of soil sample was taken in a large evaporating dish. A creamy paste was made with the soil sample by mixing of distilled water, ensuring that could be worked readily without capturing air bubbles. Shrinkage dish was taken, washed and mass of that dish was noted. Then, mercury was poured in the dish and the excess mercury was removed by pressing a plain glass plate over the top of the shrinkage dish. The plate should be flushed with top of dish ensuring that there would be no entrapped air bubbles. Then, the poured mercury in the shrinkage dish was transferred to a weighing dish and the mass was noted with a precision of 0.1 gram. The volume of shrinkage dish can be calculated by the mass of the mercury (grams) divided by specific gravity of mercury. The inner side of the shrinkage dish was coated with silicon grease. Then, the creamy paste made from the soil specimen was transferred to the dish such that 1/3rd of the dish was filled. Specimen was tapped on a firm cushion surface so that a specimen paste flows to the edges. Again, a soil specimen was added and tapped. The process was repeated until the shrinkage dish was filled. The excess soil in the top surface was struck out with the help of a straight cutting edge. Specimen soils adhering to the outside of the shrinkage dish was wiped off. The mass of that soil specimen along with the dish was noted. The mass of the weight soil was also determined. The specimen filled shrinkage dish was placed in a hot air oven at a temperature of 105°-110°C until the color of pat turned from dark to light. After that, the dry pat was cooled on the desiccator. After cooling the dry pat was removed from the dish and weighed. The dry mass of the soil was determined. In a large evaporating dish, a glass cup was placed and filled with mercury. The excess mercury was removed by firmly pressing over top of the cup with a glass plate with prongs. Excess mercury adhering to the side of the glass cup surface was wiped off. The glass cup with mercury was placed in another evaporating dish. Dry soil pat was carried out and emerged in the cup of the mercury. Glass plate with spikes was pressed resolutely over top of the cup. Displaced mercury by volume of the dry pat was collected in evaporating dish and transferred to a weighing dish. The mass of that displaced mercury was noted. The volume of dry pat was equal to mass of displaced mercury divided by mercury's specific gravity.

11.2.3.6 Compaction Characteristics of a Soil Specimen by Proctor's Test

Oven dried soil specimen finer than 4.75mm IS sieve at about 2.5kg was taken in a mixing pan. In order to increase the water content, an approximate amount of water was added to a percentage of the soil mass. Proctor compaction mold was cleaned and fixed to the base. Empty mass of the mold was taken and the collar was attached to the mold. The inner side of the mold was greased thoroughly. Soil was placed in the proctor mold at about 1/3rd of the height of the mold. 25 numbers of blows with the help of a 2.6kg rammer were given by means of compaction. The blows were provided evenly over the soil specimen surface. After compaction in the first layer, the surface was scratched with a spatula forming a grid to ensure uniformly the distribution of compaction energy to the succeeding layer. The above procedures were repeated 3 times so as to fill the mold. Collar was detached carefully without disturbing the compacted soil specimen. The final layer, i.e., the outermost surface was trimmed to the level of the mold with the help of a cutting edge so as to remove the excess soil. The compacted soil specimen along with the mold was weighed and noted down. The sample was extruded from the mold and a small representative of the soil sample was taken in air tight tin container to determine the moisture content in oven drying method. The above procedure was repeated, increasing the water content by a percentage of soil mass until the weight of the wet compacted soil decreased.

11.2.3.7 CBR

About 6kg of Soil sample (oven dried) was taken and mixed thoroughly at optimum moisture content. An empty weight of mold along with the base plate was measured. After weighing, the extension collar was attached to the mold. A space disc was inserted over the base plate. On top of the spacer disc a coarse filter paper was placed. Soil was poured into the mold in such a way that 1/3rd of the mold height is filled. After pouring, the soil was compacted by giving 25 number of blows using a 2.6kg rammer. After that, the top surface of compacted soil was tampered with the help of a spatula. These above steps were repeated 3 times, such that the mold was completely filled. Then, the collar was separated carefully. Excess soil over the mold height was cleaned by a knife. Then, the mold was turned upside down. Then, the base plate as well as the spacer disc was removed. Thereafter, the weight of the mold with the compacted soil was recorded. Then, the mold with surcharge weights was placed on the penetration test machine. The penetration piston with smallest possible load was placed at the center of the specimen. Stress-strain dial gauges were set to zero. Load at a penetration rate of 1.25mm/min was applied on piston. Load readings were recorded at penetration of 0.5, 1, 1.5, 2, 2.5, 3, 4, 5, 6, 7mm. Then, the mold was detached from the loading equipment. An archetypal of soil was taken from top of soil specimen to determine moisture content. Load versus penetration graph was plotted; as the first portion of the curve was concave downwards, so no corrections were required.

$$CBR = \left(P_T / P_S\right) * 100 \tag{11.2}$$

where P_T = test load corresponding to selected penetration
P_S = standard load for same penetration

For soaked CBR determination, the specimen making was the same as the aforementioned unsoaked procedure. For soaked CBR determination the specimen was immersed in a water tank for 96 hours. After 96 hours the specimen was taken out and placed in the penetration equipment. P_T for soaked condition was calculated and soaked CBR was determined.

11.2.3.8 Consolidation

Two porous stones were saturated by dipping them in distilled water for about 4 hours. The fittings of the consolidometer were moistened. The empty consolidation ring was weighed. Inner diameter and height of the consolidation ring were measured. Using a thin-walled Shelby tube sampler, specimen was made. Initial moisture content and specific gravity were determined. Approximately a 3-inch-long specimen was extruded from the sampler. The soil specimen was placed on the consolidation ring the excess adhering soil to the outer surface of the ring were trimmed and wiped. The consolidometer was accumulated with sample specimen and porous stones at the bottom and top of the soil specimen. One filter paper was provided in between porous stone and soil specimen. The pressure pad was centrally positioned on the porous stone which is placed at the top. The mold assembly was placed on loading frame in such a way that axial load was applied. The sample got saturated through the mold assembly, which was connected to a water reservoir. An initial load of intensity 0.05kg/cm^2 was applied. Dial gauge readings in various time intervals were recorded. The load was successively incremented as 0.1, 0.2, 0.5, 1, 2, 4 and 8kg/cm^2. After completing the last loading, load was reduced to its previous load and allowed to stand for 24 hours. Further reductions were applied until 0.1kg/cm^2 was reached. Final dial gauge readings were recorded. Then, the mass of specimen with ring was measured. The specimen was then removed from the ring carefully and was oven dried. Dry specimen was weighed and noted.

11.2.3.9 Permeability

About 1kg of soil (oven dried) was taken on a mixing pan and blended thoroughly with optimum moisture content. The mixture was kept aside for 24 hours in an air tight container. The permeameter was assembled in such a way that the 3cm collar was attached to bottom part of 0.3litres mold and 2cm collar was attached to top part. Mold assembly was reinforced over a 2.5cm end plug. Inner wall of 0.3 liter mold was greased. Then, soil specimen was transferred to the mold and the top 3cm end plug was inserted into the top collar and tamped with the hands. Then, the entire assembly now kept under a compressive machine and split collar was detached. Compressive force was put till flange of the both end plugs touched the corresponding collar. Then, the top 3cm plug and the collar was detached. On the top of the specimen, one filter paper was kept on fine wire mesh. Then, perforated base plates were fixed. Consequently, the mold assembly was turned upside down and the 2.5cm end plug and the collar were removed. The top perforated base plate was placed on the top of the soil specimen and the top cap was fixed. The permeability mold was assembled at the bottom of the tank and the tank was filled with water.

11.3 Result and Discussion

The outcomes of the mentioned experiments are shown; graphs and findings from the analysis are represented, followed by discussions about each of the results.

11.3.1 Grain Size Distribution of Soil

For soil particles of size 75 micron and bigger, dry sieve analysis is conducted for quantitatively determining the grain or particle size distribution. Standard sieves used in this case are (75μ, 150μ, 212μ, 300μ, 425μ, 600μ, 1mm and 2mm). Figure 11.1 demonstrates the grain size analysis of the soil considered for experimental study.

$$C_u = D_{60} / D_{10} \tag{11.3}$$

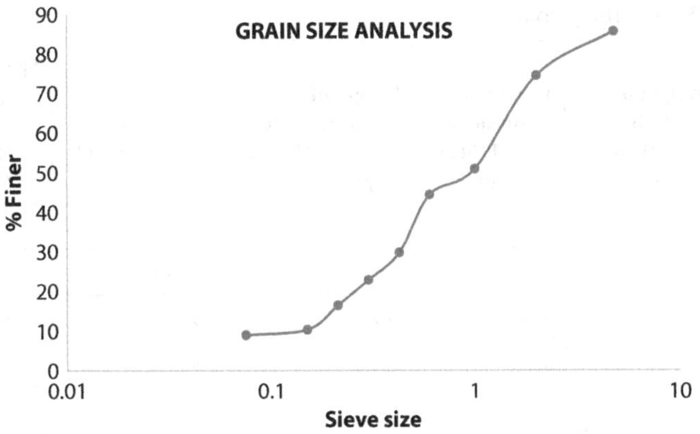

Figure 11.1 Grain size analysis.

$$C_c = D_{30}^2 / (D_{60} \times D_{10}) \tag{11.4}$$

From Cu and Cc value, we can classify the soil type whether it is well graded or poorly graded. When $C_U>4$ and $C_C=1$ to 3, it is well graded gravel. For $C_U>6$ and $C_C= 1$ to 3, that is classified as well graded sand. From the above equation we can specify the grade of the soil. Table 11.2 provides the results obtained after the grain size analysis. Similarly, Table 11.3 provides the specific gravity of the obtained soil.

Table 11.2 Grain size analysis.

Sieve size	Total wt. in gm.	Wt. retained in gm	% Wt. retained	Cumulative % wt. retained	% passing (or) % finer
4.75mm	1000	144	14.4	14.4	85.6
2mm	1000	111	11.1	25.5	74.5
1mm	1000	236	23.6	49.1	50.9
600μ	1000	65	6.5	55.6	44.4
425μ	1000	147	14.7	70.3	29.7
300μ	1000	69	6.9	77.2	22.8
212μ	1000	64	6.4	83.6	16.4
150μ	1000	61	6.1	89.7	10.3
75μ	1000	13	1.3	91.00	9
Pan	1000	90	9.0	100	0

Table 11.3 Specific gravity.

Sl. no.	Weight of the bottle (M1) in gm	Weight of bottle+dry soil (M2) in gm	Weight of bottle+water +soil (M3) in gm	Weight of bottle+ water (M4) in g	Specific gravity G = M2-M1 (M2-M1)-(M3-M4)
Expansive Soil					
1	38.580	61.71	105.54	90.73	2.78
2	38.580	63.04	107.01	90.73	2.93
3	38.580	62.86	106.87	90.73	2.98
Marble Dust					
1	38.590	48.880	97.180	90.990	2.509
2	43.610	54.510	100.280	93.620	2.570
3	22.950	26.520	64.720	62.900	2.04

Specific gravity of soil sample=2.8gm.
Hence, it is found that the specific gravity of marble dust is 2.373.

From dry sieve analysis, these are the values:

D10=150μ
D30= 425μ
D60=1000μ
Cu=D60/ D10=6.66 (C_U>6)
Cc= $(D30)^2$/D60*D10=1.20

Here, Cu>6 and Cc lies in between 1-3. So, the soil is a well graded soil. Well graded soils means different sizes of particle are present in soil mass so the voids between soil particles are very less.

11.3.2 Liquid Limit of Soil

From values of Table 11.4, Figure 11.2 is plotted. The liquid limit for virgin soil is found to be 52.20%. By adding 5%, 10%, 15% and 20% of marble dust, liquid limit is found to be 46.10%, 41.84%, 39.76% and 38.32% respectively. From the above results, it is concluded that the liquid limit decreases with the inclusion of marble dust from 5% to 20%. Soils having high liquid limit are highly compressible and thus, are not suitable for

Table 11.4 Liquid limits.

Sample	No of blows (N)	Wt. of empty container (W_1) in gm	Wt. of empty container+ wet soil (W_2) in gm	Wt. of empty container + dry soil (W_3) in gm	Water content (w) $(W) = \frac{W_2 - W_3}{W_3 - W_1}$
virgin soil					
1	20	18.76	38.91	31.92	53.11
2	25	21.84	36.01	31.15	52.20
3	29	18.93	33.46	28.54	51.19
4	33	18.63	35.82	30.11	49.73
95% virgin soil+5% marble dust					
1	15	29.6	44.78	40.82	41.24
2	34	23.07	42.16	37.11	46.97
3	18	23.02	33.40	30.72	43.53
4	23	24.29	40.63	36.06	45.98
90% virgin soil+10% marble dust					
1	38	26.80	38.65	35.23	40.56
2	25	23.21	39.82	34.92	41.84
3	27	22.81	45.66	39.05	40.7
4	14	23.07	40.75	35.49	42.35
85% virgin soil+15% marble dust					
1	13	23.10	40.69	35.63	40.38
2	20	22.83	40.59	35.51	40.06
3	29	20.16	34.67	30.56	39.52
4	42	28.55	44.22	39.83	38.91
80% virgin soil+20% marble dust					
1	34	29.05	46.54	41.82	36.96
2	26	24.97	47.03	40.92	38.3
3	21	19.78	39.63	34.12	38.42
4	19	28.54	47.98	42.55	38.75

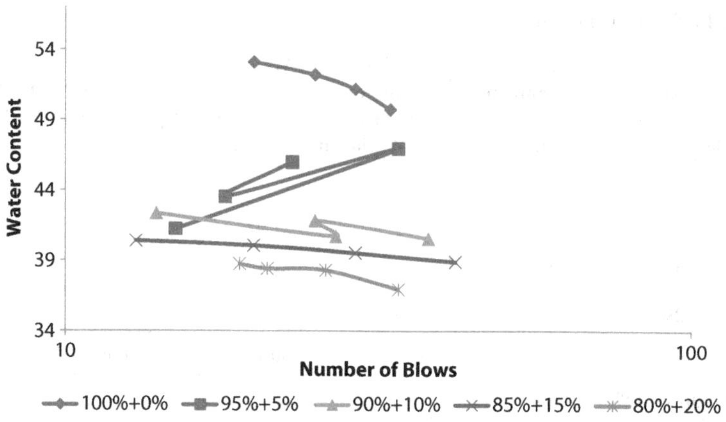

Figure 11.2 Comparison graph for various liquid limit of expansive soil with varying marble dust content.

pavement and subgrade from engineering point of view. So, the decrease in liquid limit with the inclusion of marble dust means the compressibility nature of soil decreases.

11.3.3 Plastic Limit of Soil

From Table 11.5, the plastic limit for virgin soil is found to be 17.78%. With the inclusion of 5%, 10%, 15% and 20% marble dust, the plastic limit is found to be 17.65%, 18.86%, 18.76% and 17.70% respectively. From the above results, the plasticity index is calculated.

11.3.4 Plasticity Index of Soil

$$\text{Plasticity Index} = \text{Liquid Limit} - \text{Plastic Limit} \qquad (11.5)$$

Therefore, the plasticity index for virgin soil is 34.42%; the plasticity index for 95% virgin soil+ 5% marble dust is 28.45%; the plasticity index for 90% virgin soil +10% marble dust is 22.98%; the plasticity index for 85% virgin soil +15% marble dust is 21.00% and the plasticity index for 80% virgin soil + 20% marble dust is 20.62%.The soils having high plasticity index are highly active soil; they are not suitable for both foundation and subgrade. Here, the plasticity index shows a decrease in trend with the increase in percentage inclusion of marble dust.

Table 11.5 Plastic limit.

Sample	Wt. of empty container (w_1) in gm	Wt. of empty container + wet soil (w_2) in gm	Wt. of empty container + dry soil (w_3) in gm	Water content (w)= $\frac{W_2-W_3}{W_3-W_1}$ %
virgin soil				
1	18.59	19.17	19.08	18.36
2	18.62	19.73	19.56	18.08
3	18.50	19.33	19.21	16.90
virgin soil+5% marble dust				
1	22.83	23.78	23.63	18.75
2	22.96	23.91	23.77	17.28
3	23.05	23.74	23.64	16.94
virgin soil+10% marble dust				
1	23.18	24.20	24.03	20
2	22.81	23.90	23.74	17.20
3	14.67	15.84	15.65	19.38
virgin soil+15% marble dust				
1	23.01	24.22	24.02	19.8
2	30.88	31.90	31.74	18.6
3	27.98	29.43	29.21	17.88
virgin soil+20% marble dust				
1	18.23	18.80	18.71	17.69
2	18.52	19.78	19.62	18.71
3	18.41	19.46	19.33	16.72

From the Table 11.6, the shrinkage limit for virgin soil is found to be 18.35%. By adding 5%, 10%, 15%, and 20% of marble dust the shrinkage limit is found to be 19.84%, 21.80%, 34.68% and 34.72% respectively.

When water content changes from liquid limit to shrinkage limit, the volume of soil mass also decreases from liquid limit to shrinkage limit, but below shrinkage limit the decrease in water content will not lead to decrease in volume of soil mass due to voids filled by air. Soils having more shrinkage limit are less susceptible to volumetric change. Shrinkage limit increases with the increase in the percentage of marble dust due to the increase in volume of soil mass.

Table 11.6 Shrinkage limit test.

Sample	Soil	Virgin soil+5% marble dust	Virgin soil+10% marble dust	Virgin soil +15% marble dust	Virgin soil+ 20% marble dust
Wt. of empty container	38.49	38.5	40.02	40.02	38.49
Wt. of empty container wet soil	79.31	78.79	79.32	76.59	76.42
Wt. of empty container dry soil	65.62	64.91	65.84	63.3	62.9
Wt. of empty container+ mercury	364	364	365	365	367
Wt. of mercury dish (gm)	22.36	22.36	22.36	22.36	22.36
Wt. of mercury dish+mercury (gm)	649	649	649	649	649
Wt. of mercury	626.4	626.4	626.4	626.4	626.4
Wt. of wet soil(M_1)	40.82	40.29	39.3	36.57	37.93
Wt. of dry soil(M_2)	27.13	26.41	25.82	23.28	24.41
Mercury displaced	209	210	203	247	273
Vol. of mercury V_1	23.93	23.93	23.89	23.895	24.155
Vol. of displaced mercury V_2	15.36	15.44	14.93	18.161	20.073
Shrinkage limit	18.88	21.4	17.47	32.45	38.67

11.3.5 Compaction Characteristics

Multiple compaction tests were done with the expansive soil to confirm the moisture content with corresponding dry densities. The soil was blended with 0%, 5%, 10%, 15% and 20% of marble dust. OMC and MDD were evaluated from the test. Graphs were plotted with increasing percentage of silica fume thereby representing variation of OMC and MDD respectively.

Compaction effort: Light compaction energy
Weight of hammer: 2.60 kg
Volume of mold (X): 1000 cm^3
Weight of empty mold (Y): 3633 g

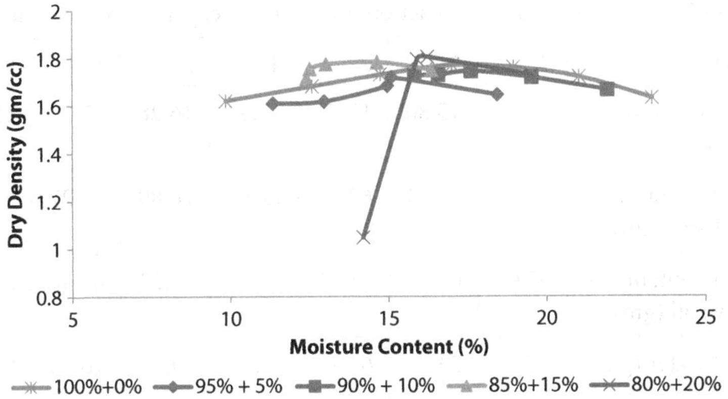

Figure 11.3 Comparison graph for various compaction characteristics of expansive soil with varying marble dust content.

Figure 11.3 illustrates the comparison graph for various compaction characteristics of expansive soil with varying marble dust content. From Table 11.7, the graph is plotted showing compaction characteristics of expansive soil by adding varying percentages of marble dust. From the graph, OMC and MDD of virgin soil is found to be 17.02% and 1.77gm/cc, OMC and MDD of 95% expansive soil + 5% marble dust is found to

Table 11.7 Standard proctor compaction test for 100% expansive soil.

Sl. no.	**1**	**2**	**3**	**4**	**5**	**6**	**7**
Water % added	6	8	10	12	8	10	12
Wt. of mold+ compacted soil	6245	6354	6450	6529	6559	6544	6516
Wt. of compacted soil (gm)	1779	1888	1984	2063	2093	2078	2050
Wt. density (gm/cc)	1.779	1.888	1.984	2.063	2.093	2.078	2.050
Container no.	1	2	3	4	5	6	7

(Continued)

Table 11.7 Standard proctor compaction test for 100% expansive soil. (*Continued*)

Sl. no.	1	2	3	4	5	6	7
Wt. of container (gm)	16.55	15.80	15.74	15.84	16.28	17.21	15.95
Wt. of container + wet soil (gm)	56.22	69.14	68.49	72.84	76.89	79.27	61.59
Wt. of container + dry soil (gm)	52.66	63.18	61.71	64.55	67.23	68.49	52.96
Wt. of water (gm)	3.56	5.96	6.78	8.29	9.66	10.78	8.63
Wt. of dry soil (gm)	36.11	47.38	45.97	48.71	50.95	51.28	37.01
Moisture content (%)	9.86	12.58	14.75	17.02	18.96	21.02	23.32
Dry density (gm/cc)	1.62	1.68	1.73	1.77	1.76	1.72	1.63

be 15.11% and 1.717gm/cc, OMC and MDD of 90% expansive soil + 10% marble dust is found to be 17.62% and 1.744gm/cc, OMC and MDD of 85% expansive soil +15% marble dust is found to be14.65% and 1.782gm/cc and OMC and MDD of 80% expansive soil + 20% marble dust is found to be 16.24% and 1.804gm/cc respectively. Therefore, an increase in trend of dry density of a soil indicates a greater strength, lesser permeability and better volume stability. The standard proctor compaction test for varying percentage of marble dust is provided in Table 11.8.

11.3.6 CBR Test

C.B.R test of the soil was determined using various percentage of marble dust varying from 0%, 5%, 10%, 15%, 20% according to specifications. The sample was prepared at OMC. Two samples were made for each concentration of marble dust, one sample tested at OMC (unsoaked) and other was tested at saturation after four days soaking.

Table 11.8 Standard proctor compaction test for varying percentage of marble dust.

Sl. no.	1	2	3	4	5	1	2	3	4	5
	95% Expansive soil + 5% Marble dust					**90% Expansive soil + 10% Marble dust**				
Water % added	10	12	14	16	18	12	14	16	18	20
Wt. of mold + compacted soil	5424	5459	5566	5610	5580	5634	5650	5685	5687	5662
Wt. of compacted soil (gm)	1791	1826	1933	1977	1947	2001	2017	2052	2054	2029
Wt. density (gm/cc)	1.791	1.826	1.933	1.977	1.947	2.001	2.017	2.052	2.054	2.029
Container no.	1	2	3	4	5	1	2	3	4	5
Wt. of container (gm)	21.58	18.14	18.17	18.28	18.92	21.54	18.14	18.12	18.39	18.92
Wt. of container + wet	38.67	35.47	28.92	39.00	35.88	32.67	33.53	36.34	33.65	31.55
Wt. of container + dry	37.12	33.48	27.52	36.28	33.24	31.14	31.34	33.61	31.14	29.28
Wt. of water (gm)	1.55	1.99	1.40	2.72	2.64	1.53	2.19	2.73	2.51	2.27
Wt. of dry soil (gm)	15.54	15.34	9.35	18.00	14.32	9.6	13.2	15.49	12.75	10.36
Moisture content (%)	11.35	12.97	14.97	15.11	18.45	15.83	16.59	17.62	19.53	21.91
Dry density (gm/cc)	1.608	1.616	1.681	1.717	1.643	1.727	1.729	1.744	1.718	1.664

(Continued)

Table 11.8 Standard proctor compaction test for varying percentage of marble dust. (*Continued*)

Sl. no.	**1**	**2**	**3**	**4**	**5**	**1**	**2**	**3**	**4**	**5**
	85% Expansive soil + 15% Marble dust					**80% Expansive soil + 20% Marble dust**				
Water % added	12	13	14	15	16	12	14	16	18	
Wt. of mold + compacted soil	5559	5607	5639	5677	5670	5597	5713	5731	5699	
Wt. of compacted soil (gm)	1926	1974	2006	2044	2037	1964	2080	2098	2066	
Weight density (gm/cc)	1.926	1.974	2.006	2.004	2.037	1.964	2.080	2.098	2.066	
Container no.	1	2	3	4	5	1	2	3	4	
Wt. of container (gm)	21.58	18.54	18.17	18.28	18.74	21.58	18.14	18.12	18.39	
Wt. of container + wet	38.15	33.15	36.03	30.57	51.84	35.90	32.20	34.51	30.57	
Wt. of container + dry	36.32	31.48	33.97	29.00	38.59	34.12	30.27	32.22	28.58	
Wt. of water (gm)	1.83	1.67	2.06	1.57	3.25	1.78	1.93	2.29	1.99	
Wt. of dry soil (gm)	14.74	13.34	15.8	10.72	19.85	12.54	12.13	14.10	10.19	
Moisture content (%)	12.41	12.51	13.03	14.65	16.37	14.19	15.91	16.24	19.52	
Dry density (gm/cc)	1.713	1.754	1.774	1.782	1.750	1.047	1.794	1.804	1.728	

11.3.6.1 Unsoaked CBR Test

The unsoaked CBR test for expansive soil was carried out following IS: 2720 Part-16 and the observations were tabulated in Table 11.9.

Capacity of proving ring: 10 KN
Value of 1 division: 1.176

From the Table 11.9 and Table 11.10, the graph (Figure 11.4) is plotted. CBR value for virgin soil is found to be 6%. The inclusion of 5%, 10%, 15% and 20% marble dust showed CBR values to be 8.15, 4.72, 6.43 and 8.58 respectively. CBR is a measure of subgrade strength. As our unsoaked CBR values are gradually increasing from 100% expansive soil to 80% expansive soil with 20% marble dust inclusion, it indicates that the soil exhibits a greater strength.

Table 11.9 Unsoaked CBR test for 100% expansive soil.

Plunger penetration (mm)	Proving ring dial gauge reading	Load intensity (PT)	Standard load intensity (PS)	CBR (%)
0	0	0		
0.5	17	19.992		
1.0	30	35.28		
1.5	44	51.744		
2.0	58	68.208		
2.5	70	82.32	1370	6
3.0	80	94.08		
4.0	91	107.016		
5.0	96	112.896	2055	5.49
6.0	101	118.776		
7.0	105	123.48		

Table 11.10 Unsoaked CBR test for varying percentage of marble dust.

Plunger penetration (mm)	Proving ring dial gauge	Load intensity (PT)	Standard load intensity (PS)	CBR (%)	Proving ring dial gauge	Load intensity (PT)	Standard load intensity (PS)	CBR (%)
	95% Expansive soil + 5% Marble dust				**90% Expansive soil + 10% Marble dust**			
0	0	0			0	0		
0.5	17.5	20.58			20	23.52		
1.0	35	41.16			30	35.28		
1.5	60	70.56			37.5	44.1		
2.0	77.5	91.14			45	52.92		
2.5	95	111.72	1370	8.15	55	64.68	1370	4.72
3.0	115	135.24			60	70.56		
3.5	125	147			70	82.32		
4.0	135	158.76			77.5	91.14	2055	4.43
5.0	140	164.64	2055	8.01	85	99.96		
6.0	192.5	226.38			90	105.84		
7.0	215	252.84						

(Continued)

Table 11.10 Unsoaked CBR test for varying percentage of marble dust. (*Continued*)

Plunger penetration (mm)	Proving ring dial gauge	Load intensity (PT)	Standard load intensity (PS)	CBR (%)	Proving ring dial gauge	Load intensity (PT)	Standard load intensity (PS)	CBR (%)
	85% Expansive soil + 15% Marble dust				**80% Expansive soil + 20% Marble dust**			
0	0	0			0	0		
0.5	22.5	26.46			35	41.16		
1.0	37.5	44.1			55	64.68		
1.5	47.5	55.86			70	82.32		
2.0	64	75.264			85	99.96		
2.5	75	88.2	1370	6.43	100	117.6	1370	8.58
3.0	84	98.784			110	129.36		
3.5	93	109.368			127.5	149.94		
4.0	101	118.776			140	164.64	2055	8.01
5.0	110	129.361	2055	6.29	150	176.4		
6.0	125	147			155	182.28		
7.0	132.5	155.82						

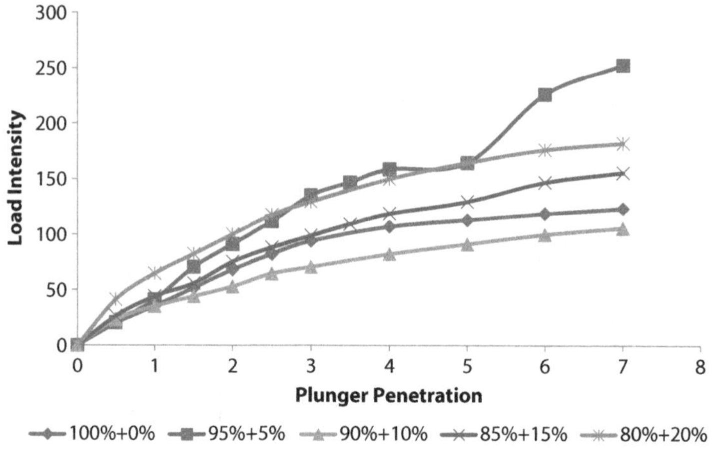

Figure 11.4 Variation in unsoaked CBR for Expansive soil with varying percentage of marble dust.

Table 11.11 Soaked CBR test for 100% expansive soil.

Plunger penetration (mm)	Proving ring dial gauge	Load intensity (PT)	Standard load intensity (PS)	CBR (%)
0	0	0		
0.5	7	8.232		
1.0	14	16.464		
1.5	21	24.696		
2.0	27	31.752		
2.5	32	37.632	1370	2.74
3.0	38	44.688		
4.0	43	50.568		
5.0	46	54.096	2055	2.63
6.0	49	57.624		
7.0	51	59.976		

11.3.6.2 Soaked CBR Test

The soaked CBR test for expansive soil was carried out and the observations were tabulated in Table 11.11.

Capacity of proving ring: 10 KN
Value of 1 division: 1.176

From the Table 11.11 and Table 11.12, the graph (Figure 11.5) is plotted. CBR value for virgin soil is found to be 2.74%. The inclusion of 5%, 10%, 15% and 20% marble dust showed CBR values to be 1.71, 1.07, 0.85 and 0.42 respectively. Here, the soaked CBR values showed a decreasing trend. Hence, this kind of soil stabilization is not suitable for heavy rainfall areas.

11.3.7 Consolidation Test

From the above Table 11.13 and Table 11.14, the graph (Figure 11.6) is plotted.

$$S = \frac{CcH}{1+eo}\log\left(\frac{Po' + \delta P'}{Po'}\right) \tag{11.6}$$

where S = Settlement

C_c = Compression Index = Slope of the e vs log P_0' graph
H = Thickness of soil sample
e_0 = Void Ratio

From the consolidation results, the settlement for virgin soil is found to be 0.309cm. By the inclusion of 5%, 10%, 15% and 20% marble dust, the settlement gradually decreased to 0.280cm, 0.251cm, 0.211cm and 0.187cm respectively.

11.3.8 Permeability Test

From the above Table 11.15, the hydraulic conductivity of virgin soil is found to be 1.38 x 10^{-2} cm/sec. By the inclusion of 5%, 10%, 15% and 20% marble dust, the hydraulic conductivity is found to be 1.26 x 10^{-2} cm/sec, 1.23 x 10^{-2} cm/sec, 1.14 x 10^{-2} cm/sec and 8.73 x 10^{-3} cm/sec respectively. A decreasing trend indicates that the material becomes less permeable with the inclusion of varying percentages of marble dust.

Table 11.12 Soaked CBR test for varying percentage of marble dust.

Plunger penetration (mm)	Proving ring dial gauge	Load intensity (PT)	Standard load intensity (PS)	CBR (%)	Proving ring dial gauge	Load intensity (PT)	Standard load intensity (PS)	CBR (%)
	95% Expansive Soil + 5% Marble dust				**90% Expansive Soil + 10% Marble dust**			
0	0	0			0	0		
0.5	15	17.64			5	5.88		
1.0	16	18.816			7.5	8.82		
1.5	17.5	20.58			10	11.76		
2.0	19	23.344			11	12.936		
2.5	20	23.52	1370	1.71	12.5	14.7	1370	1.07
3.0	21	24.696			13	15.288		
3.5	23.5	27.636			15	17.64		
4.0	27.5	32.34			16	18.816	2055	0.91
5.0	29	34.10	2055	1.65	17.5	20.58		
6.0	32.5	38.22			19	22.344		
7.0	36	42.336						

(Continued)

Table 11.12 Soaked CBR test for varying percentage of marble dust. (*Continued*)

Plunger penetration (mm)	Proving ring dial gauge	Load intensity (PT)	Standard load intensity (PS)	CBR (%)	Proving ring dial gauge	Load intensity (PT)	Standard load intensity (PS)	CBR (%)
	85% Expansive Soil + 15% Marble dust				**80% Expansive Soil + 20% Marble dust**			
0	0	0			0	0		
0.5	6	7.056			2.5	2.94		
1.0	9	10.584			2.5	2.94		
1.5	9	10.584			5	5.88		
2.0	9	10.584			5	5.88		
2.5	10	11.76	1370	0.85	5	5.88	1370	0.42
3.0	11	12.936			5	5.88		
3.5	11	12.936			7	8.23		
4.0	12	14.112			7.3	8.58	2055	0.4
5.0	13	15.288	2055	0.74	10	11.76		
6.0	15	17.64			10	11.76		
7.0	15	17.64						

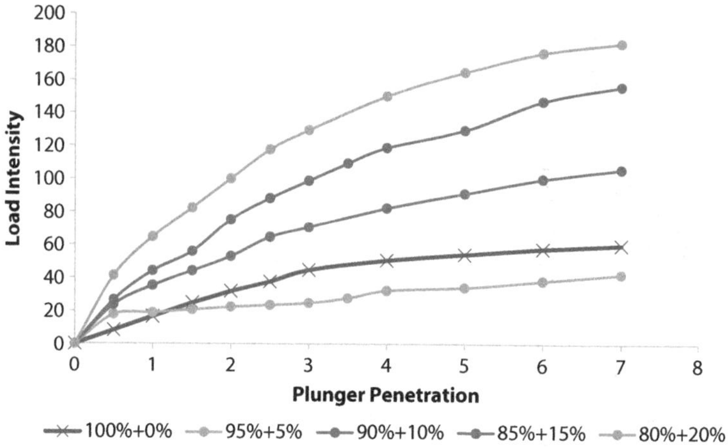

Figure 11.5 Variation in soaked CBR for Expansive soil with varying percentage of marble dust.

Table 11.13 Consolidation of 100% expansive soil.

Applied pressure (Kg/cm²)	Final dial gauge reading	Change in dial gauge reading	Thickness of soil sample H (cm)	Equivalent height of voids H-Hs (cm)	Void ratio (e) =H-Hs Hs
0	0	0	2.1	0.89	0.73528
0.1	0	0	2.1	0.89	0.73427
0.2	0.038	0.038	2.09	0.88	0.72787
0.5	0.062	0.024	2.089	0.879	0.72673
1	0.254	0.192	2.088	0.878	0.72542
2	0.784	0.53	2.087	0.877	0.72431
4	1.72	0.936	2.086	0.876	0.72407
8	2.71	0.99	2.085	0.875	0.72308

Table 11.14 Consolidation test for varying percentage of marble dust.

Applied pressure (Kg/cm²)	Final dial gauge reading	Change in dial gauge reading	Thickness of soil sample H (cm)	Equivalent height of voids H-Hs (cm)	Void ratio $(e) = \frac{H - H_s}{H_s}$	Final Dial gauge reading	Change in dial gauge reading	Thickness of soil sample H(cm)	Equivalent height of voids H-Hs(cm)	Void ratio $(e) = \frac{H - H_s}{H_s}$
	95% Expansive soil and 5% Marble dust					**90% Expansive soil and 10% Marble dust**				
0	0	0	2.1	0.86	0.69354	0	0	2.1	0.82	0.6406
0.05	0.012	0.012	2.0999	0.8599	0.69346	0.002	0.002	2.099	0.819	0.6398
0.1	0.078	0.066	2.0998	0.8598	0.69338	0.022	0.02	2.098	0.818	0.6390
0.2	0.114	0.036	2.0997	0.8597	0.69330	0.044	0.022	2.097	0.817	0.6382
0.5	0.122	0.008	2.0996	0.8596	0.69322	0.114	0.07	2.089	0.809	0.6320
1	0.24	0.118	2.0995	0.8595	0.69314	0.236	0.122	2.060	0.78	0.6095
2	1.18	0.94	2.0993	0.8593	0.69298	0.468	0.232	1.951	0.671	0.5242
4	2.16	0.98	2.0991	0.8591	0.69282	0.784	0.316	1.703	0.423	0.3304
8	3.03	0.87	2.0989	0.8589	0.69266	0.828	0.846	1.487	0.207	0.1617
	85% Expansive Soil and15% Marble Dust					**80% Expansive Soil and 20% Marble Dust**				
0	0	0	2.1	0.79	0.6030	0	0	2.1	0.68	0.4788
0.05	0.026	0.026	2.099	0.789	0.6022	0.03	0.03	2.099	0.679	0.4781
0.1	0.136	0.11	2.084	0.774	0.5908	0.144	0.114	2.082	0.662	0.4661
0.2	0.154	0.118	2.081	0.771	0.5885	0.172	0.028	2.077	0.657	0.4626
0.5	0.234	0.08	2.062	0.752	0.5740	0.276	0.104	2.048	0.628	0.4422
1	0.446	0.212	1.967	0.657	0.5015	0.534	0.258	1.910	0.49	0.3450
2	0.558	0.112	1.904	0.594	0.4534	0.61	0.076	1.863	0.443	0.3119
4	0.704	0.146	1.801	0.491	0.3648	0.796	0.186	1.714	0.294	0.2070
8	0.886	0.182	1.639	0.329	0.2511	0.984	0.188	1.529	0.109	0.1767

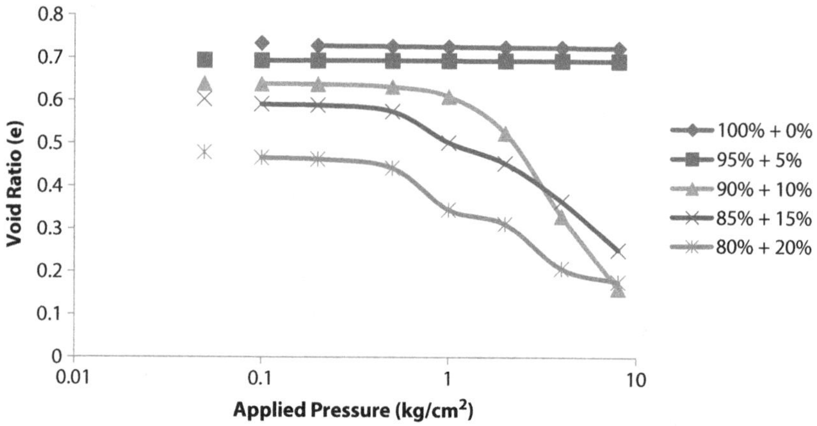

Figure 11.6 Graph of e vs log P_o' with varying percentages of marble dust.

Table 11.15 Permeability test for varying percentage of marble dust.

Sample	H_1	H_2	T (Sec)	K
100% soil +0% Marble Dust	95	60.2	180	1.38 x 10^{-2} cm/sec
95% soil +5% Marble Dust	90	64.2	180	1.26 x 10^{-2} cm/sec
90% soil +10% Marble Dust	85	64.8	180	1.23 x 10^{-2} cm/sec
85% soil +15% Marble Dust	85	62.6	180	1.14 x 10^{-2} cm/sec
80% soil +20% Marble Dust	65	51.5	180	8.73 x 10^{-3} cm/sec

11.4 Conclusion

An experimental study was taken to know the efficacy of the inclusion of waste marble dust in varying amount to expansive soil for the enhancement of engineering properties as well as the index properties of expansive soil. The conclusions derived from the above experimental investigations are summarized below.

1. The utilization of 20% marble dust was found to provide the best improvement in increasing the shrinkage limit of expansive soil, i.e., 18.35% for virgin soil to 34.72% after inclusion of 20% marble dust.

2. Liquid limit as well as plasticity index depicted a decreasing trend with the increase in percentage of marble dust inclusion to the expansive soil. The liquid limit decreased from 52.20% of virgin soil to 38.32% after the inclusion of 20% marble dust. Plasticity index decreased from 34.42% of virgin soil to 20.62% after the inclusion of 20% marble dust. Decreasing trends of liquid limit and plasticity index are due to the increase in shear resistance of expansive soil by the inclusion of marble dust.
3. The key factor to control the compacting characteristics of expansive soil and marble dust mixture is the formation of cementitious compounds. The gradual increase of marble dust in expansive soil from 0% to 20% showed the increase in maximum dry density from 1.77 to 1.804 respectively.
4. The CBR values of the specimen containing the optimum 20% of marble dust showed maximum value, i.e., 8.58% while the virgin soil had a CBR value of 6%.
5. On the basis of consolidation of test results, the inclusion of 20% marble dust have given the least settlement, i.e., 0.187cm while the virgin soil had a settlement of 0.309cm.
6. Hydraulic conductivity values gradually decreased from 1.38×10^{-2} cm/sec for virgin soil to 8.73×10^{-3} cm/sec by the inclusion of 20% marble dust. This is mainly due to the presence of marble dust as a seepage barrier.

From the above experimental results, the overall conclusion is that the marble dust as a waste from construction industries can be utilized for soil stabilization of road sub-base materials in a semi-arid climate where expansive soils are predominantly prevailing. An inclusion of 20% marble dust was found for decreasing of liquid limit, plasticity index, hydraulic conductivity, settlement to its minimum and increase of shrinkage limit, CBR values and MDD to its maximum respectively.

References

1. Samantasinghar, S., Geo-engineering properties of lime treated plastic soils (Doctoral dissertation), 2014.
2. Mohanta, N.R. and Samantaray, S., Study of Combined Effect of Metakaolin and Steel Fiber on Mechanical Properties of Concrete. *Pertanika J. Sci.*

Technol., 27, 3, 1381–1396, 2019, http://journals-jd.upm.edu.my/resources/files/Pertanika%20PAPERS /JST%.
3. Singh, P., Dash, H.M., Samantaray, S., Effect of silica fume on engineering properties of expansive soil. *Mater. Today Proc.*, 33, 5035–5040, 2020, https://doi.org/10.1016 /j.matpr.2020.02.839.
4. Kumar, M.H., Mohanta, N., Samantaray, S., Kumar, N.M., Combined effect of waste glass powder and recycled steel fibers on mechanical behavior of concrete. *SN Appl. Sci.*, 3, 1–18, 2021a, https://doi.org/10.1007/s42452-021-04422-w.
5. Kumar, M.H., Macharyulu, I.S., Ray, T., Mohanta, N.R., Jain, M., Samantaray, S., Sahoo, A., Effect of water absorption and curing period on strength and porosity of triple blended concrete. *Mater. Today Proc.*, *43*, 2162–2169, 2021b, https://doi.org/10.1016/j.matpr.2020.12.092.
6. Ray, T., Mohanta, N.R., Kumar, M.H., Samantaray, S., Study of effect of temperature on behavior of alkali activated slag concrete. *Mater. Today Proc.*, *43*, 1352–1357, 2021, https://doi.org/10.1016/j.matpr.2020.09.169.
7. Kumar, M.H., Saikrishnamacharyulu, I., Mohanta, N.R., Ashutosh, A., Mishra, P., Samantaray, S., Mechanical behaviour of high strength concrete modified with triple blend of fly ash, silica fume and steel fibers. *Mater. Today Proc.*, *65*, 933–942, 2022a, https://doi.org/10.1016/j.matpr.2022.03.528.
8. Kumar, M.H., Mohanta, N.R., Patel, N., Samantaray, S., Reddy, S.V.B., Influence of Fly ash and Metakaoline on the crack resistance and shrinkage property of concrete. *Iran. J. Sci. Technol. Trans. Civ. Eng.*, 46, 2011–2026, 2022b, https://doi.org/10.1007/s40996-021-00748-y.
9. Ene, E. and Okagbue, C., Some basic geotechnical properties of expansive soil modified using pyroclastic dust. *Eng. Geol.*, *107*, 1–2, 61–65, 2009.
10. Sebastian, B., Cryus, S., Jose, B.T., December. Effect of inclusion of coir fibre on the shear strength of marine clay, in: *Proceedings of Indian Geotechnical Conference*, Dec, 15-17, 2011.
11. Sabat, A.K. and Nanda, R.P., Effect of marble dust on strength and durability of Rice husk ash stabilised expansive soil. *Int. J. Civ. Struct. Eng.*, *1*, 4, 939–948, 2011.
12. Sarkar, R., Abbas, S.M., Shahu, J.T., Study of geotechnical behaviour of pond ash mixed with marble dust. *Int. J. Adv. Technol. Civ. Eng.*, *1*, 2, 99–106, 2012.
13. Ali, R., Khan, H., Shah, A.A., Expansive soil stabilization using marble dust and bagasse ash. *Int. J. Sci. Res.*, *3*, 6, 2812–2816, 2014.
14. Gupta, N. and Sharma, T., Experimental Study on The Effect of Marble Dust and Paddy Straw on The Strength Characteristics of Clayey Soil. *Int. J. Adv. Sci. and Technol.*, 29, 5267–5281, 2020.
15. Singh, P.S. and Yadav, R.K., Effect of marble dust on index properties of black cotton soil. *Int. J. Eng. Res. Sci. Tech.*, *3*, 158–63, 2014.
16. Bhavsar, S.N. and Patel, A.J., Analysis of swelling & shrinkage properties of expansive soil using brick dust as a stabilizer. *Int. J. Emerg. Technol. Adv. Eng.*, *4*, 303–8, 2014.

17. Saygili, A., Use of waste marble dust for stabilization of clayey soil. *Mater. Sci.*, *21*, 4, 601–606, 2015.
18. Akinwumi, I.I. and Booth, C.A., Experimental insights of using waste marble fines to modify the geotechnical properties of a lateritic soil. *J. Environ. Eng. Landscape Manage.*, *23*, 2, 121–128, 2015.
19. Bansal, H. and Sidhu, G.S., Influence of waste marble powder on characteristics of clayey soil. *Int. J. Sci. Res. (IJSR)*, *5*, 8, 87–82, 2016.
20. Neeladharan, C., Sathish, P., Nandhini, A., Priya, R., Fathima, I.S., Srimathi, J., Melvisharam, V., Stabilization of soil by using marble dust with sodium silicate as binder. *Int. J. Adv. Res. Trends Eng. Technol.*, *5*, 5, 45–49, 2018.
21. Choksi, R., Mishra, C., Patel, N., Pursuance of waste marble powder to improve soil stabilization. *Int. Res. J. Eng. Technol. (IRJET)*, *5*, 5, 1695–1698, 2018.
22. Priyanka, T.R. and Rao, S.E., Stabilization of black cotton soil with marble dust. *Int. J. Res. Appl. Sci. Eng. Technol.*, *6*, 5, 2057–2060, 2018.
23. Kumar, S.S., Teja, S.L., Needhidasan, S., Evaluation of compaction and stabilization of black cotton soil by using admixtures as marble dust and coconut shell powder. *Int. J. Pure Appl. Math.*, *119*, 17, 3111–3118, 2018.
24. Soni, M.K. and Singh, S., Statistical interpretation of marble dust and alccofine for soil stabilization. *Int. J. Innov. Technol. Explor. Eng.*, *8*, 7, 1609–1613, 2019.
25. Jha, A.K., Jain, A.K., Shivansh, Potential of marble dust to improve the physical behavior of soil, in: *Geotechnics for Transportation Infrastructure: Recent Developments*, 2019.
26. Grower, K. and Goyal, T., Experimental study of waste foundry sand and marble dust as a soil stabilizing material. *Int. Res. J. Eng. Technol. (IRJET)*, *6*, 06, 1265–1272, 2019.
27. Gupta, C. and Sharma, R.K., Influence of marble dust, fly ash and beas sand on sub grade characteristics of expansive soil. *J. Mech. Civ. Eng.*, *13*, 13–18, 2014.
28. Jain, A.K. and Jha, A.K., Geotechnical behaviour and micro-analyses of expansive soil amended with marble dust. *Soils Found.*, *60*, 4, 737–751, 2020.
29. Eltwati, A.S., Tarhuni, F., Elkaseh, A., Engineering properties of clayey soil stabilized with waste granite dust. *J. Crit. Rev.*, *7*, 16, 794–802, 2020.
30. Sivrikaya, O., Uysal, F., Yorulmaz, A., Aydin, K., The efficiency of waste marble powder in the stabilization of fine-grained soils in terms of volume changes. *Arabian J. Sci. Eng.*, *45*, 8561–8576, 2020.
31. Rai, P., Pei, H., Meng, F., Ahmad, M., Utilization of marble powder and magnesium phosphate cement for improving the engineering characteristics of soil. *Int. J. Geosynth. Ground Eng.*, *6*, 1–13, 2020.
32. Vijayan, D.S. and Parthiban, D., Effect of Solid waste based stabilizing material for strengthening of Expansive soil-A review. *Environ. Technol. Innov.*, *20*, 101108, 2020.

33. Zorluer, I. and Gucek, S., The usability of industrial wastes on soil stabilization. *Rev. Constr.*, *19*, 1, 80–89, 2020.
34. Singh, P., Boora, A., Gupta, A.K., Sub-grade characteristics of clayey soil incorporating municipal solid waste incineration ash and marble dust. *J. Eng. Des. Technol.*, *20*, 3, 712–726, 2022.
35. Tak, D., Sharma, J.K., Grover, K.S., Use of Kota stone powder to improve engineering properties of black cotton soil, in: *Problematic Soils and Geoenvironmental Concerns: Proceedings of IGC 2018*, pp. 113–126, 2021.
36. Manish, V.K.S., Use of waste foundry sand (WFS) and crushed waste glass (CWG) on stabilization of black cotton soil-review. *Int. Res. J. Eng. Technol. (IRJET)*, *8*, 01, 746–748, 2021.

Index

Printed in the USA/Agawam, MA
September 16, 2026

910661.096